Springer Proceedings in Mathematics & Statistics

Volume 125

Springer Proceedings in Mathematics & Statistics

This book series features volumes composed of selected contributions from workshops and conferences in all areas of current research in mathematics and statistics, including operation research and optimization. In addition to an overall evaluation of the interest, scientific quality, and timeliness of each proposal at the hands of the publisher, individual contributions are all refereed to the high quality standards of leading journals in the field. Thus, this series provides the research community with well-edited, authoritative reports on developments in the most exciting areas of mathematical and statistical research today.

More information about this series at http://www.springer.com/series/10533

Mihir K. Chakraborty · Andrzej Skowron
Manoranjan Maiti · Samarjit Kar
Editors

Facets of Uncertainties and Applications

ICFUA, Kolkata, India, December 2013

Springer

Editors
Mihir K. Chakraborty
Jadavpur University
Kolkata
India

Andrzej Skowron
Institute of Mathematics
University of Warsaw
Warszawa
Poland

Manoranjan Maiti
Vidyasagar University
Paschim Medinipur, West Bengal
India

Samarjit Kar
Department of Mathematics
National Institute of Technology (NIT)
Durgapur
India

ISSN 2194-1009 ISSN 2194-1017 (electronic)
Springer Proceedings in Mathematics & Statistics
ISBN 978-81-322-2300-9 ISBN 978-81-322-2301-6 (eBook)
https://doi.org/10.1007/978-81-322-2301-6

Library of Congress Control Number: 2015936163

Mathematical Subject Classification: 03Bxx, 41Axx, 68Uxx, 90Bxx, 93Exx

Springer New Delhi Heidelberg New York Dordrecht London

Printed on acid-free paper

Springer (India) Pvt. Ltd. is part of Springer Science+Business Media
(www.springer.com)

Preface

The international conference on "Facets of Uncertainties and Applications" (ICFUA 2013) was organized under the joint collaboration of the Operational Research Society of India (the Calcutta chapter) and Department of Applied Mathematics, University of Calcutta.

The conference aimed at contributing to better understanding between practitioners (both the theoreticians and researchers involved in applications) dealing with uncertainties, mainly of nonprobabilistic category. These papers, but one, presented at the conference, focus on various types of uncertainties which are essentially nonprobabilistic in nature. These types include vagueness, roughness, incompleteness, ambiguity, and such other features. Various mathematical formalisms have emerged during the past few decades to deal with such uncertainties, for example, fuzzy set theory, rough set theory, soft set theory, uncertainty theory. Papers compiled here are of two categories: invited articles presented at the plenary sessions and contributed articles read at regular sessions of the said conference. Invited articles are from experts of high standing in the field, while contributed articles are by senior and young researchers. The papers deal with the state of the art of the theories as well as their applications.

The scope of the conference included the following topics:

- Modeling different types of uncertainty (nonprobabilistic)
- Logic of uncertainty (fuzzy logic and rough logic)
- Rough sets and fuzzy sets in approximate reasoning
- Rough fuzzy hybridization and applications
- Analysis of complex systems and complex network
- Applications of fuzzy sets and rough sets in optimization and decision-making problems
- Image and speech signal processing, prediction, and control
- Robotics

- Expert systems
- Biology and medicine
- Business and management
- Noncomputational mathematics
- Complex system analysis
- Risk management
- Environment engineering
- Data mining
- Other applications

The program of the conference was organized mainly along four tracks:

- Uncertainty modeling
- Logic of uncertainty
- Hybridization of uncertainties
- Role of uncertainties in real problems

Each track contained a plenary session followed by three concurrent parallel sessions. Both the plenary and parallel sessions provided participants ample opportunity to exchange ideas on further research, research collaboration, and training.

The conference was highly interactive and intensive in nature and attracted budding researchers and young faculties working in related disciplines. The conference attracted more than 80 participants from India and abroad. The exchange among these participants has provided them with a comprehensive overview of the techniques and approaches being applied to uncertainty theory and applications.

The program committee for this conference consisted of:

- Didier Dubois, University Paul Sabatier, Toulouse
- Baoding Liu, Tsinghua University, China
- Andrzej Skowron, Warsaw University, Poland
- Roman Slowinski, Poznan University of Technology, Poland
- Dominik Slezak, University of Warsaw, Poland
- Piero Pagliani, Research group on Knowledge and Communication, Italy
- Davide Ciucci, Italy
- Manoranjan Maiti, Vidyasagar University, India
- Amit Konar, Jadavpur University, India
- Mohua Banerjee, IIT Kanpur, India

The conference was supported by the Department of Applied Mathematics, University of Calcutta; Board of Research in Nuclear Science (BRNS), Department of Atomic Energy (DAE), India; Department of Science and Technology (DST), West Bengal; Department of Higher Education, West Bengal; and Indian Statistical Institute (ISI), Kolkata. We are grateful to these organizations for their very generous support.

We thank all the authors for kindly submitting their articles to the conference proceedings. We are very thankful to all the reviewers for their constructive comments and suggestions for the finalization of the papers and to the editorial board of Springer for supporting the publication of the present volume.

Mihir K. Chakraborty
Andrzej Skowron
Manoranjan Maiti
Samarjit Kar

Contents

Part III Hybridization of Uncertainties

Part IV Roll of Uncertainties

About the Editors

Mihir Kumar Chakraborty, Ph.D., is visiting professor at the School of Cognitive Sciences, Jadavpur University, and director of Sivatosh Mookerjee Centre of Sciences, Kolkata. Earlier, he was professor of Pure Mathematics at the University of Calcutta. Professor Chakraborty had also been visiting professor at the Centre for Soft Computing Research, Indian Statistical Institute, Kolkata; Institute for Logic Language and Cognition, Sun Yat-Sen University, Guangzhou, China; Institut de Recherche en Informatique de Toulouse (IRIT), University Paul Sabataire, Toulouse, France; University of Paris VIII, France; University of Wollongong, Australia; University of Regina, Canada; NIAS, Bangalore, India; and VisvaBharati, Santiniketan, India. A recipient of Deutscher Akademischer Austuasch Dienst (DAAD) Fellowship, IISc Fellowship, and Fellowship of West Bengal Academy of Sciences, Professor Chakraborty is member of the Council and Research Project Committee (RPC) of Indian Council for Philosophical Research (ICPR), and guest professor at South West University of Chongqing, China. He has about 150 research papers to his credit in several international journals and edited volumes, co-authored one book *A Geometry of Approximation* (Springer), authored three books in vernacular Bengali on philosophy of mathematics, and coedited several scientific publications. A member of the editorial board of several international journals and a book series *Logic in Asia: Studia Logic Library* (Springer), Professor Chakraborty's area of research include nonstandard logics, rough set theory, fuzzy set theory, reasoning in uncertainty and vagueness, logic of diagrams, topology/functional analysis, and philosophy of mathematics. He has also supervised 16 Ph.D. students. Professor Chakraborty is founder of Calcutta Logic Circle, Association for Logic in India, Indian Society for Fuzzy Mathematics and Information Processing (ISFUMIP), and member of the advisory board of the International Rough Set Society and Indian Rough Set Society.

Andrzej Skowron, European Coordination Committee for Artificial Intelligence (ECCAI) Fellow, received his Ph.D. and D.Sc. (Habilitation) from the University of Warsaw, Poland. In 1991, he received the Scientific Title of Professor. He is full professor in the Faculty of Mathematics, Computer Science and Mechanics at the

University of Warsaw, and honorary professor of Chongqing University of Posts and Telecommunications, China. He is the author of more than 400 scientific publications and editor of many books. His areas of expertise include reasoning with incomplete information, approximate reasoning, soft computing methods and applications, rough sets, rough mereology, granular computing, intelligent systems, knowledge discovery and data mining, decision support systems, adaptive and autonomous systems, perception-based computing, and interactive computational systems. He has supervised more than 20 Ph.D. theses and was the editor-in-chief of the journal *Fundamenta Informaticae* during 1995–2009. Professor Skowronis also on the editorial boards of many international journals. During 1996–2000, he was the president of the International Rough Set Society. He delivered numerous invited talks at international conferences including plenary talk at the 16th IFIP World Computer Congress (Beijing, 2000) and served as program chair at more than 200 international conferences. He was involved in numerous research and commercial projects.

Manoranjan Maiti, Ph.D., has earlier worked at Indian Institute of Tropical Meteorology, Poona; Structural Engineering Division, Vikram Sarabhai Space Centre, ISRO, Trivandrum; Department of Mathematics, Calcutta University Post Graduate Centre (presently, Tripura University), Agartala; Department of Applied Mathematics, Vidyasagar University, West Bengal. He was also dean, Faculty Council of Science, for a period of 10 years and vice-chancellor (pro-tempore) of Vidyasagar University, for a short period. Twenty six students have been awarded Ph.D. degree in mathematics under his guidance at Vidyasagar University and NIT Durgapur, West Bengal, as well as several students are pursuing Ph.D. under him. He has published more than 250 research papers in several international journals. He was associate editor of Applied Mathematical Modelling. His fields of interest include inventory control system, supply chain, fuzzy optimization, transportation, etc.

Samarjit Kar completed his Ph.D., in mathematics from Vidyasagar University, West Bengal. He is associate professor at the Department of Mathematics, National Institute of Technology, Durgapur, India. With over 15 years of experience in teaching, Professor Kar is also a visiting professor at the Department of Mathematical Sciences, Tsinghua University, China. Moreover he has visited several universities and institutes in India and abroad. He has co-authored more than 120 technical articles in international journals, contributed volumes and conference proceedings. An author of 2 textbooks and edited 5 contributed books, Professor Kar has guided 10 Ph.D. students. He is the associate editor of Journal of *Uncertainty Analysis and Application* (Springer) and is presently associated with an ongoing project, “Hybrid modelling of uncertainty analysis in environmental risk assessments” under BRNS, Department of Atomic Energy (DAE), Govt. of India. His research interests include operations research and optimization, soft computing, uncertainty theory and financial modelling.

Part I
Uncertainty Modelling

Rough Sets and Other Mathematics: Ten Research Programs

Piero Pagliani

Abstract Since its inception, interesting connections between Rough Set Theory and different mathematical and logical topics have been investigated. This paper is a survey of some less known although highly interesting connections, which extend from Rough Set Theory to other mathematical and logical fields. The survey is primarily thought of as a guide for new directions to be explored.

Keywords Rough sets · Algebraic logic · Topology

1 Information from Data and Information as Metaphor

As is well known, the starting point of Rough Set Theory is an *Indiscernible space* $\langle U, E\rangle$, where U is a set and $E \subseteq U \times U$ is an equivalence relation such that $\langle x, y\rangle \in E$ states that items x and y take exactly the same attribute-values according to an evaluation recorded in an *Information System*.

Given any relational structure $\langle U, R\rangle$, with $R \subseteq U \times U$, and $X \subseteq U$, the set $R(X) = \{y : \exists x \in X(\langle x, y\rangle \in R)\}$ will by named the *R-neighborhood of X*. If $X = \{a\}$ we shall write $R(a)$. Thus, by means of $E-$neighborhoods, from any indiscernibility space the following operators are defined on $\wp(U)$:

$$(lE)(X) = \{x : E(x) \subseteq X\} = \{x : \forall y(\langle x, y\rangle \in E \Rightarrow y \in X)\} \tag{1}$$

$$(uE)(X) = \{x : E(x) \cap X \neq \emptyset\} = \{x : \exists y(\langle x, y\rangle \in E \wedge y \in X)\} \tag{2}$$

$(lE)(X)$ is called the *lower approximation* of X (*via* E), while $(uE)(X)$ is called the *upper approximation* of X (*via* E), and $\langle U, (uE), (lE)\rangle$ is called an *approximation space*. Any equivalence class modulo E is a neighborhood $E(a)$ for some $a \in U$, and represents a "basic property," that is, a unique array of attribute-values, hence a subset of U *definable* by means of the given evaluation. Moreover, forany X, $(lE)(X)$ and

P. Pagliani (✉)
Rome, Italy
e-mail: pier.pagliani@gmail.com

M.K. Chakraborty et al. (eds.), *Facets of Uncertainties and Applications*,
Springer Proceedings in Mathematics & Statistics 125,
DOI 10.1007/978-81-322-2301-6_1

$(uE)(X)$ output either an E–equivalence class or a union of equivalence classes, which represents nonelementary definable set.

The well-known properties of (lE) and (uE) depend on the fact that E is an equivalence relation. Particularly, it is possible to show that $\{(lE)(X) : X \subseteq U\} = \{(uE)(X) : X \subseteq U\}$. This set will be denoted by $Df(U)$. It collects the *definable subsets* of U. If $X \in Df(U)$, then $X = (lE)(X) = (uE)(X)$: a definable set does not need to be approximated.

All the concepts not introduced in the paper can be found in [16][1]

Observation 1 *The second part of definitions* (1) *and* (2) *displays the dual logical constructions* $(\forall, \Rightarrow)$ *and* $(\exists, \wedge)$. *They are the backbone to a number of mathematical concepts. Notably,* $(\forall, \Rightarrow)$ *is the logical core of* interior *and* necessity *operators, while* $(\exists, \wedge)$ *is that of* closure *and* possibility *operators.*

We remind that an operator ϕ on a lattice $\mathbf{L}$ is an interior (resp. closure) operator if it is (i) decreasing: $\phi(x) \leq x$ (resp. increasing: $x \leq \phi(x)$), (ii) monotone: $x \leq y$ implies $\phi(x) \leq \phi(y)$ and (iii) idempotent: $\phi(\phi(x)) = \phi(x)$. An interior (resp. closure) operator ϕ is *topological* if it is (iv) multiplicative: $\phi(x \wedge y) = \phi(x) \wedge \phi(y)$ (resp. additive: $\phi(x \vee y) = \phi(x) \vee \phi(y)$) and (v) co-normal: $\phi(1) = 1$ (resp. normal: $\phi(0) = 0$).

Indeed, (lE) and (uE) are interior and, respectively, closure topological operators. This was one of the first results of Pawlak's approximation spaces and it can be stated under different points of views:

Facts 1 1. (lE) is an interior, $\mathbb{I}$, and (uE) a closure, $\mathbb{C}$, operator of a topological space with a basis of clopen - closed and open - subsets.
2. $\mathbf{AS}(U) = \langle Df(U), \cap, \cup, -, \emptyset, U\rangle$ is a subalgebra of the Boolean on $\wp(U)$.
3. $\langle \wp(U), \cap, \cup, -, \emptyset, U, (uE), (lE)\rangle$ is a topological Boolean algebra.
4. $\langle \wp(U), \cap, \cup, -, \emptyset, U, (uE), (lE)\rangle$ is a model for S5 modal logic, where (lE) stands for the necessity operator $\Box$, and (uE) for the possibility operator $\Diamond$.

By extension, $\mathbf{AS}(U)$ will also be called an approximation space. Since the two approximation operators are defined by means of E-neighborhoods, straightforward generalizations were proposed in the Rough Set literature since its inception, by considering other types of binary relations, R. However, problems arise if definitions (1) and (2) are merely traced. For instance, $(lR)(X)$ could fail to be decreasing, which is "strange" for a *lower* approximation. In view of these problems, usually generalized approximation operators do not mechanically trace the original definitions (see for instance [6]). Anyway, in view of Observation 1, when definitions (1) and (2) are used, properties of generalized upper and lower approximations can be easily derived from the literature on modal logics with Kripke models (see [16], Chap. 4.13). The following issue arises:

---

[1]Except for [7], this book is the only work of the author's that will be cited. The story of the results can be found in the mentioned chapters.

ISSUE A. INFORMATIONAL INTERPRETATION OF KRIPKEAN MODAL LOGICS, THROUGH ROUGH SETS: *For any binary relation R, give (lR)* and *(uR) a meaningful informational interpretation. That is, a concretely justified interpretation of R, such as that of relative accessibility relation (as defined in* [15]*). Conversely, give a modal interpretation to generalized approximation operators.*

2 Algebras of Rough Sets

Given an approximation space $\mathbf{AS}(U)$, a *rough set* is an equivalence class modulo (lE) and (uE) on the powerset $\wp(U)$. Thus, the rough set of X can be identified by the ordered pair $\langle (uE)(X), (lE)(X)\rangle$, called *decreasing representation*, or $\langle (lE)(X), -(uE)(X)\rangle$, called *disjoint representation*. The symbol $rs(X)$ will denote both of these representations. However, not all the ordered pairs of decreasing (disjoint) elements of $\mathbf{AS}(U)$ represent a rough set. In fact, if $S \subseteq U$ is a singleton equivalence class, then for any $X \subseteq U$ the following equivalent conditions hold (cf. [16], Chap. 7):

$$\text{(a) } S \subseteq (uE)(X) \textit{ iff } S \subseteq (lE)(X); \text{ (b) } S \subseteq (lE)(X) \textit{ or } S \subseteq -(uE)(X) \tag{3}$$

The informational explanation of this fact is that singletons represent completely defined objects. Thus, U divides into two parts: an *exact* part, given by the union B of all singleton equivalence classes, and an *uncertain* part, given by its complement $P = U \cap -B$. Indeed, a clause equivalent to conditions (3) is $(uE)(X) \cap B = (lE)(X) \cap B$. It states a *local* property: on B there is no roughness because lower and upper approximations coincide, which is the characteristic of definable sets. Consequently, the set of all and only the rough sets of an approximation space $\mathbf{AS}(U)$ is definable as follows.
In decreasing representation:

$$Dc_{\equiv_{JB}}(\mathbf{AS}(U)) = \{\langle A_1, A_2\rangle \in \mathbf{AS}(U)^2 : A_2 \Rightarrow A_1 = U, A_1 \Rightarrow A_2 \equiv_{JB} U\} \tag{4}$$

where for all $X, Y \in \mathbf{AS}(U)$, $X \Rightarrow Y$ is $-X \cup Y$ and $X \equiv_{JB} Y$ if and only if $B \Rightarrow X = B \Rightarrow Y$. So, the first clause just means $A_2 \subseteq A_1$, while conditions (3) follow from the second clause.
For the disjoint representation we have:

$$Dj_{\equiv_{JB}}(\mathbf{AS}(U)) = \{\langle A_1, A_2\rangle \in \mathbf{AS}(U)^2 : A_1 \cap A_2 = \emptyset, A_1 \cup A_2 \equiv_{JB} U\}. \tag{5}$$

With $D_{\equiv_{JB}}(\mathbf{AS}(U))$ we denote either of these collections. Now, notice that $\equiv_{JB}$ is a (Boolean) congruence on $\mathbf{AS}(U)$. In general, given any Heyting algebra $\mathbf{H}$ and a Boolean congruence $\equiv$ on it (i.e. $\mathbf{H}/_{\equiv}$ is a Boolean algebra), the operations in the following table are definable on the set $Dj_{\equiv}(\mathbf{H})$. If $\mathbf{H}$ is a Boolean algebra, corresponding operations are definable on the set $Dc_{\equiv}(\mathbf{H})$. It is understood that

$a = \langle a_1, a_2\rangle, b = \langle b_1, b_2\rangle$ and the operations between elements of the pairs are those of **H**:

Symbol	$Dc_{\equiv}(\mathbf{H})$	$Dj_{\equiv}(\mathbf{H})$	Name
$0, 1$	$\langle 0, 0\rangle, \langle 1, 1\rangle$	$\langle 0, 1\rangle, \langle 1, 0\rangle$	Bottom, resp. Top
$\sim a$	$\langle \neg a_2, \neg a_1\rangle$	$\langle a_2, a_1\rangle$	Strong negation
$a \longrightarrow b$	$\langle a_2 \Rightarrow b_1, a_2 \Rightarrow b_2\rangle$	$\langle a_1 \Rightarrow b_1, a_1 \wedge b_2\rangle$	Weak implication
$a \wedge b$	$\langle a_1 \wedge b_1, a_2 \wedge b_2\rangle$	$\langle a_1 \wedge b_1, a_2 \vee b_2\rangle$	Inf
$a \vee b$	$\langle a_1 \vee b_1, a_2 \vee b_2\rangle$	$\langle a_1 \vee b_1, a_2 \wedge b_2\rangle$	Sup

Derived operations:

Symbol	Definition	Name
$\lrcorner a$	$a \longrightarrow 0$	Weak negation
$a \supset b$	$\sim \lrcorner \sim a \vee b \vee (\lrcorner a \wedge \lrcorner \sim b)$	Pre-relative pseudocomplementation
$\neg a$	$a \supset 0 = \sim \lrcorner \sim a$	Pre-pseudocomplementation
$a \subset b$	$\sim (\sim a \supset \sim b)$	Pre-relative co-pseudocomplementation
$a \overset{c}{\Longrightarrow} b$	$\sim \lrcorner (a \supset b)$	Pre-relative pseudosupplementation
$a \overset{c}{\Longleftarrow} b$	$\lrcorner \sim (a \subset b)$	Pre-relative co-pseudosupplementation
!	$1 \overset{c}{\Longrightarrow} a$	Pre-pseudosupplementation
¡	$0 \overset{c}{\Longleftarrow} a$	Pre-copseudosupplementation

Facts 2 (cf. [16], Chaps. 7, 8 and 9.6) (1) $\sim\sim a = a$; (2) $\sim (a \wedge b) = \sim a \vee \sim b$; (3) $\sim (a \vee b) = \sim a \wedge \sim b$; (4) $\lrcorner a = \langle \neg a_2, \neg a_2\rangle$in $Dc_{\equiv}(\mathbf{H})$ and $\langle \neg a_1, a_1\rangle$ in $Dj_{\equiv}(\mathbf{H})$, (5) $\neg a = \langle \neg a_1, \neg a_1\rangle$ in $Dc_{\equiv}(\mathbf{H})$ and $\langle a_2, \neg a_2\rangle$ in $Dj_{\equiv}(\mathbf{H})$. If **H** is a Boolean algebra: (6) $\supset$ is a *relative pseudocomplementation* in the lattices $\langle Dc_{\equiv}(\mathbf{H}), \leq\rangle$ and $\langle Dj_{\equiv}(\mathbf{H}), \leq\rangle$, where $a \leq b$ if and only if $a \wedge b = a$. Hence, for all a, b, c of these lattices, $c \wedge a \leq b$ iff $c \leq a \supset b$. As a consequence, $\neg$ is a *pseudocomplementation.* (7) $\subset$ is a *relative co-pseudocomplementation,* that is, $c \vee a \geq b$ iff $c \geq a \subset b$. Since $\lrcorner a = a \subset 1$, $\lrcorner$ is a *co-pseudocomplementation;* (8) $¡a = \sim \neg a = \lrcorner \sim a = \neg\neg a = \lrcorner \neg a$; (9) $!a = \sim \lrcorner a = \neg \sim a = \lrcorner \lrcorner a = \neg \lrcorner a$.

In what follows we set $D_1 = \phi_1 = ¡$, $D_2 = \phi_2 = !$, $e_0 = 0$, $e_2 = 1$ and $e_1 = \langle U, B\rangle$ if $D_{\equiv_{JB}}$ is $Dc_{\equiv_{JB}}$, while $e_1 = \langle B, \emptyset\rangle$ if $D_{\equiv_{JB}}$ is $Dj_{\equiv_{JB}}$.
Since any approximation space $\mathbf{AS}(U)$ is a Boolean algebra one can prove:

Facts 3 (cf. [16], Chaps. 6–10) Let $B \in \mathbf{AS}(U)$, then:

1. $\langle D_{\equiv_{JB}}(\mathbf{AS}(U)), \wedge, \vee, \longrightarrow, \sim, \lrcorner, 0, 1\rangle$ is a *semi-simple Nelson algebra.*
2. $\langle D_{\equiv_{JB}}(\mathbf{AS}(U)), \wedge, \vee, \sim, \phi_1, \phi_2, 0, 1\rangle$, is a *three-valued Łukasiewicz algebra.*
3. $\langle D_{\equiv_{JB}}(\mathbf{AS}(U)), \wedge, \vee, \neg, \supset, 0, 1\rangle$, is a *Heyting algebra.*
4. $\langle D_{\equiv_{JB}}(\mathbf{AS}(U)), \wedge, \vee, \lrcorner, \subset, 0, 1\rangle$, is a *co-Heyting algebra.*
5. $\langle D_{\equiv_{JU}}(\mathbf{AS}(U)), \wedge, \vee, \sim, 0, 1\rangle$ is a *Boolean algebra* isomorphic to $\mathbf{AS}(U)$.
6. $\langle D_{\equiv_{J\emptyset}}(\mathbf{AS}(U)), \wedge, \vee, \neg, \supset, D_1, D_2, e_0, e_1, e_2\rangle$, is a *Post algebra of order three.*

We have enough material for a number of interesting observations.

Observation 2 *Given a topological space on a set* U *and* $X \subseteq U$*, the boundary* $\mathbb{B}(X)$ *of* X *is given by* $\mathbb{C}(X) \cap -\mathbb{I}(X)$. *In [10] William Lawvere pointed out that in co-Heyting algebras the topological notion of a boundary is definable as* $\partial(x) = x \wedge \lrcorner x$, *and observed that for all* x, y:

(1) $\partial(x \wedge y) = (\partial(x) \wedge y) \vee (x \wedge \partial(y))$; *(2)* $\partial(x \wedge y) \vee \partial(x \vee y) = \partial(x) \vee \partial(y)$.

The first equation is essentially the usual Leibniz rule for differentiation of a product. Lawvere emphasizes that though its validity for boundaries of closed sets is supported by our space intuition, nevertheless it is virtually unknown in general topology literature. Moreover, given an element x *of a co-Heyting algebra, Lawvere calls* $\lrcorner\lrcorner x$ *the* regular core *of* x. *In the context of Continuum Physics, he claimed that a part* x *may be considered a* sub-body *(or shortly a* body*) if and only if* $\lrcorner\lrcorner x = x$ *and noticed that any element* x *is the join of its core and its boundary:* $x = \lrcorner\lrcorner x \vee \partial(x)$.

In view of Lawvere's observations and Fact 3.(4), the notion of a co-Heyting boundary was exploited by the author in the context of Rough Set analysis. Given an Approximation Space $\mathbf{AS}(U)$, $X \subseteq U$ and $a = \langle (uE)(X), (lE)(X) \rangle$, $a \wedge \lrcorner a$ (or, equivalently, $a \wedge \sim a$), is $\langle \mathbb{B}(X), \emptyset \rangle$. In order to obtain the rough set of $\mathbb{B}(X)$ it is sufficient to compute $\neg\neg(a \wedge \lrcorner a)$. Moreover, $\lrcorner\lrcorner a = \langle (lE)(X), (lE)(X) \rangle$, which is the rough set of $(lE)(X)$. But $(lE)(X)$ is the *internal* or *necessary* part of X, (in a literal sense when $\mathbf{AS}(U)$ is interpreted as an S5 modal space). This part is stable because (lE) is idempotent. This means that $\delta(\lrcorner\lrcorner a) = 0$; that is, the boundary of $(lE)(X)$ is empty.

In a private communication, Lawvere said that to his knowledge this was the first nontrivial, albeit simple, example of his algebraic characterization of topological boundaries. A new issue arises, thus:

ISSUE B. MORE GENERAL ALGEBRAIC CHARACTERIZATION OF TOPOLOGICAL BOUNDARIES THROUGH GENERALIZED ROUGH SETS: *Exhibit more general examples of rough set systems in which Lawvere's algebraic descriptions of a "body", a "core" and a "boundary" can be expressed.*

Rough set systems induced by pre or partial orders $\mathbf{P} = \langle U, \leq \rangle$ are natural candidates, because the set $F(\mathbf{P})$ of order filters of $\mathbf{P}$ is a Heyting algebra $\mathbf{H}(\mathbf{P})$ (for $X \subseteq U$ the *order filter* $\uparrow X$, or $\uparrow x$ if $X = \{x\}$, is nothing but the $\leq$ −neighborhood of X). But, given a Heyting algebra $\mathbf{H}$, and a Boolean congruence $\equiv$ on it, $\mathbf{N}_{\equiv}(\mathbf{H}) = \langle Dj_{\equiv}(\mathbf{H}), \longrightarrow, \sim, \lrcorner, \wedge, \vee, 0, 1 \rangle$ is a Nelson algebra, which is a model for *Constructive Logic with Strong Negation*, $\mathcal{CLSN}$ (from this result one obtains Fact 3.(1)). But when is $\mathbf{N}_{\equiv}(\mathbf{H})$ a Heyting algebra? When a bi-Heyting algebra? Is it possible a characterization of these cases and, moreover, a rough set, hence informational, interpretation as it is done in [7] for finite algebras and particular infinite cases (see Facts 6)?

Observation 3 *An operator* J *on a Heyting algebra* **H**, *is a* Lawvere-Tierney operator *if it is idempotent, increasing and multiplicative. The operator* J^B *of definitions*

(4) and (5) is such an operator. A Lawvere-Tierney operator J on the dual Heyting algebra $\mathbf{H}(\mathbf{P})$ *of a preorder* $\mathbf{P} = \langle U, \leq \rangle$ *defines an association between elements p of U and subfilters of $\uparrow p$, in the following manner:*

$$J_{[p]} = \{\uparrow p \cap X : p \in J(X), X \in \mathbf{H}(\mathbf{P})\}. \tag{6}$$

The family $\Gamma = \{J_{[p]} : p \in U\}$ *is called a* Grothendieck topology *and the system* $\langle \mathbf{P}, \Gamma \rangle$ *an* ordered site *(see [16], Chap. 7.3). The logical importance of ordered sites is the following.* $\mathbf{H}(\mathbf{P})$ *is the set of possible evaluations* $[\![A]\!]$ *from intuitionistic formulas A to the Kripke model* $\langle \mathbf{P}, \models \rangle$. *Given an element $x \in U$, a formula A is said to be* locally valid *on x,* $x \models \langle l \rangle A$, *if* $[\![A]\!] \in J_{[x]}$ *for some Grothendieck topology Γ on* $\mathbf{P}$.

This formalizes our intuition that *locally* on B sets are not rough but exact (see [16], Chap. 7).

ISSUE C: ROUGH SET SYSTEMS AND LOGIC: *Find a faithful logic for rough set systems. The problem is that, for instance, three-valued Łukasiewicz logic, which is often proposed as the logic of rough set systems, is not able to grasp the distinction between the exact behaviour on B and the inexact behaviour on P. In fact, this logic encompasses the cases in which $B = U$, $B = \emptyset$ and $B \neq U$, $B \neq \emptyset$. Maybe, Labeled Deduction Systems could be useful (see* [4] *and subsequent works).*

Observation 4 *From Facts* 2.(8)–(9), *one derives* $\sim \neg\neg = \lrcorner\lrcorner \sim$ *and* $\sim \lrcorner\lrcorner = \neg\neg \sim$, *which suggest that the double negations $\neg\neg$ and $\lrcorner\lrcorner$ behave in modal* ($\sim \Box = \Diamond \sim$, $\sim \Diamond = \Box \sim$) *and topological* ($-\mathbb{I} = \mathbb{C}-$, $-\mathbb{C} = \mathbb{I}-$) *manners.*

The rough set explanation of this fact is given by the following equations:

$$\neg\neg rs(X) = rs((uE)(X)); \quad \lrcorner\lrcorner rs(X) = rs((lE)(X)) \tag{7}$$

From Facts 2.(8)-(9), $\neg\neg = \lrcorner\neg$ and $\lrcorner\lrcorner = \neg\lrcorner$. Thus they are particular cases of two more general operators definable in bi-Heyting algebras. Indeed, let us define the following sequences in a σ-complete bi-Heyting algebra $\mathbf{BH}$: (i) $\Box_0 = \Diamond_0 = Id$; (ii) $\Box_{n+1} = \neg\lrcorner\Box_n$, $\Diamond_{n+1} = \lrcorner\neg\Diamond_n$; (iii) $\blacksquare(a) = \bigwedge_{i=1}^{n} \Box_i(a)$; (iv) $\blacklozenge(a) = \bigvee_{i=1}^{n} \Diamond_i(a)$, $\forall a \in \mathbf{BH}$. In [19] it is shown that for any a, $\blacksquare(a)$ is the largest complemented element of $\mathbf{BH}$ below a, while $\blacklozenge(a)$ is the smallest complemented element above a. From Facts 3.(3)-(4), a rough set system is a bi-Heyting algebra where $\blacksquare = \Box_1$ and $\blacklozenge = \Diamond_1$ (from (7), because (lE) and (uE) are idempotent or directly from idempotency of $\neg\neg$ and $\lrcorner\lrcorner$).

This property is related to the following laws: (DM1) Let $\mathbf{H}$ be a Heyting algebra. $\mathbf{H}$ satisfies the De Morgan law for $\neg$, if $\forall x, y$, $\neg(x \wedge y) = \neg x \vee \neg y$. (DM2) Let $\mathbf{CH}$ be a co-Heyting algebra. $\mathbf{CH}$ satisfies the De Morgan law for $\lrcorner$, if $\forall x, y$, $\lrcorner(x \vee y) = \lrcorner x \wedge \lrcorner y$. It can be shown that in bi-Heyting algebras the law for $\neg$ implies $\blacksquare(a) = \neg\lrcorner a$ and that the law for $\lrcorner$ implies $\blacklozenge(a) = \lrcorner\neg a$ (the reverse of the implications does not hold). Actually, both laws hold in rough set systems. The proof is in the following list, as well as some consequences:

Facts 4 1. (DM1) is equivalent to the fact that $Reg(\mathbf{H}) = \{x \in \mathbf{H} : x = \neg\neg x\}$ is a sublattice of **H** (see [8]). Dually for (DM2) and $coReg(\mathbf{CH}) = \{x \in \mathbf{CH} : x = \lrcorner\lrcorner x\}$.

2. Since $\neg\neg$ is a Lawvere-Tierney operator, it is multiplicative but, in general, not additive. This means that generally $Reg(\mathbf{H})$ is not a sublattice of **H**. Dually, $coReg(\mathbf{CH})$ is not a sublattice of a generic co-Heyting algebra **CH**.
3. But in rough set systems, from (7) and Facts 1.(1), $\neg\neg$ is a topological closure, hence additive, while $\lrcorner\lrcorner$ is a topological interior, hence multiplicative. It follows that both $Reg(D_{\equiv_{JB}}(\mathbf{AS}(U)))$ and $coReg(D_{\equiv_{JB}}(\mathbf{AS}(U)))$ are sublattices of $D_{\equiv_{JB}}(\mathbf{AS}(U))$.
4. The set of the complemented elements of a lattice **L** is called the *center of* **L**, $Ctr(\mathbf{L})$. One can prove that $Reg(D_{\equiv_{JB}}(\mathbf{AS}(U))) = coReg(D_{\equiv_{JB}}(\mathbf{AS}(U))) = Ctr(D_{\equiv_{JB}}(\mathbf{AS}(U)))$. Of course, if $a \in Ctr(D_{\equiv_{JB}}(\mathbf{AS}(U)))$, then $\delta(a) = 0$.

ISSUE D: INFORMATIONAL INTERPRETATION OF THE SITUATION WHERE THE SEQUENCES $\Box_n$ AND $\Diamond_n$ DO NOT STABILIZE AT STEP 2: *Longer steps for stabilization reflect the fact that new boundaries must be included after each application of closure and smaller internal parts must be grasped after any interior application. Informational interpretations of this situation should be provided. In Sect.* 4 *a first answer is suggested.*

Facts 5 1. In a Heyting algebra **H**, an element x is called *dense* if $\neg\neg x = 1$. If **H** has a least dense element d, then $Reg(\mathbf{H})$ is isomorphic to $\mathbf{H}/_{\equiv_{Jd}}$. It can be proved ([16] Chap. 7) that in $Dj_{\equiv_{JB}}(\mathbf{AS}(U))$ the least dense element is $\langle B, \emptyset\rangle$, while in $Dc_{\equiv_{JB}}(\mathbf{AS}(U))$ is $\langle U, B\rangle$.

2. Thus, one can prove that $\langle D_{\equiv_{JB}}(\mathbf{AS}(U)), \wedge, \vee, !, e_0, e_1, e_2\rangle$, is a P_2*-lattice of order three,* if $B \neq U$. Here, $e_0 \leq e_1 \leq e_2$ is the chain of values. Moreover, if A is a classical tautology (with $\sim$, $\neg$ or $\lrcorner$ as negation), then $e_1 \leq [\![A]\!] \leq e_2$, while if A is a classical contradiction, then $e_0 \leq [\![A]\!] \leq {\sim}e_1$ (see [16] Chap. 9.6). Thus, e_1 is a local classical top, and $\sim e_1$ is a local classical bottom.
3. Eventually, $\langle D_{\equiv_{JB}}(\mathbf{AS}(U)), \wedge, \vee, \overset{c}{\Longrightarrow}, \overset{c}{\Longleftarrow}, \supset, \subset, !, ¡, 0, 1\rangle$, is a *P-algebra.* In this case $a \overset{c}{\Longrightarrow} b$ is the largest element e of the center such that $e \wedge a \leq b$, while $a \overset{c}{\Longleftarrow} b$ is the least element e of the center such that $e \vee a \geq b$.

The last result leads us to a new issue:

ISSUE E: ROUGH SET SYSTEMS AND TOPOS THEORY: *In* [3] *it is shown that a lattice* **L** *has a stalk (etalé) space representation when for each* $s \in \mathbf{L}$, *the mapping* $\varphi_s : Ctr(\mathbf{L}) \longrightarrow \mathbf{L}$; $\varphi_s(e) = e \wedge s$ *is residuated, or, equivalently, if* $\eta_s : Ctr(\mathbf{L}) \longrightarrow \mathbf{L}$; $\eta_s(e) = e \vee s$ *is residuated. But from Fact 5.(3)* $\overset{c}{\Longrightarrow}$ *and* $\overset{c}{\Longleftarrow}$ *are the required residuations, respectively. Moreover, the definition of a bi-Heyting algebra in* [19] *is given in terms of a topos* $\mathcal{E}$, *a Boolean topos* $\mathcal{B}$, *and a surjective geometric morphism* $\Gamma : \mathcal{E} \longrightarrow \mathcal{B}$, *such that the canonical map* $\delta : \Omega_{\mathcal{E}} \longrightarrow \Omega_{\mathcal{B}}$ *has a lax adjoint. Finally, also the construction of rough set systems through Grothendieck (Lawvere-Tierney) topologies is a signal that these systems are some sort of topos. Therefore,* *a unitary and coherent description of (generalized) rough sets from the point of view of etalé spaces and topos theory would be welcome (some hints: the representation of a rough set system in terms of the dual of a set of disjoint chains of max length 2, or as a product of a Boolean algebra and a Post algebra of order three (see* [16] *Chaps. 10.4 and 8.3); the fact that the prime ideals of a $P-$algebra lie in disjoint maximal chains (*[3]*)).*

We have seen that rough set systems induced by pre or partial orders $\mathbf{P}$ are well-behaving because the dual $\mathbf{H}(\mathbf{P})$ of a preorder $\mathbf{P}$ is a Heyting algebra. Not only they are useful in data-mining (cf. [5]), but in this case the construction of rough set systems assumes an unexpected amazing meaning (see [7]) (on the topic, see also [14]). In fact, if $\mathbf{P}$ is a *partial order* upper bounded by a set M of maximal elements (always if it is finite), then for all $m \in M$, $\uparrow m$ is a singleton definable set. In view of the previous discussion, the corresponding rough set system is given by $Dj_{\equiv_{JM}}(\mathbf{H}(\mathbf{P}))$. One has, thus:

Facts 6 1. M is the least dense element of $\mathbf{H}(\mathbf{P})$ (i.e. $\neg\neg M = U$).
2. $\equiv_{JM}$ is the Glivenko congruence: i.e. $X \equiv_{JM} Y$ iff $\neg X = \neg Y$.
3. $Dj_{\equiv_{JM}}(\mathbf{H}(\mathbf{P}))$ belongs to the subvariety of Nelson algebras of pairs $\langle a, b\rangle$ determined by the equation $\neg a = \neg\neg b$.
4. $Dj_{\equiv_{JM}}(\mathbf{H}(\mathbf{P}))$ is a model for the logic E_0, which is $\mathcal{CLSN}$ plus the following modal axioms: $({\sim}A \longrightarrow \perp) \longrightarrow \mathbf{T}(A)$ and $(A \longrightarrow \perp) \longrightarrow {\sim}\mathbf{T}(A)$. Logic E_0 was introduced in [11] where it is proved $\vdash_{\mathcal{CL}} A$ if and only if $\vdash_{\mathcal{E}_0} \mathbf{T}(A)$, also in the predicative case, thus extending the well-known Gödel-Glivenko theorem stating $\vdash_{\mathcal{CL}} A$ if and only if $\vdash_{\mathcal{INT}} \neg\neg A$, for A any classical *propositional* formula, $\mathcal{CL}$ a classical logic system and $\mathcal{INT}$ an intuitionistic system.

We arrive at a new issue:
ISSUE F: ROUGH SET SYSTEMS AND SUBSTRUCTURAL LOGICS: *$\mathcal{CLSN}$ is a substructural logic* ([24], *but see also* [23]). *Is there any rough set-based informational interpretation of this fact? (for some suggestions see* [12, 13], *and* [25]*).*

3 Purely Relational Approximation Operators

Consider a structure $\mathbf{P} = \langle U, M, R\rangle$, with U, M sets and $R \subseteq U \times M$. We interpret U and M as sets of objects and, respectively, properties, so that $\langle g, m\rangle \in R$ means that object g fulfills property m. $\mathbf{P}$ will be called a *property system*. Let us define the following functions, where $R^{\smile}$ is the reverse of R (see [22], cf. [16], Chap. 2.):

- $\langle e\rangle : \wp(M) \longmapsto \wp(U)$; $\langle e\rangle(Y) = \{a \in U : \exists b(b \in Y \wedge b \in R(a))\}$;
- $[e] : \wp(M) \longmapsto \wp(U)$; $[e](Y) = \{a \in U : \forall b(b \in R(a) \Rightarrow b \in Y)\}$;
- $\langle i\rangle : \wp(U) \longmapsto \wp(M)$; $\langle i\rangle(X) = \{b \in M : \exists a(a \in X \wedge a \in R^{\smile}(b))\}$;
- $[i] : \wp(U) \longmapsto \wp(M)$; $[i](X) = \{b \in M : \forall a(a \in R^{\smile}(b) \Rightarrow a \in X)\}$.

A function is decorated by 'e' when its application gives an *extension*, i.e., a set of objects, and it is decorated by 'i' when it outputs an *intension*. From Observation 1, it is clear why two of them are $\Diamond$-shaped (*possibility*), and two are $\Box$-shaped (*necessity*). These functions fulfill a strategic property: $\langle\langle i\rangle, [e]\rangle$ and $\langle\langle e\rangle, [i]\rangle$ are *Galois adjunctions*: $\langle i\rangle(X) \subseteq Y$ iff $X \subseteq [e](Y)$, $\langle e\rangle(Y) \subseteq X$ iff $Y \subseteq [i](X)$, for all $X \subseteq G, Y \subseteq M$. Exploiting this fact one immediately obtains that $\langle i\rangle[e]$ and $\langle e\rangle[i]$ are pre-topological interior operators, while $[i]\langle e\rangle$ and $[e]\langle i\rangle$ are pre-topological closure operators, on M and U, respectively. For this reason we set, for all $X \subseteq U, Y \subseteq M$:

$$\text{(a) } int(X) = \langle e\rangle([i](X)); \text{ (b) } cl(X) = [e](\langle i\rangle(X)).$$
$$\text{(c) } \mathcal{A}(Y) = [i](\langle e\rangle(Y)); \text{ (d) } \mathcal{C}(Y) = \langle i\rangle([e](Y)).$$

$\mathcal{A}$ and $\mathcal{C}$ are the "formal" counterparts of cl and, respectively, int. One has:

$$int(X) \subseteq X \subseteq cl(X), \quad any\ X \subseteq U. \tag{8}$$

If $R(U) = M$ and $R^{\smile}(M) = U$, then int is co-normal and cl is normal (in this case we shall say that the property system is *normal*). It can be proved that (lR) and (uE) are special cases of int, respectively, cl.

ISSUE G. ROUGH SET SYSTEMS FROM ADJOINT OPERATORS: *What are the logico-algebraic properties of the set of ordered pairs of the form $\langle cl(X), int(X)\rangle$ or $\langle int(X), -cl(X)\rangle$? (Some hints from* [1] *or* [2]).

Observation 5 *It is worth noticing that the above machinery can be rephrased in the framework of* Chu spaces. *Since they provide models for Linear Logic (see [17, 18]), one could add this ingredient to Issue F for a more comprehensive description of the "substructural picture".*

4 Approximation by Means of Neighborhoods

Consider a structure $\mathbf{N} = \langle U, \wp(U), R\rangle$, with $R \subseteq U \times \wp(U)$. It can be considered a concrete instance of a neighborhood system. If $u' \in N \in R(u)$, we say that u' is a *neighbor* and N a *neighborhood* of u. We call $\mathcal{N}(U) = \{R(u) : u \in U\}$ a *neighborhood system*. Let us define the following operators on $\wp(U)$:

$$\text{(a) } G(X) = \{u : X \in R(u)\}; \text{ (b) } (X) = -G(-X) = \{u : -X \notin R(u)\}.$$

Consider the following conditions on $\mathcal{N}(U)$, for any $x \in U, A, N, N' \subseteq U$:
1. $U \in R(x)$; **0**. $\emptyset \notin R(x)$; **Id**. if $x \in G(A)$ then $G(A) \in R(x)$;
N1. $x \in N$, for all $N \in R(x)$; **N2**. if $N \in R(x)$ and $N \subseteq N'$, then $N' \in R(x)$;
N3. if $N, N' \in R(x)$, then $N \cap N' \in R(x)$.

They induce the following properties of the operators G and F:

Condition	Equivalent properties of G	Equivalent properties of F
1	$G(U) = U$	$F(\emptyset) = \emptyset$
0	$G(\emptyset) = \emptyset$	$F(U) = U$
Id	$G(X) \subseteq G(G(X))$	$F(F(X)) \subseteq F(X)$
N1	$G(X) \subseteq X$	$X \subseteq F(X)$
	$X \subseteq Y \Rightarrow G(X) \subseteq G(Y)$	$X \subseteq Y \Rightarrow F(X) \subseteq F(Y)$
N2	$G(X \cap Y) \subseteq G(X) \cap G(Y)$	$F(X \cup Y) \supseteq F(X) \cup F(Y)$
N3	$G(X \cap Y) \supseteq G(X) \cap G(Y)$	$F(X \cup Y) \subseteq F(X) \cup F(Y)$

But **N** is a property system, too. So it is possible to define int and cl. One can prove that $int = G$ and $cl = F$ if conditions **Id**, **N1** and **N2** are satisfied. Moreover, if **N** is normal, then **1** and **0** are satisfied, too. Neighborhood systems satisfying these conditions will be classified as $\mathcal{N}_{2Id}$. A topology is a $\mathcal{N}_{2Id}$ neighborhood system which fulfills **N3** in addition.

Now, we compare formal and concrete pre-topological spaces by exploiting the *formal semi-cover* relation $\blacktriangleleft$ introduced in [20]. Let $b \in M$ and $Y, Y' \subseteq M$:

$$(basis)\ b \blacktriangleleft Y \text{ iff } b \in \mathcal{A}(Y), \qquad (step)\ Y \blacktriangleleft Y' \text{ iff } \forall y \in Y, y \blacktriangleleft Y'.$$

Moreover, we assume M to be a monoid with a binary operation "$\cdot$" and unity 1. The operation "$\cdot$" is a formal counterpart of intersection. Then we put for $X, Y \subseteq M$:

$$\text{(a) } X \cdot Y = \{x \cdot y : x \in X \;\&\; y \in Y\}; \quad \text{(b) } X \bullet Y = \mathcal{A}(X \cdot Y). \tag{9}$$

Let us put $Sat_{\mathcal{A}}(M) = \{X \subseteq M : \mathcal{A}(X) = X\}$ and let $\perp$ be any subset of M. Then we say that a pre-topological formal system $\langle M, \cdot, 1, \perp, \blacktriangleleft\rangle$ is topological if $\langle Sat_{\mathcal{A}}(M), \bullet, \vee, M, \mathcal{A}(\perp)\rangle$ is a complete lattice with complete distributivity and ordering $\subseteq$. It can be proved that a pre-topological formal system is topological if the following (*left*) and (*right*) properties hold:

$$(left)\ \frac{b \blacktriangleleft Y}{b \cdot b' \blacktriangleleft Y}; \quad (right)\ \frac{b \blacktriangleleft Y \quad b \blacktriangleleft Y'}{b \blacktriangleleft Y \cdot Y'}.$$

Since **N** is a property system, we can define the operation $\mathcal{A}$ on $\wp(U)$ obtaining the pre-topological system, $\tau = \langle \wp(U), \cap, U, \emptyset, \blacktriangleleft\rangle$ which is in between a formal and a concrete system. Thus, the question is (see [16], Chap. 14.2): given a relational structure $\langle U, \wp(U), R\rangle$, is there any connection between the properties **1**, **0**, **Id**, **N1**, **N2**, **N3** and **N4**, of $\mathcal{N}(U)$ and the properties (*left*) and (*right*) of τ? There are two answers, for the present: (A): If $\mathcal{N}(U)$ fulfills **N3**, then (*right*) holds in τ (the converse does not hold). (B) $\mathcal{N}(U)$ fulfills **N2** if and only if (*left*) holds.

At this point a couple of issues arises:

ISSUE H. G, F, $\blacktriangleleft$, int AND cl IN PARTNERSHIP: *Since* $G = int$ *and* $F = cl$ when conditions **Id**, **N1** and **N2** hold in $\mathcal{N}(U)$, the above results provide us with a glimpse of the relationships between those operators and the properties of $\blacktriangleleft$. However, it is worthwhile a thorough investigation of the connections which link the couples of "concrete" operators (int, cl) and (G, F) and the abstract relation $\blacktriangleleft$, along with their informational interpretations.

ISSUE I. *$Sat_{\mathcal{A}}(M)$ can be made into a logical model in which $\blacktriangleleft$ plays the role of the sequent relation $\vdash$. It happens that in such a system, $(right)$ corresponds to the* contraction rule *and* (*left*) *to the* weakening rule (cf. [20, 21]). *Thus an amazing task would be to put this fact and those of Issue F and Observation 5 into a sound comprehensive picture.*

If one collects different observations about the same sets of objects and properties, a composite system $\langle U, M, \{R_i\}_{1\leq i\leq n}\rangle$ is obtained. If $M = U$, we call it a *Dynamic System* (see [16], Chap. 12). In this case, one can ask what are the lower and the upper approximations of a subset of U according to a certain number of observations. Let us then set the following operators $\varkappa^m$ and ε^m, for $1 \leq m \leq n$:

1. (*Contraction*): We say that $x \in \varkappa^m(A)$, if $R_i(x) \subseteq A$ for at least m indices.
2. (*Expansion*): We say that $x \in \varepsilon^m(A)$, if $R_i(x) \cap A \neq \emptyset$ for at least $n + 1 - m$ indices.

In [16], Chap. 12.6, it is explained how to compute these operators. However an issue arises about them:

ISSUE J: ALGEBRAIC AND TOPOLOGICAL PROPERTIES OF DYNAMIC OPERATORS: *What are the topological and algebraic properties of $\varkappa^m$ and ε^m? Do graded operators int^m, cl^m, G^m and F^m make any sense?*

A few results are available, and just for simple cases (namely if all $R \in \{R_i\}_{1\leq i\leq n}$ are preorders, then $\varkappa^1$ is a pretopological interior operator and ε^n is a pretopological closure operator—see [16], Chap. 12). This topic is connected with multiple source approximation spaces (see [9]).

5 Conclusions

The above connections are not exhaustive. Rough Set Theory is productive of new unexpected intersections and partnerships with surprising fields. We just mention that it suggested a semantic interpretation of the *Logic of conjectures and assertions* (see G. Bellin's page *profs.sci.univr.it/~bellin/papers.html*) and a tool for *spatial reasoning* (see the works by T. Bittner and J. S. Stell at *www.comp.leeds.ac.uk/jsg* and the works on spatial reasoning by I. Düntsch and E. Orłowska). Eventually, a lot of work is still required to understand the logico-algebraic properties of approximations of relations (for some results in simple cases, see [16], Chap. 15.18).

References

1. Cattaneo, G.: Rough Set Theory II. The Approximation Space Approach. with Applications to Roughness and Fuzzy Set Theory. http://www.fislab.disco.unimib.it/dl.php?file=/lib/exe/fetch.php/book/
2. Cattaneo, G., Ciucci, D.: Lattices with Interior and Closure Operators and Abstract Approximation Spaces. Transactions on Rough Sets X. Springer (2009)
3. Crown, G.D., Harding, J., Janowitz, M.F.: Boolean products of lattices. Order **13**, 175–205 (1996)
4. Gabbay, D.M.: Labelled Deductive Systems. Oxford University Press (1997)
5. Greco, S., Matarazzo, B., Slowinski, R.: Rough sets theory for multi-criteria decision analysis. Euro. J. Op. Res. **129**(1), 1–47 (2001)
6. Huang, A., Zhu, W.: Topological characterizations for three covering approximation operators. In: Rough Sets, Fuzzy Sets, Data Mining and Granular Computing, LNCS, 8170, pp. 277–284, Springer (2013)
7. Järvinen, J., Pagliani, P., Radeleczki, S.: Information completeness in Nelson algebras of rough sets induced by quasiorders. Studia Logica (2012). doi:10.1007/s11225-012-9421-z
8. Johnstone, P.T.: Conditions related to De Morgan's law. In: Fourman, M.P., Mulvey C.J., Scott, D.S., (eds.), Applications of Sheaves (Durham 1977), LNM 753, pp. 479–491, Springer-Verlag (1979)
9. Khan, M.A., Banerjee, M.: A Study of Multiple-Source Approximation Systems. Transactions on rough sets. XII. Springer (2010)
10. Lawvere, F.W.: Intrinsic co-heyting boundaries and the Leibniz rule in certain toposes. In: Carboni, A., Pedicchio, M.C., Rosolini, G. (eds.), Category Theory (Como 1990), LNM 1488, pp. 279–297, Springer (1991)
11. Miglioli, P.A., Moscato, U., Ornaghi, M., Usberti, U.: A constructivism based on classical truth. Notre Dame J. Formal Logic **30**(1), 67–90 (1989)
12. Mundici, D.: The C^*-algebras of three-valued logic. In: Ferro, R., et al. (eds.) Logic Colloquium'88, pp. 61–67. North-Holland, Amstedam (1989)
13. Mundici, D.: Ulam's games, Łukasiewicz logic and AF C^*-algebras. Fundamenta Informaticae **18**, 151–161 (1993)
14. Nagarajan, E.K.R., Umadevi, D.: A method of representing rough sets system determined by quasi orders. Order **30**(1), 313–337 (2013)
15. Orłowska, E.: Kripke models with relative accessibility and their applications to inferences from incomplete information. In: Mirkowska, G., Rasiowa, H. (eds.), Mathematical Problems in Computation Theory, p. 329–339, Banach Center Publications, 21 PWN—Polish Scientific Publishers, Warsaw (1988)
16. Pagliani, P., Chakraborty, M.K.: A geometry of Approximation. Trends in Logic, 27. Springer (2008)
17. V. de Paiva, V.: A dialectica-like model of linear logic. In: Proceedings of the Conference on Category Theory and Computer Science, LNCS, 389, pp. 341–356. Springer (1989)
18. Pratt, V.: Chu spaces as a semantic bridge between linear logic and mathematics. Theor. Comput. Sci. **294**, 439–471 (1998)
19. Reyes, G.E., Zolfaghari, N.: Bi-Heyting algebras, toposes and modalities. J. Philos. Logic **25**, 25–43 (1996)
20. Sambin, G.: Intuitionistic formal spaces—a first communication. In: Skordev, D. (ed.) Mathematical Logic and Its Applications, pp. 187–204. Plenum, New York (1987)
21. Sambin, G.: The semantics of pretopologies. In: Schroeder-Heister, P., Dosen, K. (eds.) Substructural Logics, pp. 293–307. Clarendon, Oxford (1993)

22. Sambin, G., Gebellato, S.: A preview of the basic picture: a new perspective on formal topology. In: Proceedings of the TYPES '98. International Workshop on Types for Proofs and Programs, LNCS 1657, pp. 194–207, Springer (1999)
23. Sen, J., Chakraborty, M.K.: A study of interconnections between rough and 3-valued Łukasiewicz logics. Fundamenta Informaticae **51**, 311–324 (2002)
24. Spinks, M., Veroff, R.: Constructive Logic with Strong Negation is a Substructural Logic—I and II. Studia Logica, 88, pp. 325–348 and 89, 401–425 (2008)
25. Wansing, H.: The Logic of Information Structures. Springer, Berlin (1993)

Dealing with Uncertainty: From Rough Sets to Interactive Rough-Granular Computing

Andrzej Jankowski, Andrzej Skowron and Roman Swiniarski

If you thought that science was certain-well, that is just an error on your part.

—Richard P. Feynman,
The Nobel Prize in Physics (1965)

Abstract We discuss an approach for dealing with uncertainty in complex intelligent systems. The approach is based on interactive computations over complex objects called here complex granules (c-granules, for short). C-granules are defined relative to a given agent. Any c-granule of a given agent specifies a perceived structure of local environment of physical objects, called hunks. There are three kinds of such hunks: (i) hunks in the agent external environment creating the hard_suit of c-granule, (ii) internal hunks of agent, creating the soft_suit of c-granule, some of which can be represented by agent as infogranules, and (iii) hunks creating the link_suit of c-granule and playing the role of links between hunks from the hard_suit and soft_suit. This structure is used for recording by means of infogranules the results of interactions of hunks from the local environment. We begin from the discussion on dealing with uncertainty in the rough set approach and next we move toward interactive computations on c-granules. In particular, from our considerations

The original version of this chapter was revised: The grant numbers approved by Polish National Science Centre (NCN), mentioned in the acknowledgement section of this chapter has been corrected. The correction to this chapter is available at https://doi.org/10.1007/978-81-322-2301-6_26

R. Swiniarski: Deceased

A. Jankowski
The Dziubanski Foundation of Knowledge Technology, Nowogrodzka 31, 00-511 Warsaw, Poland
e-mail: Andrzej.AdgaM@gmail.com

A. Skowron (✉)
Institute of Mathematics, Warsaw University, Banacha 2, 02-097 Warsaw, Poland
e-mail: skowron@mimuw.edu.pl

A. Skowron
Systems Research Institute PAS, Newelska 6, 01-447 Warsaw, Poland

M.K. Chakraborty et al. (eds.), *Facets of Uncertainties and Applications*, Springer Proceedings in Mathematics & Statistics 125,
https://doi.org/10.1007/978-81-322-2301-6_2

it follows that the fundamental issues of intelligent systems based on interactive computations concern the efficiency management in controlling of computations performed by such systems. Our approach is a step toward realization of the Wisdom Technology (WisTech) program. The approach was developed over years of work, based on the work on different real-life projects.

Keywords Information granule · Physical object · Interaction · Complex granule · Granular computing · Rough set · Complex vague concept approximation · Adaptive judgment · Efficiency management

1 Introduction

There are quite many well-known different approaches for dealing with uncertainty (e.g., [13, 16, 17, 23, 24, 43, 44]). We emphasize some basic issues related to uncertainty in: (i) object perception, (ii) concept perception as well as (iii) reasoning about concepts. In real-life applications, the objects and concepts we are dealing with are complex. Moreover, they are often vague what causes additional problems in coping with them.

We start from the rough set approach proposed by Professor Pawlak [23, 24, 27] as a tool for dealing with imperfect knowledge, in particular with vague concepts. Rough set theory has attracted the attention of many researchers and practitioners all over the world. We discuss uncertainty issues in object and concept perception in the rough set framework.

Granular Computing (GC) is now an active area of research [29]. Objects we are dealing with in GC are *information granules* (or *infogranules*, for short). Such granules are obtained as the result of information granulation [47]:

> Information granulation can be viewed as a human way of achieving data compression and it plays a key role in implementation of the strategy of divide-and-conquer in human problem-solving.

The concept of granulation is rooted in the concept of a linguistic variable introduced by Professor Lotfi Zadeh in 1973. Information granules are constructed starting from some elementary ones. More compound granules are composed of finer granules that are drawn together by distinguishability, similarity, and functionality [45].

Understanding of interactions of objects on which are performed computations is fundamental for modeling of complex systems [3]. For example, in [21] this is expressed in the following way:

> [...] interaction is a critical issue in the understanding of complex systems of any sorts: as such, it has emerged in several well-established scientific areas other than computer science, like biology, physics, social and organizational sciences.

When we move to dealing with perception of interacting complex objects in observed situations one should consider that due to resource bounds only some parts of complex objects may be perceived at a given moment of time. These parts are perceived as values of compound attributes computed on the basis of the delivered (e.g., by control of the agent) parameters of sensors and recorded in relevant information (decision) systems as the results of sensory measurements. Hence, uncertainty in identification of the environment state often causes that results of interactions with and within the environment cannot be predicted with certainty. As a consequence, e.g., results of performed actions may be different than the predicted ones.

In this paper, we outline an extension of Interactive Rough-Granular Computing (IRGC) approach (see, e.g., [29, 38, 41, 42]) by introducing *complex granules* (*c-granules*, for short) making it possible to model interactive computations performed by an agent. In such computations, interactions among physical objects and interactions of these physical objects with information granules possessed by the agent are represented.

In IRGC, the rough set approach in combination with other soft computing approaches are used for inducing approximations of complex vague concepts.

Different problems related to dealing with uncertainty in IRGC are outlined in the paper.

Let us mention here that our discussion on IRGC based on c-granules is strongly related to the following sentences:

> As far as the laws of mathematics refer to reality, they are not certain; and as far as they are certain, they do not refer to reality. (Albert Einstein, [2])

> Constructing the physical part of the theory and unifying it with the mathematical part should be considered as one of the main goals of statistical learning theory. (Vladimir Vapnik, [43] p. 721)

This paper covers some issues presented in the invited talk at ICFUA 2013.

In Sect. 2, we discuss some basic problems related to dealing with uncertainty in the rough set approach. Section 3 outlines the approach to IRGC based on c-granules and reports some issues concerning uncertainty in IRGC. In particular, due to uncertainty, e.g., in identification of the global environment state, development of the efficiency management techniques for controlling by agent computations performed over c-granules for achieving goals is crucial for intelligent systems based on IRGC.

2 Rough Sets and Uncertainty

2.1 Uncertainty in Object Perception

The rough set philosophy [23, 24, 27] is founded on the assumption that with every object of the universe of discourse, we associate some information (data, knowledge) called the object signature. Objects characterized by the same information are indiscernible (similar) in view of the available information about them. The *indiscernibility relation* generated in this way is the mathematical basis of rough set theory. This understanding of indiscernibility is related to the idea of Gottfried Wilhelm Leibniz that objects are indiscernible if and only if all available functionals take

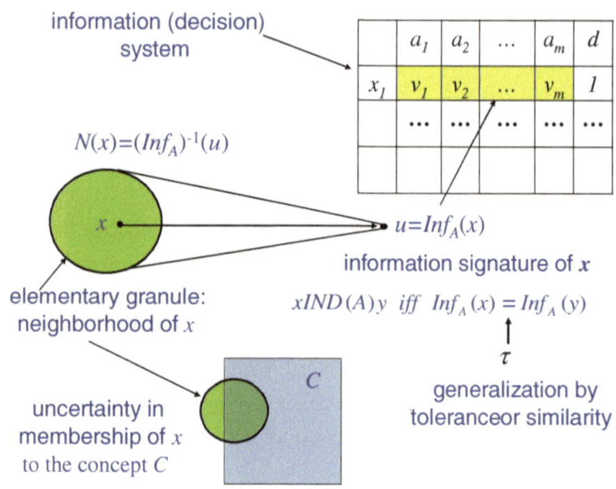

Fig. 1 Elementary granules in rough sets defined by signatures of objects and their partial inclusion in concepts (sets)

on them identical values [15]. However, in the rough set approach, indiscernibility is defined relative to a given set of functionals (attributes).

Any set of all indiscernible (similar) objects is called an *elementary granule*, and forms a basic granule (atom) of knowledge about the universe. In Fig. 1, we illustrate the elementary granules defined by the *indiscernibility relation* $IND(A)$ on the basis of the *object signatures* $Inf_A(x)$, where

$$Inf_A(x) = \{(a, a(x)) : a \in A\} \tag{1}$$

for $x \in U$ [27].

If, e.g., the results of measurements are uncertain, one should consider more compound elementary granules (see, e.g., Fig. 2).

Many research papers on rough sets are dedicated to different issues related to *uncertainty of object perception* [39], e.g., to uncertainty caused by: (i) missing attribute values, (ii) imperfect measurement of attribute values, (iii) noise, (iv) unknown relevant structure or context of objects in hierarchical learning, and (v) unknown relevant attributes for approximation (feature selection and constructive induction problems). All the above aspects concerning uncertainty in object

Fig. 2 Examples of elementary granules with centers and different uncertainty in membership measurement: binary case, monotonic discrete multivalued case, monotonic continuous fuzzy case, non-monotonic discrete case

perception have an impact on definitions of indiscernibility (discernibility) relations, elementary granule, granules constructed from them, and approximations of concepts.

2.2 Uncertainty in Concept Perception

In this section, we start from an explanation how uncertainty in object perception influences the concept perception. Next, we consider the impact of imperfect information about concepts of their perception.

Any union of elementary granules is referred to as *crisp* (precise) set. If a set is not crisp then it is called *rough* (imprecise, vague).

Note that due to the computational complexity of searching for relevant crisp sets in solving problems related to concept approximation, the searching is usually restricted to a feasible subfamily of the family of all possible unions of elementary sets, e.g., consisting of conjunctions of descriptors only [25, 27].

Each rough set has *borderline cases*, i.e., objects which cannot be classified with certainty as members of either the set or its complement. Obviously, crisp sets have no borderline elements at all. This means that borderline cases cannot be properly classified by employing available knowledge.

Thus, the assumption that objects can be "seen" only through the information available about them leads to the view that knowledge has granular structure. Due to the granularity of knowledge, some objects of interest cannot be discerned and appear as the same (or similar). As a consequence, vague concepts in contrast to precise concepts, cannot be characterized in terms of information about their elements. Therefore, in the proposed approach, we assume that any vague concept is replaced by a pair of precise concepts—called the *lower and the upper approximation of the vague concept*. The lower approximation consists of all objects which definitely belong to the concept and the upper approximation contains all objects which possibly belong to the concept. The difference between the upper and the lower approximation constitutes the *boundary region* of the vague concept. Approximation operations are two basic operations in rough set theory. Hence, rough set theory expresses vagueness not by means of membership, but by employing a boundary region of a set. If the boundary region of a set is empty it means that the set is crisp, otherwise the set is rough (inexact). A nonempty boundary region of a set means that our knowledge about the set is not sufficient to define the set precisely.

In the literature, one can find more details on different aspects of rough set approximations of *vague concepts*. In particular, discussion on vague (imprecise) concepts in philosophy includes the following characteristic features of them [12]: (i) the presence of borderline cases, (ii) boundary regions of vague concepts are not crisp, and (iii) vague concepts are susceptible to sorites paradoxes. The rough set approach is consistent with this view. For example, one should consider that the set of attributes and the set of objects and/or attributes are changing. Hence, the boundary region is drifting and it is only possible to construct temporary crisp definitions of boundary region.

The original approach by Pawlak [23, 24, 27] was based on indiscernibility defined by equivalence relations. Any such indiscernibility relation defines a partition of the universe of objects. Over the years many generalizations of this approach were introduced and the most of them are based on coverings rather than partitions. In particular, one can consider similarity (tolerance)-based rough set approach, binary relation-based rough sets, neighborhood and covering rough sets, dominance-based rough set approach, hybridization of rough sets and fuzzy sets, and many others [26, 39]. One should note that dealing with coverings requires solving several new algorithmic problems such as selection of family of definable sets or resolving problems with selection of relevant definition of approximation of sets among many possible ones. For a given problem (e.g., classification problem), it is necessary to discover the relevant covering (or partial covering) for the target classification task. In the literature, there are numerous papers dedicated to theoretical aspects of the *covering rough set approach*. However, still more work should be done on algorithmic problems concerning discovery of the relevant covering.

Another issue investigated in the rough set approach concerns (rough) *inclusion measures* [26]. In particular, approximation spaces with rough inclusion measures have been investigated [26, 35]. This approach was further extended to rough mereological approach [30, 31]. More general cases of approximation spaces with rough inclusion were also discussed in the literature including approximation spaces in GC [38]. It is worthwhile mentioning here the approach for ontology approximation used in hierarchical learning of complex vague concepts (see, e.g., [1, 39]). Different rough inclusion measures and based on them quality measures are used for inducing from dataset decision rules, dependencies of attributes, concept description, clusters, or classifiers. They are based, e.g., on the positive region, different kinds of entropy or relative entropy. In the case of rough set-based classifiers, often are used different versions of the *minimum length description principle* [32, 33]. In searching for the high quality classifiers, quality measures aggregating two components are used. The first one is related to the data model quality and the second one to the model description length. The aggregation of such uncertainty measures is optimized for obtaining the high quality classifiers [27, 39]. Let us also note that many known similarity indices can be defined by rough inclusion measures [4].

Due to uncertainty in perception of concepts, the rough set approach is used for developing methods for inducing approximations of concepts in the form of classifiers or clusters. This direction is strongly related to *inductive reasoning* and also to more general reasoning called *adaptive judgment* [6–9, 11]. The general idea is as follows. From a given decision table, a set of granules in the form of decision rules is induced together with arguments *for* and *against* for each decision rule and decision class. For any new object with known signature, one can select rules matching this object. Note that the left-hand sides of decision rules are described by formulas making it possible to check for new objects if they satisfy them assuming that the signatures of these objects are known. In this way, one can consider two semantics of formulas: on a sample of objects U and on its extension $U^* \supseteq U$. Definitely, one should consider a risk of such generalization in the decision rule inducing. Next, a conflict resolution should be applied for resolving conflicts between rules matching the new object and voting for different decisions. The whole procedure can be generalized for the case of approximation of more compound information granules than concepts.

It is worthwhile mentioning that in the rough set approach were also developed approaches for inducing approximate reasoning schemes [36, 39].

3 IRGC and Uncertainty

Solving under uncertainty problems concerning interactive computations require to consider issues such as [6]: (i) changing attention in time relative to parts of complex objects which cannot be perceived as the whole at a given moment of time, (ii) values of compound attributes are computed on the basis of the delivered (e.g., by the agent control) parameters of sensors and recorded (in relevant information systems) results of sensory measurements, and (iii) interaction with the environment may cause different results of actions than the predicted ones.

3.1 Complex Granules and Computations Over Complex Granules

Any c-granule of a given agent specifies a perceived structure of local environment of portions of matter (physical objects), called hunks [5]. There are three kinds of such hunks: (i) hunks in the agent external environment creating the hard_suit of c-granule, (ii) internal hunks of agent, creating the soft_suit of c-granule, some of which can be represented by agent as *information granules* (*infogranules*, for short), and (iii) hunks creating the link_suit of c-granule and playing the role of links between hunks from the hard_suit and soft_suit. This structure is used in recording by means of infogranules, the results of interactions of hunks from the local environment [6, 10].

Any atomic infogranule g of c-granule can be treated as a hunk h_g with states encoded by objects such as numbers or words. Objects encoding the states of h_g are possible values of g (or h_g). More formally, one can treat h_g as a collection of hunks consisting of values of h_g. In the interaction of h_g with the local environment of c-granule, the hunk encoding the relevant state is selected from h_g. More compound infogranules are obtained by relevant aggregation of already defined infogranules. This is a generalization of a notion of infogranules considered e.g., in [29, 36], where the values are assumed to be given while here they are obtained as the result of interaction processes in the local environment of c-granule.

One can distinguish two functionalities of each c-granule.

The first one, corresponding to the c-granule syntax frame, consists of the specification of the local environment structure. This structure, defined in the soft_suit part of c-granule and represented by infogranules, consists of representations of hunks and links, defined in link_suit part of c-granule, between the representations of hunks and corresponding to them parts of the physical world—creating the hard_suit of c-granule. Roughly speaking, the structure of the local environment of c-granule describes structure of glasses of c-granule through which interactions of physical objects in the local environment of c-granule may be perceived. We assume that any c-granule has the ability to check if its local environment has the required structure.

The second one, corresponding to the c-granule semantics, is making it possible to record properties of processes such as degrees of satisfiability of features or values of compound attributes. The processes are running in the perceived local environment with the structure of interacting hunks predefined for the c-granule by agent. Roughly speaking, the second part of c-granule is making it possible to record and process the relevant results of the perceived interactions, observed through the glasses of c-granule, in the local environment and next to provide them to other c-granules.

The above two functionalities of c-granule are making it possible to perceive by c-granule of interactions in its local environment of hunks.

We also assume that to any c-granule may be assigned pair(s) of the form (pre-condition, post-condition). It is assumed that if c-granule is perceiving a structure satisfying the precondition (including interaction initiation), then the results of interactions of physical objects (perceived by means of properties of hunks from the local environment of c-granule) recorded by infogranules from the c-granule are expected (by control of agent) to satisfy the post-condition of the c-granule. One should remember that due to uncertainty, e.g., unpredictable interactions in the environment, the real results of interactions (which may be recorded only a posteriori) may differ from the expected ones. Roughly speaking, the pre-condition and the post-condition of c-granule are used to describe the expected changes of properties of the local environment of c-granule caused by interactions in this environment. The changes may be interpreted as the result of performing an action realized by c-granule. Actions (sensors or plans) represented by an agent in link_suits of c-granules are used by the agent for exploration and/or exploitation of the environment on the way to achieve the targets. Again, due to the bounds of the agent perception abilities, usually only a partial information about the interactions in the physical world may be available for agents. Hence, in particular, the results of performed actions by agents cannot be predicted with certainty. This causes the necessity of adaptation of preconditions and post-conditions by c-granule or agents.

The above-described extension of c-granules by preconditions and post-conditions may be interpreted as a delegation by an agent of some control functionalities to c-granules and may take more advanced forms. For example, more compound c-granules may have more pairs (precondition, post-condition) which lead to a possibility of realization by c-granule packages or plans of actions together with some autonomy embedded by agent into c-granules concerning some control functionalities, e.g., in the process of selection-relevant actions for realization.

Any agent operates in its local world of c-granules by generating (or elimination) some c-granules and measuring the results of perceived interactions. The agent is

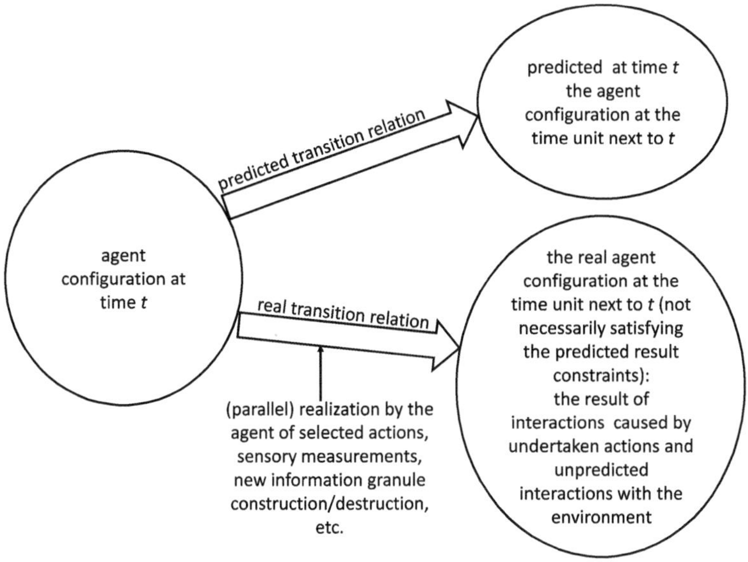

Fig. 3 Two transition relations

aiming to control computations performed by c-granules from this local world for achieving the target goals.

The *transition relation* is usually defined between configurations of agent ag at succeeding moments of time. Any configuration of ag at time t (relative to the time clock of ag) consists of all c-granules being at time t under the agent control. It is worthwhile mentioning that the configuration at the time next to t cannot be defined *a priori* (at time t). Due to uncertainty, in particular, unpredictable interactions with the environment, the agent ag can only predict the next configuration and the real one, resulting due to interactions, can be perceived by ag at time $t' > t$ when the results of interactions can be perceived by the agent using the relevant c-granules. Hence, we obtain two transition relations: the *predicted transition relation* and the *real transition relation* (see Fig. 3).

Note that the introduced model of interactive computations based on c-granules differs from the Turing model of computations. The results of computations based on c-granules depend on interactions of physical objects and linked to them information granules (also represented by means of physical objects).

Agents and societies of agents may also be represented as (generalized) c-granules. For more details on IRGC based on c-granules also on the agent architecture as well as on societies of agents and communication languages, the reader is referred to [6].

3.2 Agent Interactions and Communication Languages

Languages of agents consist of special c-granules called *(agent) expressions*. A soft_suit of expression (treated as a c-granule) includes an infogranule corresponding to syntax of the expression as well as the specification of the local environment for the expression. The expression value (e.g., the satisfiability degree when expression is a formula) is computed using the functionalities of the expression (treated as a c-granule) concerning perception of interactions of hunks in its local environment. Note that some hunks of expression may belong to its soft_suit, e.g., may belong to the agent memory or its "brain".

The agents can crete new names or expressions, e.g., for new structured objects or their indiscernibility (similarity) classes. Expressions from languages of agents consist of partial descriptions of situations (or their indiscernibility or similarity classes) perceived by agents using c-granules as well as description of approximate reasoning schemes about the situations and their changes caused by actions and/or plans. The situations may be represented in hierarchical modeling by structured objects (e.g., relational structures over attribute value vectors and/or indiscernibility (similarity classes) of such structures [40]).

From the point of view of dealing with uncertainty, it is important to observe that any expression usually represents classes of hunks [5] rather than a single hunk. This follows from the fact that the agents have bounded abilities on discerning of perceived objects. Also more compound expressions, e.g., representing different behavioral patterns may be indiscernible relative to the set of attributes used by the agent. Hence, it follows that the agents perceive in the same way objects belonging to the same indiscernibility and/or similarity class. This is an important feature allowing agents to use generalization. For example, a new, unseen so far, situation may be matched and classified to the perceived indiscernibility classes what allows agents to use strategies of generalization.

In reasoning about the situation changes [37], one should take into account that the predicted actions and/or plans may depend not only on the changes recognized in the past situations but also on the performed actions and plans in the past. This is strongly related to the idea of perception pointed out in [19]:

> The main idea of this book is that perceiving is a way of acting. It is something we do. Think of a blind person tap-tapping his or her way around a cluttered space, perceiving that space by touch, not all at once, but through time, by skillful probing and movement. This is or ought to be, our paradigm of what perceiving is.

Many challenging issues are related to the origin and evolution of communication languages of agents (see, e.g., [20]). Here, we present only a few preliminary comments on these issues. We assume that the agents can perceive behavioral patterns of other agents or their groups, and based on this they can try to exchange some messages [18]. It is worthwhile mentioning that at the beginning, agents do not have common understanding of the meaning of such messages. In the consequence, this leads to misunderstanding, uncomfortable situation for agents. However, after series of trials in a dialogues they have a chance to set up common meaning of some behavioral patterns. In other words, they start to create common c-granules which use fixed in dialogues links to other hunks or infogranules. For example, at the beginning the messages could be linked to warning situations or to identifications of some sources required for satisfiability of some agent needs. This kind of simple messages could be passed by very simple behavioral pattern. Next, based on these very simple behavioral patterns the agents can develop more compound messages related to c-granules corresponding to common plans of cooperation of group of agents or/and competition with other groups of agents. This very general framework could be implemented in many ways using different AI paradigms. Especially, many models from Natural Computing could be quite helpful (e.g., modification of cellular automata or evolutionary programming). However, our proposal is to implement this general scheme by agents (using c-granules) built up on the hierarchies of interactive information (decision) systems linked to configurations of hunks. The approach based on rough sets is quite convenient for implementation by computers well prepared for manipulation of data tables.

Let us consider a simple example illustrating how the names may originate in the environment where agents are interacting. Let us assume that an agent possesses a metaphorically understood "brain" with the states represented by configurations of hunks. The brain of agent *ag* is involved in interaction processes (IP) with the local environment and is perceiving a hunk h in this environment. In effect, the brain of agent *ag* launches an interaction process IP' with the environment. IP' introduces a hunk h' which is an image of h and constitutes a compressed form of h. The hunk h' can be considered as a name for the hunk h in language of agent *ag*. The frequent perception by another agent of hunk structures constituting of co-occurrence of agent *ag* and hunks h, h' may lead to accepting h' as the name for h by this agent (see Fig. 4).

The perception by an agent of hunk structures constituting expressions in its language and their aggregations leads to the creation of grammar rules. The agent learns the usage rules of its language through interaction with the environment.

4 Adaptive Judgment: Toward Efficiency Management in Interactive Computations Over Complex Granules

The reasoning making it possible to derive relevant c-granules used for obtaining solutions of the target tasks is called *adaptive judgment*. *Intuitive judgment* and *rational judgment* are distinguished as different kinds of judgment [11]. Deduction and induction as well as abduction or analogy-based reasoning as well as reasoning for efficiency management are involved in adaptive judgment. Among the tasks for adaptive judgment are the following ones supporting reasoning under uncertainty toward: searching for relevant approximation spaces, discovery of new features, selection of relevant features, rule induction, discovery of inclusion measures, strategies for conflict resolution, adaptation of measures based on the minimum description length principle, reasoning about changes, perception (action and sensory) attributes selection, adaptation of quality measures over computations relative to agents, adaptation of object structures, discovery of relevant context, strategies for knowledge representation and interaction with knowledge bases, ontology acquisition and approximation, learning in dialogue of inclusion measures between information granules from different languages (e.g., the formal language of the system and the user natural language), strategies for adaptation of existing models, strategies for development and evolution of communication language among agents in distributed environments, strategies for efficiency management, e.g., risk management in distributed computational systems. Definitely, in the language used by agents for dealing with adaptive judgment (i.e., intuitive and rational) some deductive systems known from logic may be applied for reasoning about knowledge relative to closed worlds. This may happen, e.g., if the agent languages are based on classical mathematical logic. However, if we move to interactions in open worlds then new specific rules or patterns relative to a given agent or group of agents in such worlds should be discovered. The process of inducing such rules or patterns is influenced by uncertainty because they are induced by agents under uncertain and/or imperfect knowledge about the environment. Hence, considering only the absolute truth becomes unsatisfactory.

It is worthwhile mentioning that we propose to base adaptive judgment about interactive computations on complex granules not only on risk management (in particular, on risk assessment) but on a more general approach based on efficiency management using properly adopted well-known techniques such as SWOT analysis, Cost–Benefit Analysis (CBA) and others [6]. In the efficiency analysis, one should consider a variety of complex vague concepts and relations between them

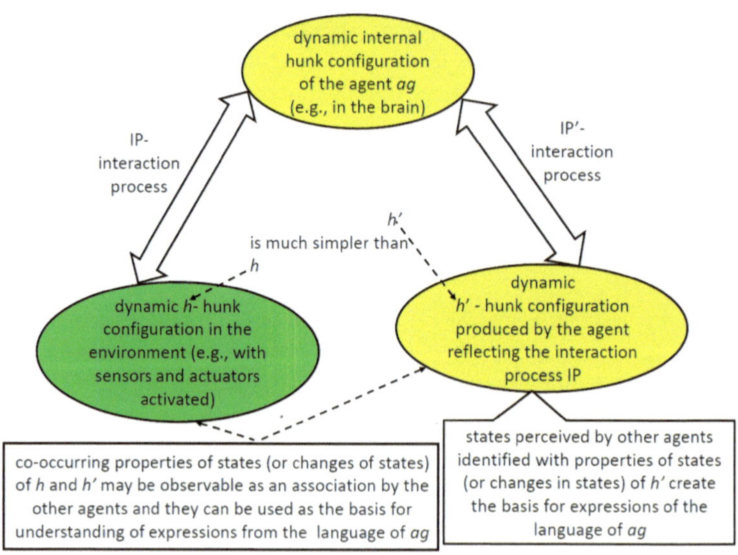

Fig. 4 Creating names

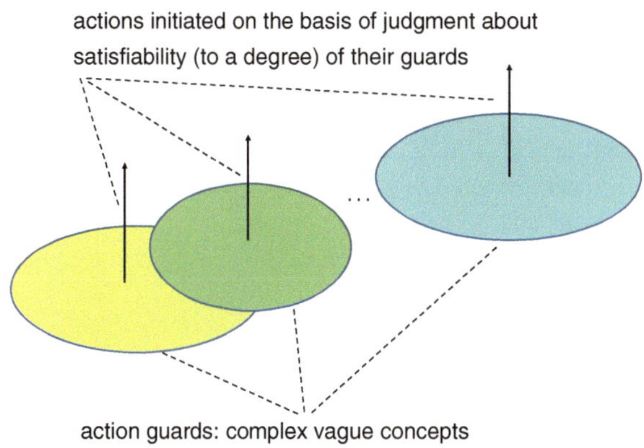

Fig. 5 Games based on complex vague concepts

as well as reasoning schemes over such concepts and relations related, e.g., to the bow-tie diagram well known in the risk management area. To make such objects as vague concepts, relations among them as well as reasoning schemes over vague concepts "understandable" for agent control language, the relevant adaptive approximate methods for such objects should be developed. For example, the ontology approximation methodology was applied successfully in different real-life projects (see [1] and also the references in this paper).

One can consider the above-mentioned tasks of approximation of vague complex concepts initiating actions as the complex game discovery task (see Fig. 5) from data and domain knowledge. The agents are using the discovered games for achieving their targets in the environment. The discovery process can be based on hierarchical learning supported by domain knowledge [1, 6]. Such games are evolving in time (drifting in time) together with data and knowledge about the approximated concepts. The relevant adaptive strategies for adapting the games to changes perceived by agents are required. These adaptive strategies are used to control the behavior of agents toward achieving by them targets. Note that also these strategies should be learned from data and domain knowledge.

One can observe that some of the discussed tasks such as conflict resolution among classifiers initiating actions voting for decisions or efficiency management require an extension beyond the approaches based on ontologies. This extension requires usage of relevant fragments of natural language. In such fragments, one can express reasoning performed by humans based on concepts and relations from a given ontology. The challenge is how artificial agents can learn to perform approximate reasoning consistent to a satisfactory degree with reasoning performed by humans in those fragments of natural language. This challenge is related to the following sentences formulated by Pearl [28]:

> Traditional statistics is strong in devising ways of describing data and inferring distributional parameters from sample. Causal inference requires two additional ingredients:
>
> 1. a science-friendly language for articulating causal knowledge, and

2. a mathematical machinery for processing that knowledge, combining it with data and drawing new causal conclusions about a phenomenon.

The analogous idea was also formulated by Lotfi Zadeh in the framework of computing with words (see, e.g., [46–50]).

Issues related to interactions among objects in the physical and mental worlds as well as adaptive judgment belong to the fundamental issues in Wisdom Technology (WisTech) [6–9] based on the following meta-equation:

$$\text{WISDOM} = \text{INTERACTIONS} + \text{ADAPTIVE JUDGEMENT} + \text{KNOWLEDGE}. \tag{2}$$

5 Conclusions

We discussed some basic issues related to dealing with uncertainty in the rough set approach and in IRGC. The outlined research on the nature of interactive computations is crucial for understanding complex systems. Our approach is based on complex granules (c-granules) on which agents are performing interactive computations. More compound granules represent agents and societies of agents. Computations over c-granules are controlled by the agent control. We emphasized the role of risk management and other techniques from management theory in IRGC. In our research, we plan to further develop the foundations of interactive computations based on c-granules toward tools for modeling and analysis of computations in Natural Computing [34], Wisdom Web of Things [51], or Cyber-Physical Systems [14].

Acknowledgements This work was supported by the Polish National Science Centre (NCN) grants DEC-2011/01/D /ST6/ 06981 and DEC-2012/05/B/ST6/03215 as well as by the Polish National Centre for Research and Development (NCBiR) under the grant O ROB/0010/03/001.

References

1. Bazan, J.: Hierarchical classifiers for complex spatio-temporal concepts. Trans. Rough Sets IX: J. Subline LNCS **5390**, 474–750 (2008)
2. Einstein, A.: Geometrie und Erfahrung (Geometry and Experience). Julius Springer, Berlin (1921)
3. Goldin, D., Smolka, S., Wegner, P. (eds.): Interactive Computation: The New Paradigm. Springer, Berlin (2006)
4. Gomolińska, A., Wolski, M.: Rough inclusion functions and similarity indices. In: Concurrency, Specification and Programming, CS&P 2013. Warszawa, Poland (2013) (to appear)
5. Heller, M.: The Ontology of Physical Objects. Four Dimensional Hunks of Matter. Cambridge Studies in Philosophy, Cambridge University Press (1990)
6. Jankowski, A.: Complex Systems Engineering: Conclusions from Practical Experience. Springer, Heidelberg (2015)

7. Jankowski, A., Skowron, A.: A WisTech paradigm for intelligent systems. Trans. Rough Sets VI: J. Subline 94–132
8. Jankowski, A., Skowron, A.: Logic for artificial intelligence: The Rasiowa—Pawlak school perspective. In: Ehrenfeucht, A., Marek, V., Srebrny, M. (eds.) Andrzej Mostowski and Foundational Studies, pp. 106–143. IOS Press, Amsterdam (2008)
9. Jankowski, A., Skowron, A.: Wisdom technology: a rough-granular approach. In: Marciniak, M., Mykowiecka, A. (eds.) Bolc Festschrift. Lectures Notes in Computer Science, vol. 5070, pp. 3–41. Springer, Heidelberg (2009)
10. Jankowski, A., Skowron, A., Swiniarski, R.: Interactive complex granules. In: Szczuka, M., Czaja, L., Kacprzak, M. (eds.) Proceedings of the 22nd International Workshop on Concurrency, Specification and Programming (CS&P'2013), Warsaw, Poland, 25–27 September 2013, vol. 1032, pp. 206–218 (2013). (CEUR Workshop Proceedings, RWTH Aachen University)
11. Kahneman, D.: Maps of bounded rationality: psychology for behavioral economics. Am. Econ. Rev. **93**, 1449–1475 (2002)
12. Keefe, R.: Theories of Vagueness. Cambridge Studies in Philosophy. Cambridge (2000)
13. Klir, G.J.: Uncertainty and Information: Foundations of Generalized Information Theory. Wiley, Hoboken (2006)
14. Lamnabhi-Lagarrigue, F., Di Benedetto, M.D., Schoitsch, E.: Introduction to the special theme cyber-physical systems. Ercim News **94**, 6–7 (2014)
15. Leibniz, G.W.: Discourse on metaphysics. In: Leibniz, G.W. (ed.) Philosophical Essays, pp. 35–68 (1686)
16. Marti, K., Ermoliev, Y., Makowski, M. (eds.): Coping with Uncertainty: Robust Solutions. Springer, Berlin (2010)
17. Marti, K., Ermoliev, Y., Makowski, M., Pflug, G. (eds.): Coping with Uncertainty: Modeling and Policy Issues. Springer, Berlin (2006)
18. Minsky, M.: Emotion Machine: Commonsense Thinking, Artificial Intelligence, and the Future of the Mind. Simon & Schuster, New York (2006)
19. Noë, A.: Action in Perception. MIT Press (2004)
20. Nolfi, S., Mirolli, M. (eds.): Evolution of Communication and Language in Embodied Agents. Springer (2010)
21. Omicini, A., Ricci, A., Viroli, M.: The multidisciplinary patterns of interaction from sciences to computer science. In: Goldin et al. (eds.) Interactive Computation: The New Paradigm, pp. 395–414. Springer, Berlin (2006)
22. Pal, S.K., Polkowski, L., Skowron, A. (eds.): Rough-Neural Computing: Techniques for Computing with Words. Cognitive Technologies. Springer, Heidelberg (2004)
23. Pawlak, Z.: Rough sets. Int. J. Comput. Inf. Sci. **11**, 341–356 (1982)
24. Pawlak, Z.: Rough Sets: Theoretical Aspects of Reasoning About Data. System Theory, Knowledge Engineering and Problem Solving, vol. 9. Kluwer Academic Publishers, Dordrecht, The Netherlands (1991)
25. Pawlak, Z., Skowron, A.: Rough sets and Boolean reasoning. Inf. Sci. **177**(1), 41–73 (2007)
26. Pawlak, Z., Skowron, A.: Rough sets: some extensions. Inf. Sci. **177**(28–40), 1 (2007)
27. Pawlak, Z., Skowron, A.: Rudiments of rough sets. Inf. Sci. **177**(1), 3–27 (2007)
28. Pearl, J.: Causal inference in statistics: an overview. Stat. Surv. **3**, 96–146 (2009)
29. Pedrycz, W., Skowron, S., Kreinovich, V. (eds.): Handbook of Granular Computing. Wiley, Hoboken (2008)
30. Polkowski, L. (ed.): Approximate Reasoning by Parts. An Introduction to Rough Mereology. Intelligent Systems Reference Library, vol. 20. Springer, Heidelberg (2011)
31. Polkowski, L., Skowron, A.: Rough mereology: a new paradigm for approximate reasoning. Int. J. Approx. Reason. **15**(4), 333–365 (1996)
32. Rissanen, J.: Modeling by shortes data description. Automatica **14**, 465–471 (1978)
33. Rissanen, J.: Minimum-description-length principle. In: Kotz, S., Johnson, N. (eds.) Encyclopedia of Statistical Sciences, pp. 523–527. Wiley, New York (1985)
34. Rozenberg, G., Bäck, T., Kok, J. (eds.): Handbook of Natural Computing. Springer (2012)

35. Skowron, A., Stepaniuk, J.: Tolerance approximation spaces. Fundamenta Informaticae **27**(2–3), 245–253 (1996)
36. Skowron, A., Stepaniuk, J.: Information granules and rough-neural computing. In: Pal et al. (eds.) Rough-Neural Computing: Techniques for Computing with Words. Cognitive Technologies, pp. 43-84. Springer, Heidelberg (2004)
37. Skowron, A., Stepaniuk, J., Jankowski, A., Bazan, J.G., Swiniarski, R.: Rough set based reasoning about changes. Fundamenta Informaticae **119**(3–4), 421–437 (2012)
38. Skowron, A., Stepaniuk, J., Swiniarski, R.: Modeling rough granular computing based on approximation spaces. Inf. Sci. **184**, 20–43 (2012)
39. Skowron, A., Suraj, Z. (eds.): Rough Sets and Intelligent Systems. Professor Zdzislaw Pawlak in Memoriam. Series Intelligent Systems Reference Library. Springer (2013)
40. Skowron, A., Szczuka, M.: Toward interactive computations: a rough-granular approach. In: Koronacki, J., et al. (eds.) Advances in Machine Learning II: Dedicated to the Memory of Professor Ryszard S. Michalski. Studies in Computational Intelligence, vol. 263, pp. 23–42. Springer, Heidelberg (2009)
41. Skowron, A., Wasilewski, P.: Information systems in modeling interactive + computations on granules. Theor. Comput. Sci. **412**(42), 5939–5959 (2011)
42. Skowron, A., Wasilewski, P.: Interactive information systems: toward perception based computing. Theor. Comput. Sci. **454**, 240–260 (2012)
43. Vapnik, V.: Statistical Learning Theory. Wiley, New York (1998)
44. Zadeh, L.A.: Fuzzy sets. Inf. Control **8**, 338–353 (1965)
45. Zadeh, L.A.: Toward a theory of fuzzy information granulation and its centrality in human reasoning and fuzzy logic. Fuzzy Sets Syst. **90**, 111–127 (1997)
46. Zadeh, L.A.: From computing with numbers to computing with words—from manipulation of measurements to manipulation of perceptions. IEEE Trans. Circuits Syst. **45**, 105–119 (1999)
47. Zadeh, L.A.: A new direction in AI: toward a computational theory of perceptions. AI Mag. **22**(1), 73–84 (2001)
48. Zadeh, L.A.: Foreword. In: Pal et al. (eds.) Rough-Neural Computing: Techniques for Computing with Words. Cognitive Technologies, pp. IX-XI. Springer, Heidelberg (2004)
49. Zadeh, L.A.: Generalized theory of uncertainty (GTU)-principal concepts and ideas. Comput. Stat. Data Anal. **51**, 15–46 (2006)
50. Zadeh, L.A. (ed.): Computing with Words: Principal Concepts and Ideas. Studies in Fuzziness and Soft Computing, vol. 277. Springer, Heidelberg (2012)
51. Zhong, N., Ma, J., Huang, R., Liu, J., Yao, Y., Zhang, Y., Chen, J.: Research challenges and perspectives on wisdom web of things (W2T). J. Supercomput. **64**, 862–882 (2013)

An Evolutionary Approach to Secondary Membership Function Selection in Generalized Type-2 Fuzzy Sets

Reshma Kar, Amit Konar, Aruna Chakraborty and Pratyusha Rakshit

Abstract Lack of knowledge about secondary membership function acts as an impediment to using generalized type-2 fuzzy sets in real-world problems. This chapter shows a new direction to compute secondary memberships in the settings of a strategic optimization problem. It employs three strategies to design an optimization objective as a function of secondary memberships and employs differential evolution algorithm to determine secondary memberships as the optimal solution to the optimization problem. The proposed method of secondary membership function evaluation has successfully been applied to an emotion recognition problem.

Keywords Generalized type-2 fuzzy sets · Primary membership · Secondary membership · Differential evolution · Emotion recognition

1 Introduction

Fuzzy sets have widely been used over the last few decades for uncertainty management of real-world systems. Classical fuzzy set, also called type-1 fuzzy set, represents a vague concept by a membership function. For example, the fuzzy concept about "height is tall" is represented by a type-1 membership function which

R. Kar · A. Konar (✉) · P. Rakshit
Department of Electronics and Telecommunication Engineering,
Jadavpur University, Kolkata, India
e-mail: konaramit@yahoo.co.in

R. Kar
e-mail: rkar317@gmail.com

P. Rakshit
e-mail: pratyushar1@gmail.com

A. Chakraborty
Department of Computer Science and Engineering, St. Thomas College
of Engineering and Technology, Kolkata, India
e-mail: aruna_stcet@rediffmail.com

M.K. Chakraborty et al. (eds.), *Facets of Uncertainties and Applications*,
Springer Proceedings in Mathematics & Statistics 125,
DOI 10.1007/978-81-322-2301-6_3

monotonically increases with "height" and saturates at a certain height, say 6 ft. Since a number of monotonically increasing functions of "height" can be constructed to represent the type-1 membership function, identifying the best selection of type-1 membership function to represent the given concept remained an unsolved problem.

One approach to alleviate the above problem is to consider the union of all possible type-1 membership functions describing a concept. This union of type-1 membership function together is often referred to as the Footprint of Uncertainty (FOU) in the nomenclature of Interval Type-2 Fuzzy Set (IT2FS) [1]. An IT2FS thus can better capture the uncertainty of different sources. IT2FS is nowadays widely been used in scientific and engineering systems as a fundamental model to describe the uncertainty of the measurement variables used therein. The type-1 membership function resources used in the construction of FOU carries intrapersonal level uncertainty of a source at each value of the fuzzy linguistic variable, for example, "height," in the context of "height is tall." Moreover, at each distinct position of the linguistic variable, the FOU provides a frame work for interpersonal uncertainty of several sources.

Mendel and his research team took initiatives [2] to describe the degree of the interpersonal uncertainty for each type-1 membership function. The certainty referred to above at a given value of the linguistic variable "height" on the primary assignment of "height is tall" is denoted by a two-dimensional membership function of "height" and its type-1 membership (also called primary membership). The degree of certainty thus obtained for each distinct value of the linguistic variable and its corresponding primary membership is often referred to as the secondary membership. The resulting system comprising a number of primary membership functions obtained from n sources with an estimate of the secondary membership for each distinct value of the linguistic variable is called generalized type-2 fuzzy set (GT2FS).

It is apparent that GT2FS can better capture the uncertainty of a real-world problem than its IT2FS counterpart. However, one of the primary bottlenecks to use GT2FS is to provide the secondary memberships. Since users are unable to provide the secondary membership values corresponding to their primary assignments, an automatic approach to determine the secondary membership from their primary estimates is an open problem in the fuzzy literature [3].

This chapter aims at evaluating secondary memberships in GT2FS [4] by determining consistency in primary membership assignment by a given source with respect to the composite opinion of all the users' primary membership functions. The above objective is realized by constructing an objective function that attempts to minimize the sharp change in secondary membership grade. The constraint of the problem lies in the fact that the defuzzified signal obtained from the primary membership function should be close to that of the secondary membership distribution of the same source. It is obvious because both characterize the same real world parameter.

Any derivative-free evolutionary optimization algorithm can be employed to minimize the given objective function to determine the secondary membership grade for individual primary membership function. In this paper, we employed differential evolution (DE) algorithm for its proven merits in global optimization [5] and our experience of using it in solving nonconvex optimization problems [6]. Some of the attractive features of DE, justifying its selection in the secondary membership evaluation, include simplicity of its structure leading to ease of coding, very few control parameters, and faster convergence with respect to other swarm/evolutionary algorithms.

A case study is undertaken here to evaluate the comparative performance of the proposed approach of obtaining the secondary membership by optimizing the given objective function with another existing approach [3, 7]. The case study in the present context refers to the problem of emotion recognition from an unknown facial expression. Experiments undertaken to compare the relative performance of the proposed approach with the method given in [3] reveal that the proposed technique outperforms the method in [3] with respect to the classification accuracy.

The chapter is divided into six sections. Section 2 provides fundamental definitions associated with type-2 fuzzy sets, which will be required in the rest of the paper. Section 3 introduces the differential evolution algorithm. Section 4 deals with secondary membership evaluation procedure for a given primary membership function. Experimental details and performance analysis are undertaken in Sect. 5. Conclusions are listed in Sect. 6.

2 Preliminaries on Type-2 Fuzzy Sets

In this section, we define some terminologies related to type-1 (T1) and type-2 (T2) fuzzy sets. These definitions will be used throughout the paper.

Definition 1 (*Type-1 fuzzy set*) A defined on a universe of discourse X, is given by a two-dimensional membership function, and also called type-1 membership function. The (primary) *membership function*, symbolized by $\mu_A(x)$ is a crisp number in [0, 1] for a generic element $x \in X$. Usually, the fuzzy set A is expressed as a two tuple [8] given by

$$A = \{(x, \mu_A(x)) | \forall x \in X\}. \tag{1}$$

An alternative representation of the fuzzy set A is also found in the literature as given in (2).

$$A = \int_{x \in X} \mu_A(x)|x \tag{2}$$

where $\int$ denotes union of all admissible x.

Definition 2 (*A type-2 fuzzy set*) $\tilde{A}$ is represented by a three dimensional membership function, also called type-2 membership function, which itself is fuzzy. The *type-2 membership function* is usually specified by $\mu_{\tilde{A}}(x, u)$, where $x \in X$, and $u \in J_x \subseteq [0, 1]$ [3]. Usually, the fuzzy set $\tilde{A}$ is expressed as a two tuple:

$$\tilde{A} = \{((x, u), \mu_{\tilde{A}}(x, u)) | x \in X, u \in J_x \subseteq [0, 1]\} \tag{3}$$

where $\mu_{\tilde{A}}(x, u) \in [0, 1]$. An optional form of representation of the type-2 fuzzy set is given in (4).

$$\tilde{A} = \int_{x \in X} \int_{u \in J_x} \mu_{\tilde{A}}(x, u) | (x, u), \; J_x \subseteq [0, 1] \tag{4}$$

$$= \int_{x \in X} \left[\int_{u \in J_x} f_x(u)/u \right] \Big/ x, \; J_x \subseteq [0, 1] \tag{5}$$

where $f_x(u) = \mu_{\tilde{A}}(x, u) \in [0, 1]$. The $\int \int$ denotes union over all admissible x and u [2].

Definition 3 At each point of x, say $x = x'$, the two-dimensional plane containing axes u and $\mu(x', u)$ is named as the *vertical slice* of $\mu_{\tilde{A}}(x, u)$. A *secondary membership function* is a vertical slice of $\mu_{\tilde{A}}(x, u)$. Symbolically, it is represented by $\mu_{\tilde{A}}(x, u)$ *at* $x = x'$ for $x' \in X$ and $\forall u \in J_{x'} \subseteq [0, 1]$.

$$\mu_{\tilde{A}}(x = x', u) = \int_{u \in J_{x'}} f_{x'}(u) | u, \; J_{x'} \subseteq [0, 1] \tag{6}$$

where $0 \leq f_{x'}(u) \leq 1$. The amplitude of a secondary membership function is called secondary grade (of membership). In (6) $J_{x'}$ signifies the primary membership of x'.

3 An Overview of Differential Evolution Algorithm

Differential evolution (DE) starts with a population of *NP* *D*-dimensional parameter vectors representing the candidate solutions. We represent the ith vector of the population at the current generation as

$$\vec{Z}_i(G) = \left[z_{i,1}(G), z_{i,2}(G), z_{i,3}(G), \ldots, z_{i,D}(G) \right] \tag{7}$$

The initial population (at $G = 0$) should cover the entire search space as much as possible by uniformly randomizing individuals within the search space constrained by the prescribed minimum and maximum bounds: $\vec{Z}_{\min} = \left[z_{\min-1}, z_{\min-2}, z_{\min-3}, \ldots, z_{\min-D} \right]$ and $\vec{Z}_{\max} = \left[z_{\max-1}, z_{\max-2}, z_{\max-3}, \ldots, z_{\max-D} \right]$.

Hence, we may initialize the jth component of the ith vector as

$$z_{i,j}(0) = z_{j-\min} + rand_{i,j}\,(0, 1) \times (z_{j-\max} - z_{j-\min}) \tag{8}$$

where $rand_{i,j}(0, 1)$ is a uniformly distributed random number lying between 0 and 1.

3.1 Mutation

After initialization, DE creates a donor vector $\vec{V}_i(G)$ corresponding to each target vector $\vec{Z}_i(G)$ in the current generation through mutation. Although DE has five most frequently referred mutation strategies available online at (http://www.icsi.berkeley.edu/storn/code.html), we here prefer to use DE/rand/1/bin, the most popular and widely used version of DE, particularly for its simplicity and computational accuracy without sacrificing the speed.

$$\text{DE/rand/1}: \vec{V}_i(G) = \vec{Z}_{r_1^i}(G) + F(\vec{Z}_{r_2^i}(G) - \vec{Z}_{r_3^i}(G)) \tag{9}$$

The indices r_1^i, r_2^i, and r_3^i are mutually exclusive integers randomly chosen from the range [1, *NP*], and all are different from the base index *i*. The scaling factor *F* is a positive control parameter for scaling the difference vectors. $\vec{Z}_{best}(G)$ is the best individual vector with the best fitness in the population at generation *G*.

3.2 Crossover

In case of binomial crossover [5, 9], a trial vector $\vec{U}_i(G) = [u_{i,1}(G), u_{i,2}(G), u_{i,3}(G), \ldots, u_{i,D}(G)]$ is generated for each pair of donor vector $\vec{V}_i(G)$ and target vector $\vec{Z}_i(G)$ by the following operation

$$u_{i,j}(G) = \begin{cases} v_{i,j}(G) \; if \; rand_{ij} \le Cr \; or \; j = j_{rand} \\ z_{i,j}(G) \; otherwise \end{cases} \tag{10}$$

where $rand_{i,j}(0, 1) \in [0, 1]$ is a uniformly distributed random number lying in [0, 1]. $j_{rand} \in [1, D]$ is a randomly chosen index, which ensures that $\vec{U}_i(G)$ gets at least one component from $\vec{V}_i(G)$. "*Cr*" is called the crossover rate and appears as a control parameter of DE.

3.3 Selection

To keep the population size constant over subsequent generations, the next step of the algorithm calls for selection to determine whether the target or the trial vector survives to the next generation, i.e., at $G = G + 1$. The selection operation is described as

$$\begin{aligned} \vec{Z}_i(G+1) &= \vec{U}_i(G) \; if \; f(\vec{U}_i(G)) \le f(\vec{Z}_i(G)) \\ &= \vec{Z}_i(G) \; if \; f(\vec{U}_i(G)) > f(\vec{Z}_i(G)) \end{aligned} \tag{11}$$

where $f(\vec{x})$ is the function to be minimized.

4 Secondary Membership Evaluation in the Settings of an Optimization Problem

In this section, we discuss type-2 membership evaluation [2, 3, 7]. Although theoretically very sound, the application of type-2 fuzzy set remains confined over the last two decades because of the users' insufficient information about correctly assigning the secondary memberships. This paper, however, surmounts this problem by extracting type-2 membership function from its type-1 counterpart by DE algorithm. A brief outline to the construction of secondary membership function is given in this section.

Let the primary membership functions for a linguistic variable x from n sources are represented as $\mu_A^1(x), \mu_A^2(x), \ldots, \mu_A^n(x)$. Here $\mu_A^i(x)$ is the primary membership in [0, 1] of the linguistic variable x to be a member of set A, and $\mu(x, \mu_A^i(x))$ be the secondary membership of the measured variable x in [0, 1] for the ith source. The following strategies are used to construct an objective function, which is minimized to obtain the solution of the problem.

Assumption 1 Let $\mu_{A'}^i(x)$ represents the average primary membership function excluding $\mu_A^i(x)$. Then $\mu_{A'}^i(x)$ representing a special form of fuzzy aggregation is given by

$$\mu_{A'}^i(x) = \frac{\sum_{j=1, j\neq i}^{n} \mu_A^j(x)}{n-1}, \forall x \tag{12}$$

i.e., at each position of $x = x_j$, the above membership aggregation is employed to evaluate a new composite membership profile $\mu_{A'}^i(x)$. In the proposed approach, the secondary membership of the ith subject, $\mu(x, \mu_A^i(x))$, at any point x, is evaluated based on the absolute difference between $\mu_A^i(x)$ and $\mu_{A'}^i(x)$. This is based on the supposition that a small value of $\left|\mu_A^i(x) - \mu_{A'}^i(x)\right|, \forall x$ symbolizes higher level of correctness in assigning the primary membership $\mu_A^i(x)$ at each measurement point x. On the other hand, if $\left|\mu_A^i(x) - \mu_{A'}^i(x)\right|, \forall x$ is too high, it indicates a lower degree of certainty in the primary membership assignment. Based on this strategy, the secondary membership of the ith subject at any point x is calculated as

$$\mu(x, \mu_A^i(x)) = \exp(-k_x^i \,|\, \mu_A^i(x) - \mu_{A'}^i(x)|) \tag{13}$$

Here, k_x^i is a parameter to be determined at each distinct value of linguistic variable x.

Assumption 2 The unknown secondary membership at two values of x separated by a small positive δ should have a small difference. This is required to avoid sharp changes in the secondary grade [7].

Assumption 3 Type-1 defuzzification over a given primary membership function, $\mu_A^i(x)$ should return the same value as obtained by type-2 defuzzification corresponding to the same primary membership function for any ith source [3, 7]. This assumption holds because the two modalities of defuzzification, representing the same real-world parameter, should return close values.

Using strategy 2, we construct a performance index J_i to compute secondary membership for the ith subject.

$$J_i = \sum_{x=x_1}^{x_{R-1}} \left| \mu((x+\delta), \mu_A^{\prime i}(x+\delta)) - \mu(x, \mu_A^{\prime i}(x)) \right| \tag{14}$$

This term is employed to prevent abrupt changes in the membership function. In (19) x_1 and x_R are the smallest and the largest values of a given linguistic feature considered over R sampled points of $\mu_A^{\prime i}(x)$. In (14), $\delta = (x_R - x_1)/(R-1)$ and $x_k = x_1 + (k-1).\,\delta$ for $k = [1, R]$.

The defuzzified signal obtained by the centroid method [8] from the primary membership function of the ith subject is given by

$$\bar{c}_i = \frac{\sum_{\forall x} x.\mu_A^{\prime i}(x)}{\sum_{\forall x} \mu_A^{\prime i}(x)} \tag{15}$$

Further, the type-2 centroidal defuzzified signal [3] obtained from the ith primary and secondary membership functions here is defined as

$$\bar{\bar{c}}_i = \frac{\sum_{\forall x} x.\mu_A^{\prime i}(x).\mu(x, \mu_A^{\prime i}(x))}{\sum_{\forall x} \mu_A^{\prime i}(x).\mu(x, \mu_A^{\prime i}(x))} \tag{16}$$

The products of primary and secondary memberships are used in (16) to refine the primary memberships by the degree of certainty of the corresponding secondary values. Then the constraint satisfying the Assumption 3 [3] is given by $\left|\bar{c}_i - \bar{\bar{c}}_i\right| = 0$.

We now represent the optimization problem by adding the basic cost function with the constraint. Thus, the constrained optimization problem in the present context is given by,

$$J_i = \sum_{x=x_1}^{x_{R-1}} \left| \mu((x+\delta), \mu_A^{\prime i}(x+\delta)) - \mu(x, \mu_A^{\prime i}(x)) \right| + \lambda \left|\bar{c}_i - \bar{\bar{c}}_i\right| \tag{17}$$

where λ (>0) is the Lagrangian multiplier. Using Assumption 1 [4], the present constraint optimization problem is transformed to

$$J_i = \sum_{x=x_1}^{x_{R-1}} \left| \exp(-k_{x+\delta}^i \left| \mu_A^{\prime i}(x+\delta) - \mu_{A'}^{\prime i}(x+\delta) \right|) - \exp(-k_x^i \left| \mu_A^{\prime i}(x) - \mu_{A'}^{\prime i}(x) \right|) \right| + \lambda \left|\bar{c}_i - \bar{\bar{c}}_i\right| \tag{18}$$

The secondary membership evaluation problem, now transforms to minimization of J_i by selecting λ and k_x at every point x. Note that, for each subject, we have to define (13), (15), (16), and (18) and find the optimal secondary membership functions.

A pseudocode to compute the secondary membership function of a type-2 fuzzy set from its primary counterpart using DE is given below.

Pseudo Code:

Input: Primary membership distribution $\mu_A^i(x)$, for n subjects where i= [1, n], the smallest value x_1 and the largest value x_R of linguistic variable x.

Output: Secondary membership distribution $\mu(x, \mu_A^i(x))$ corresponding to the primary membership $\mu_A^i(x)$ for each of the i-th subject.

Begin
For subject i=1 to n
Repeat

1. Obtain the averaged primary membership function $\mu_{A'}^i(x)$ (corresponding to $\mu_A^i(x)$) from the primary membership functions $\mu_A^j(x)$ of n sources, only disregarding the primary membership distribution, $\mu_A^i(x)$ of the i-the subject, i.e., j= [1, n], $j \neq i$ using (12).
2. Evaluate $\overline{c}_i$ for the selected i-th primary membership distribution $\mu_A^i(x)$ using (15).
3. Call DE ($\mu_A^i(x)$, $\mu_{A'}^i(x)$, $\overline{c}_i$).

End For.
End.

Procedure DE ($\mu_A^i(x)$, $\mu_{A'}^i(x)$, $\overline{c}_i$)

Begin

1. Initialize a population of NP solutions $\vec{Z}_j(G)$, with j= [1, NP]. Here $\vec{Z}_j(G) = \{k^i{}_{j,x_1}, k^i{}_{j,x_2}, \ldots, k^i{}_{j,x_R}, \lambda_j\}$ for R sample points of the linguistic variable x.
2. Obtain the secondary membership distribution encoded in $\vec{Z}_j(G)$, corresponding to $\mu_A^i(x)$ for j= [1, NP] using (13).
3. Evaluate $\overline{\overline{c}}_{i,j}$ for the selected i-th primary membership distribution $\mu_A^i(x)$ using (16) for each $\vec{Z}_j(G)$ for j= [1, NP].
4. **While** termination criterion is not satisfied
   Begin
   a. Create trial vector using mutation and crossover schemes of DE as given in (9) and (10).
   b. Evaluate fitness using (18).
   c. **If** the trial vector is better than the target vector
      Then replace the target by the trial in the next generation.
      End If
   End While
5. **Return** $\mu(x, \mu_A^i(x)), \forall x$ after the DE converges.

Two illustrative plots of secondary membership function for a given primary are given in Figs. 1 and 2.

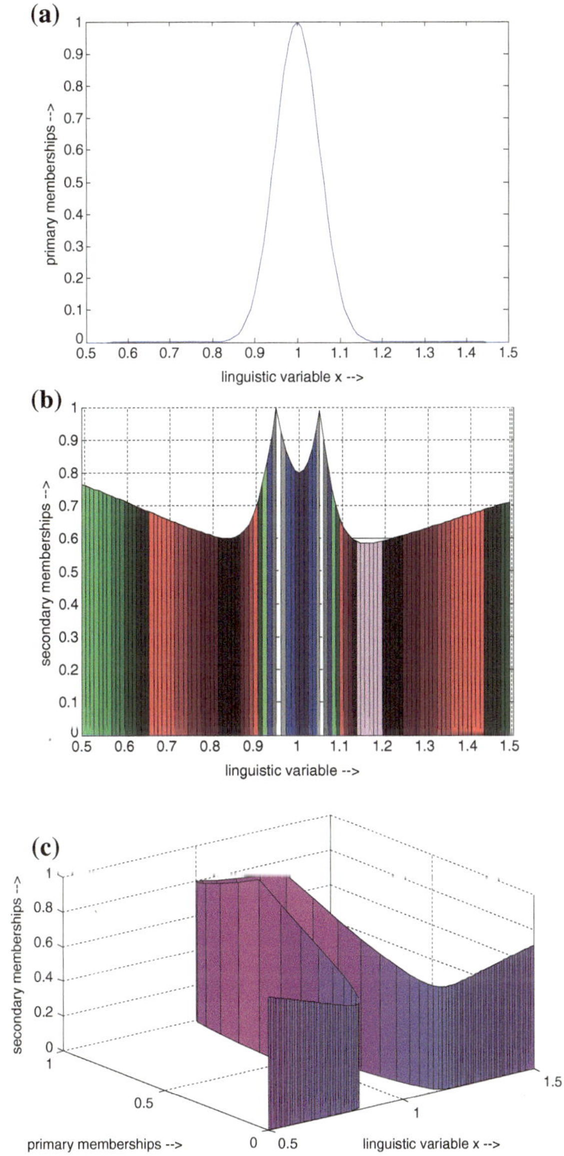

Fig. 1 **a** The primary membership function for a given linguistic variable, **b** 2-D view and **c** 3-D view of its corresponding secondary membership function obtained by minimizing J_i

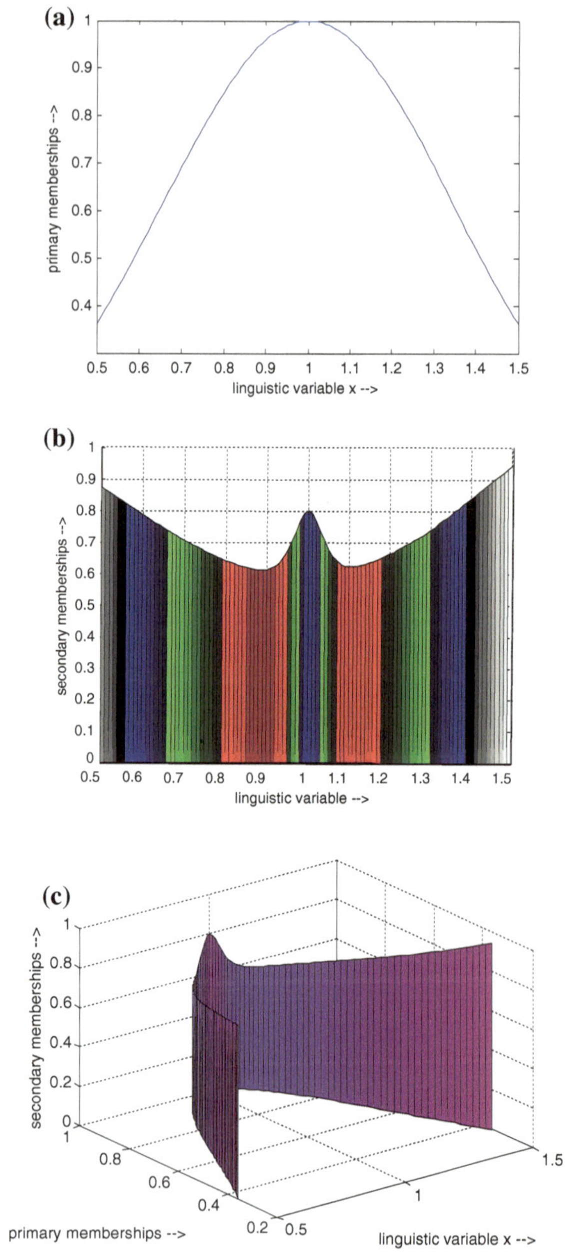

Fig. 2 **a** The primary membership function for a given linguistic variable, **b** 2-D view and **c** 3-D view of its corresponding secondary membership function obtained by minimizing J_i

5 Experiments

5.1 Experimental Environment

The secondary membership evaluation problem of GT2FS is carried out on a simulated environment on Intel Core 2 Duo processor architecture with clock speed of 2 GHz. We used the best possible parameter setting for the DE algorithm. An optimal set of parameters is chosen after experimenting with many possibilities. The crossover rate *Cr* is set to 0.9 and the scale factor *F* to 0.8. For the algorithm, the population size *NP* is taken to be ten times the dimension *D* of the problem. We repeat the mutation, crossover, and selection operations of DE until the terminating condition for convergence is reached. We stop the algorithm when the number of Function Evaluations (FEs) exceeds 10^6.

5.2 Optimization of Objective Function and Constraint to Evaluate Secondary Membership

We have considered a group of 20 primary membership functions for the linguistic variable "speed" (sp) in the context that "speed is medium (med)" as illustrated in Fig. 3a. Figure 3b elucidates an example of evaluating the secondary membership distribution $\mu(sp, \mu^5_{med}(sp))$ for the 5th subject based on its primary membership distribution $\mu^5_{med}(sp)$ and $\mu^5_{med}(sp)$ (obtained by taking average of all $\mu^i_{med}(sp)$, $i = [1, 20]$ except at $i = 5$ at all distinct values of the variable "speed"). It is observed from Fig. 3c that there exist three peaks in the secondary membership distribution at the "speed" value of 55, 70, and 140 miles/h. It is apparent from Fig. 3b that $\mu^5_{med}(sp = 55) = \mu^5_{med}(sp = 55) = 0.9$, $\mu^5_{med}(sp = 70) = \mu^5_{med}(sp = 70) = 0.6$, and $\mu^5_{med}(sp = 140) = \mu^5_{med}(sp = 140) = 0$, indicating the highest level of certainty in assigning the primary membership at these three distinct value of "speed". The lowest value of $\mu(sp = 40, \mu^5_{med}(sp = 40))$ is implied by the maximum absolute difference of $\left|\mu^5_{med}(sp = 40) - \mu^5_{med}(sp = 40)\right| = |0.15 - 0.8| = 0.65$, which in turn signifies the high level of uncertainty in the primary membership assignment. In Fig. 4, we have also plotted the objective function value of the best solution for evaluation of $\mu(sp, \mu^5_{med}(sp))$ obtained in DE-based simulations with FEs. Here $J_1 = \sum_{x=x_1}^{x_{R-1}} \left|\mu((x+\delta), \mu^i_A(x+\delta)) - \mu(x, \mu^i_A(x))\right|$, $J_2 = \lambda \left|\bar{c}_i - \bar{\bar{c}}_i\right|$ and $J = J_1 + J_2$.

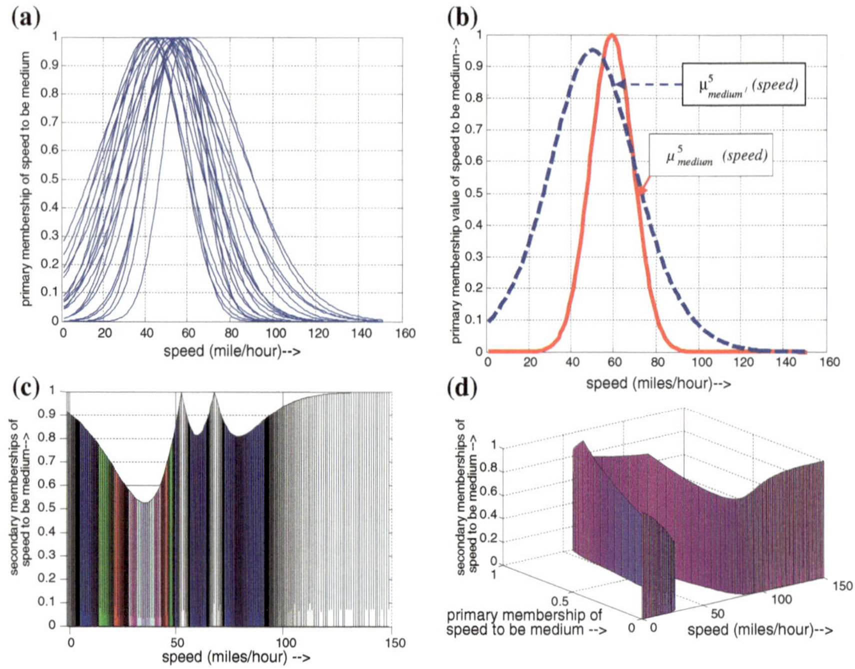

Fig. 3 **a** Group of 20 primary membership curves. **b** $\mu^5_{medium}(speed)$ and $\mu^5_{medium}(speed)$. **c** Two-dimensional view of secondary membership. **d** Three dimensional view of secondary membership

Fig. 4 Best objective function value with function evaluation

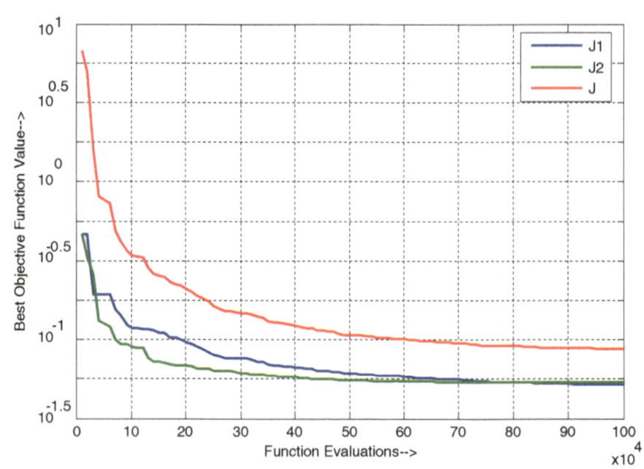

5.3 Case Study in Emotion Recognition

In this section, a case study of GT2FS based reasoning for emotion classification [3, 7, 10, 11] is undertaken to illustrate the importance of the proposed approach. Here we consider $k = 5$ distinct emotion classes: anger, fear, disgust, happiness, and relaxation. The experiment is conducted with two sets of subjects: (a) the training dataset of $n = 20$ subjects is considered for designing the fuzzy face-space and (b) the testing data set of 10 unknown subjects is considered to validate the result of the proposed emotion classification scheme. $m = 5$ facial features, (Left Eye Opening (EO_L), Right Eye Opening (EO_R), Distance between the Lower Eyelid to the Eyebrow for the Left Eye (LEE_L), Distance between the Lower Eyelid to Eyebrow for the Right Eye (LEE_R), and the Maximum Mouth opening (MO)) have been used here to design the type-2 fuzzy face-space.

Let f_i be the measurement of the ith feature for a subject with an unknown emotion class, c. Now, by consulting the n primary membership functions that were generated from n-subjects in the training data for a given emotion class, c, we obtain n primary membership values (generated from n-subjects in the training data) corresponding to f_i for emotion class c as given by $\mu^1_{\tilde{A}c}(f_i), \mu^2_{\tilde{A}c}(f_i), \ldots, \mu^n_{\tilde{A}c}(f_i)$. Let the secondary membership values for each primary membership value respectively be $\mu(f_i, \mu^1_{\tilde{A}c}(f_i)), \mu(f_i, \mu^2_{\tilde{A}c}(f_i)), \ldots, \mu(f_i, \mu^n_{\tilde{A}c}(f_i))$. In order to reduce the intralevel uncertainty, we obtained the modified primary membership value for the jth training subject for ith feature of class c as

$$^{\text{mod}}\mu^j_{\tilde{A}c}(f_i) = \mu^j_{\tilde{A}c}(f_i) \times \mu(f_i, \mu^j_{\tilde{A}c}(f_i)), \forall j = [1, n] \quad (19)$$

Now for all $f_i, i = [1, m]$ and for all $c = [1, k]$, we evaluate

$$^{\text{mod}}\underline{\mu}_{\tilde{A}c}(f_i) = \min(^{\text{mod}}\mu^1_{\tilde{A}c}(f_i), {}^{\text{mod}}\mu^2_{\tilde{A}c}(f_i), \ldots, {}^{\text{mod}}\mu^n_{\tilde{A}c}(f_i)) \quad (20)$$

$$^{\text{mod}}\overline{\mu}_{\tilde{A}c}(f_i) = \max(^{\text{mod}}\mu^1_{\tilde{A}c}(f_i), {}^{\text{mod}}\mu^2_{\tilde{A}c}(f_i), \ldots, {}^{\text{mod}}\mu^n_{\tilde{A}c}(f_i)) \quad (21)$$

Now for m different facial features $f_i^/, i = [1, m]$ of an unknown emotion, we can obtain $\left[^{\text{mod}}\underline{\mu}_{\tilde{A}c}(f_1^/), {}^{\text{mod}}\overline{\mu}_{\tilde{A}c}(f_1^/)\right], \left[^{\text{mod}}\underline{\mu}_{\tilde{A}c}(f_2^/), {}^{\text{mod}}\overline{\mu}_{\tilde{A}c}(f_2^/)\right], \ldots, \left[^{\text{mod}}\underline{\mu}_{\tilde{A}c}(f_m^/), {}^{\text{mod}}\overline{\mu}_{\tilde{A}c}(f_m^/)\right]$ under the emotion class c. Thus we can say that the unknown subject is experiencing the emotion class c at least by $S_c^{\min} = \min\left(^{\text{mod}}\underline{\mu}_{\tilde{A}c}(f_1^/), {}^{\text{mod}}\underline{\mu}_{\tilde{A}c}(f_2^/), \ldots, {}^{\text{mod}}\underline{\mu}_{\tilde{A}c}(f_m^/)\right)$ and at most to the extent $S_c^{\max} = \min\left(^{\text{mod}}\overline{\mu}_{\tilde{A}c}(f_1^/), {}^{\text{mod}}\overline{\mu}_{\tilde{A}c}(f_2^/), \ldots, {}^{\text{mod}}\overline{\mu}_{\tilde{A}c}(f_m^/)\right)$. Then the degree of support of that unknown facial expression in

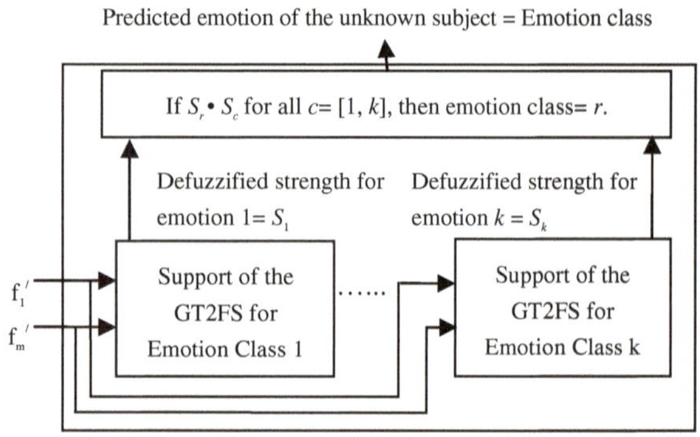

Fig. 5 GT2FS based emotion classification

the emotional class c is given by $S_c = (S_c^{\min} + S_c^{\max})/2$ [12]. Now to predict the emotion of an unknown subject, we determine S_c for each of the k classes, i.e., $c = [1, k]$ and thus determine the emotion class r, for which $S_r \geq S_c$ for all $c = [1, k]$. A complete scheme for GT2FS-based emotion recognition, considering the support of k-emotion classes is given in Fig. 5.

5.4 Relative Performance Analysis

The emotion recognition problem addressed here attempts to determine the emotion of an unknown subject from his/her facial expression. The features obtained from Fig. 6 are enlisted in Table 1. Table 2 provides the results of individual range in

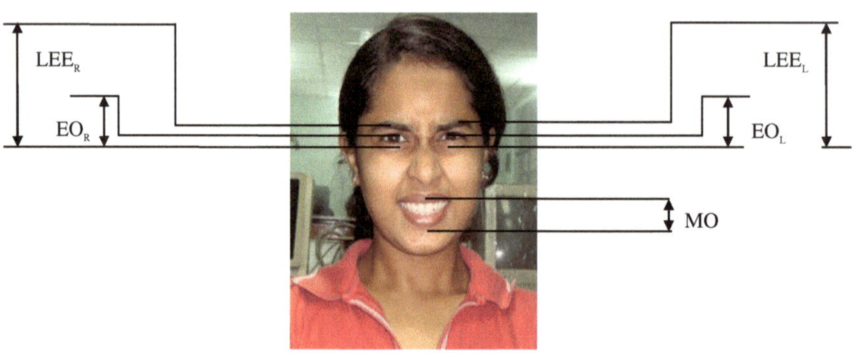

Fig. 6 Extracted features of facial image of an unknown subject

Table 1 Extracted feature values from Fig. 5

EO_L	EO_R	LEE_L	LEE_R	MO
0.069	0.065	0.116	0.127	0.156

Table 2 Calculated feature ranges and center value for each emotion using the proposed approach of evaluating secondary membership

Emotion	Range of modified primary membership for features					Range S_c^j after fuzzy meet operation (**center**)
	EO_L	EO_R	MO	LEE_L	LEE_R	
Anger	0.0361–0.6450	0.0362–0.6310	0.0312–0.8125	0.0425–0.8658	0.0382–0.8856	0.0312–0.6450 (**0.338125**)
Disgust	0.0635–0.6625	0.0613–0.6741	0.0020–0.9254	0.0149–0.8251	0.0131–0.7855	0.0020–0.6625 (0.332250)
Fear	0.0137–0.3340	0.0155–0.3410	0.0325–0.8275	0.0539–0.8700	0.0512–0.8517	0.0137–0.3340 (0.173875)
Happiness	0.0597–0.6475	0.0633–0.7145	0.0591–0.8303	0.0085–0.7757	0.0925–0.8278	0.0085–0.6475 (0.328000)
Relaxed	0.0517–0.6025	0.0552–0.5857	0.0000–0.5125	0.0000–0.7512	0.0000–0.8251	0.0000–0.5125 (0.2562500)

modified primary membership (using (19)) for each feature of the unknown emotion experimented under different emotional conditions. This uses our proposed approach to evaluate the secondary membership. The results of computing fuzzy meet operation over the range of individual features taken from facial expressions of the subject under the same emotional condition are given in the last column of Table 2. The average of the ranges along with its center value is also given in the same column. It is observed that the center has the largest value (=0.338125) for the emotion: anger. So, we can conclude that the subject in Fig. 5 shows emotion anger.

Table 3 enlists the results for the same unknown facial expression but using the approach as proposed in [3] to determine the secondary membership grade. Table III shows that the emotion is misclassified as disgust with the largest center value (=0.452). Table 4 shows the classification accuracy of our proposed approach as well as the method proposed in [3] using three facial image databases, i.e., Japanese

Table 3 Calculated feature ranges and center value for each emotion using the approach of evaluating secondary membership as given in [7]

Emotion	Range of modified primary membership for features					Range S_c^j after fuzzy meet operation (**center**)
	EO_L	EO_R	MO	LEE_L	LEE_R	
Anger	0.0661–0.7742	0.0632–0.7713	0.0360–0.9231	0.0433–0.7735	0.0401–0.7541	0.0360–0.7541 (0.395050)
Disgust	0.0290–0.8750	0.0351–0.8182	0.0349–0.9115	0.0430–0.9378	0.0327–0.9452	0.0290–0.8750 (**0.452000**)
Fear	0.0541–0.6326	0.0553–0.6531	0.0525–0.8156	0.0522–0.9325	0.0546–0.9022	0.0522–0.6326 (0.342420)
Happiness	0.0267–0.7550	0.0278–0.7796	0.0414–0.8471	0.0177–0.8666	0.0182–0.8828	0.0177–0.7550 (0.386380)
Relaxed	0.0654–0.7244	0.0641–0.7214	0.0000–0.7000	0.0005–0.8752	0.0005–0.8653	0.0000–0.7000 (0.350000)

Table 4 Percentage accuracy over three databases

GT2FS-based emotion recognition	JAFFE (%)	Indian Women (Jadavpur University) (%)	Cohn-Kanade (%)	Average accuracy (of last 3 columns) (%)
Our proposed approach	98.5	100	98	98.833
Approach given in [7]	91	94.5	91.5	92.333

Female Face Database (JAFFE), Indian Women Face Database (Jadavpur University), and Cohn-Kanade database (Fig. 7). Table 4 and comparison of Tables 2 and 3 indicate that our proposed approach has outperformed the method proposed in [7] to identify the class of an unknown emotion from the extracted features.

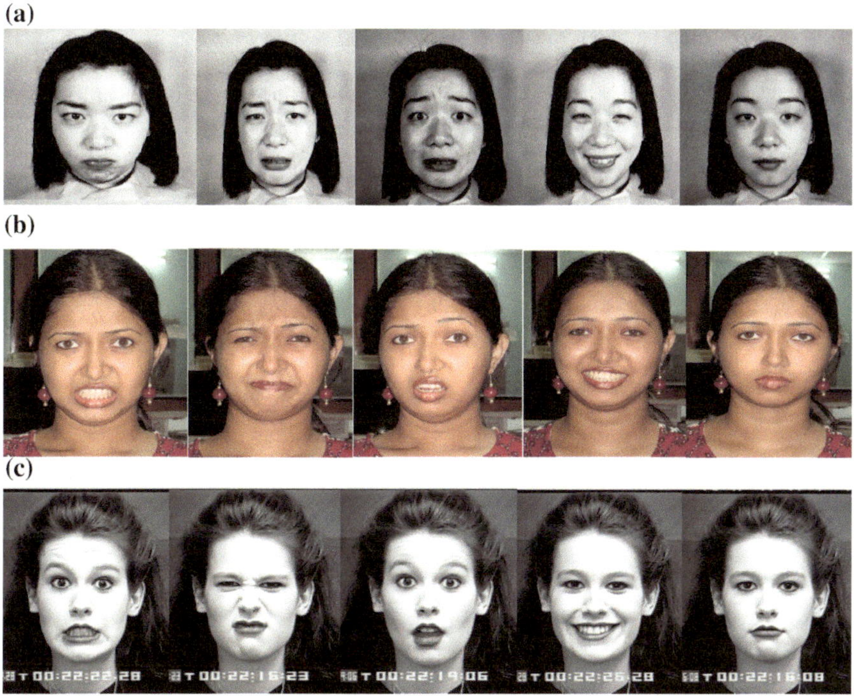

Fig. 7 Experiment done on different databases: **a** JAFFE, **b** Indian Women Database (Jadavpur University), and **c** Cohn-Kanade database

6 Conclusion

The chapter proposed a novel approach to evaluate the secondary membership grade from the primary membership distribution of a fuzzy linguistic variable and attempted to demonstrate the advantage of the proposed approach in emotion classification from facial expression. Experiments reveal that the classification accuracy in emotion recognition by our method of evaluating secondary membership is 98.833 %, while the existing approach [3] employed for the same purpose offers a classification accuracy of 92.333 %. This 6.5 % increase in classification accuracy is obtained without any significant loss in computational time. This indicates that our proposed GT2FS design method can effectively deal with both intra- and interpersonal level uncertainty associated with the assignment of primary membership values.

References

1. Mendel, J.M., John, R.I., Liu, F.: Interval type-2 fuzzy logic systems made simple. IEEE Trans. Fuzzy Syst. **14**(6), 808–821 (2006)
2. Mendel, J.M.: Type-2 fuzzy sets made simple. IEEE Trans. Fuzzy Syst. **10**(2), 117–127 (2002)
3. Halder, A., Konar, A., Mandal, R., Chakraborty, A., Bhowmik, P., Pal, N.R., Nagar, A.K.: General and interval type-2 fuzzy face-space approach to emotion recognition. IEEE Trans. Syst. Man Cybern.: Syst. **43**(3), 587–605 (2013)
4. Rakshit, P., Chakraborty, A., Konar, A., Nagar, A.K.: Secondary membership evaluation in generalized type-2 fuzzy sets by evolutionary optimization algorithm. In: IEEE International Conference on Fuzzy Systems, pp. 1–8 (2013)
5. Storn, R., Price, K.V.: Differential evolution-a simple and efficient heuristic for global optimization over continuous spaces. J. Glob. Optim. **11**(4), 341–359 (1997)
6. Rakshit, P., Konar, A., Das, S., Jain, L.C., Nagar, A.K.: Uncertainty management in differential evolution induced multi-objective optimization in presence of measurement noise. IEEE Trans. Syst. Man Cybern.: Syst. **PP**(99), 1–16 (2013)
7. Halder, A., Mandal, R., Chakraborty, A., Konar, A., Janarthanan, R.: Application of general type-2 fuzzy set in emotion recognition from facial expression. Swarm, Evolutionary and Memetic Computing. Lecture Notes in Computer Science, vol. 7076, pp. 460–468 (2011)
8. Zimmermann, H.J.: Fuzzy Set Theory and Its Applications. Springer, Netherlands (2001)
9. Rakshit, P., Konar, A., Bhowmik, P., Goswami, I., Das, S., Jain, L.C., Nagar, A.K.: Realization of an adaptive memetic algorithm using differential evolution and q-learning: a case study in multirobot path planning. IEEE Trans. Syst. Man Cybern.: Syst. **43**(4), 814–831 (2013)
10. Halder, A., Rakshit, P., Chakraborty, S., Konar, A., Kim, E., Nagar, A.K.: Reducing uncertainty in interval type-2 fuzzy sets for qualitative improvement in emotion recognition from facial expressions. In: IEEE International Conference on Fuzzy Systems, pp. 1–8 (2012)
11. Chakraborty, A., Konar, A., Kumar Chakraborty, U., Chatterjee, A.: Emotion recognition from facial expressions and its control using fuzzy logic. IEEE Trans. Syst. Man Cybern. Part A **39**(4), 726–743 (2009)
12. Mendel, J.M.: On the importance of interval sets in type-2 fuzzy logic systems. In: Proceedings of Joint 9th IFSA World Congress and 20th NAFIPS International Conference, pp. 1647–1652, Vancouver, BC, Canada, 25–28 July 2001

Specificity Based Defuzzification in Approximate Reasoning

Asim Pal and Swapan Raha

Abstract In this paper, an attempt is made to introduce a new defuzzification scheme to be used in case there are a number of clipped fuzzy sets in the output of a fuzzy system. Approximate reasoning with this defuzzification scheme is proposed. A number of defuzzification methods existing in literature are reviewed here. A comparative study with our method has been made. The results are illustrated with a DC shunt motor for better understanding.

Keywords Fuzzy set · Specificity measure · Defuzzification · Approximate reasoning

1 Introduction

In fuzzy logic, defuzzification is the process of producing a quantifiable result from a number of fuzzy sets in the output of a fuzzy system. A typical fuzzy logic system has the following components—a fuzzification procedure, a knowledge-base, an inference mechanism, and a defuzzification procedure. The performance of a fuzzy system depends heavily on fuzzification and the defuzzification strategies—as in this case, the overall performance of the system under study is determined by the controlling signal if receives.

Defuzzification procedure converts a fuzzy set into a crisp set. For a fuzzy set A defined over the universe U, the $\alpha - cut$ set A_α, where $0 \leq \alpha \leq 1$ is a crisp set $A_\alpha = \{x | \mu_A(x) \geq \alpha\}$. There may be situations when the output of a fuzzy process needs to be a single scalar quantity as opposed to a set of possible quantities. Thus, defuzzification is a procedure by which, we can obtain a typical member

This research has been partially supported by the UGC SAP (DRS) Phase-II programme of the Department of Mathematics, Visva-Bharati. Authors appreciate the financial support of the University Grants Commission.

A. Pal · S. Raha (✉)
Department of Mathematics, Visva-Bharati, Santiniketan, India
e-mail: swapan.raha@visva-bharati.ac.in

M.K. Chakraborty et al. (eds.), *Facets of Uncertainties and Applications*,
Springer Proceedings in Mathematics & Statistics 125,
DOI 10.1007/978-81-322-2301-6_4

from the universe of discourse of a linguistic variable if the best representative of the fuzzy set representing the value that the linguistic variable can take. Although, there are a number of definitions for defuzzification of a fuzzy set, there is no systematic procedure for the choice of a good defuzzification strategy. However, we will concentrate on the following basic criteria against which one can measure the methods viz., continuity (small change in the input should not produce a large change in the output), unambiguity (defuzzification always result in a unique value), plausibility (the defuzzified value should always correspond to the center of the support region), and simplicity (easy to compute).

In this paper, we propose a defuzzification method which is such that if the output fuzzy set has at least two convex subregions, then the center of gravity of the most specific subregion may be used as the defuzzified value of the output. Accordingly, from a collection of output fuzzy sets, we select one with maximum specificity and compute the centroid of the region determined by the point of the Universe at which the centroid corresponds will be the defuzzified value. For example, rules designed to decide how much pressure to apply might result in "Decrease Pressure by 15 % or, Increase Pressure by 72 %" when fired for a typical data resulting in a fuzzy output against the desire for a typical value of pressure. Defuzzification is interpreted as a procedure for making the membership degrees of the elements of a fuzzy set into a real value—a specific decision. The simplest but least useful defuzzification method is to choose the element with the highest membership and ignore the others. The problem with this approach is that it loses information. The rules that called for decreasing or maintaining pressure might as well have not been there in this case. A common and useful defuzzification technique is to find the center of gravity of the region it defines. First, the results of the rules fired must be added together in some way. Most typical fuzzy set membership functions have the graph of a triangle. Now, if this triangle was to be cut in a straight horizontal line somewhere between the top and the bottom, and the top portion was to be removed, the remaining portion forms a trapezoid. The first step of defuzzification typically "chops off" parts of the graph to form a trapezoid (or other shapes if the initial shapes were not triangles). For example, if the output has "Decrease Pressure 15 %", then this triangle will be reduced by 15 % the way up from the bottom. In the most common technique, all these trapezoids are then superimposed one upon another, forming a single geometric shape. Then, the centroid of this shape is calculated. The Projection of the centroid on the domain of definition gives the defuzzified value.

Choice of appropriate/suitable defuzzification procedure is an important task in the design of a fuzzy system. Here, we study six different defuzzification procedures, propose a new defuzzification procedure and call the same "Specificity based defuzzification." The paper is organized into five sections. After a brief introduction in Sect. 1, a review of the different existing methods of defuzzification is presented in Sect. 2. Section 3 studies the proposed specificity-based defuzzification method. In a subsection, the use of the proposed method in approximate reasoning methodology has been presented. One concrete example based on actual data is presented in Sect. 4 followed by the result of a case study on DC motor. The paper is concluded in Sect. 5 followed by a list of references.

2 Defuzzification Procedure

We can define a set of m rules as

$$if\ x_1\ is\ LX_1^{(k)}\ and \cdots and\ x_n\ is\ LX_n^{(k)}\ then\ Y\ is\ LU^{(k)}\ k = 1, \ldots, m.$$

The outcome of firing these rules with physical, crisp input values of $x_1^\star, \ldots, x_n^\star$ will either be m clipped fuzzy sets Fig. 1 denoted by

$$\overline{CLU}^{(1)}, \ldots, \overline{CLU}^{(m)}$$

or m scaled fuzzy sets denoted by

$$\overline{SLU}^{(1)}, \ldots, \overline{SLU}^{(m)}.$$

Here, we discuss the six most frequently used defuzzification methods [1]:

- Center of Area/Gravity defuzzification (COA/COG),
- Center of Sums (COS) defuzzification,
- Center of Largest-Area (COLA) defuzzification,
- First of Maxima (FOM) defuzzification,
- Middle of Maxima (MOM) defuzzification,
- Height defuzzification.

2.1 *Center-of-Area/Gravity Defuzzification (COA/COG)*

The Center-of-Area method or Center-of-Gravity method is a well-known defuzzification method. In the discrete case, let $U = \{u_1, \ldots, u_n\}$ be the universe of discourse of a linguistic variable and let A be the fuzzy set of possible values that the linguistic

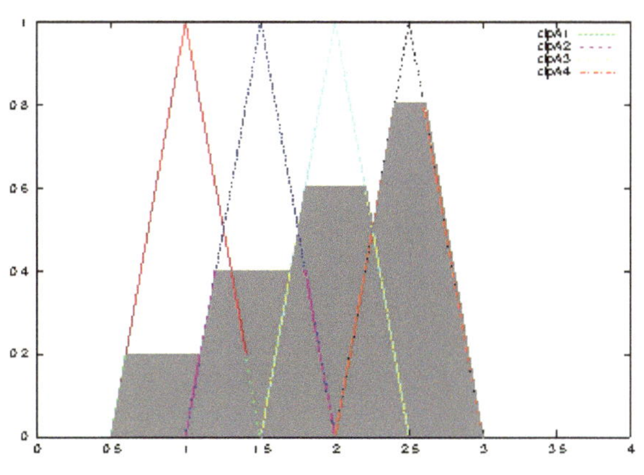

Fig. 1 Clipped fuzzy sets

variable can assume. Then, the defuzzification value will be given in (1) for the discrete and continuous cases respectively,

$$u^{*}=\frac{\sum_{i=1}^{n} u_i \cdot \mu_U(u_i)}{\sum_{i=1}^{n} \mu_U(u_i)} \text{ and } u^{*}=\frac{\int_U u \cdot \mu_U(u)}{\int_U \mu_U(u)} \tag{1}$$

where $\int$ is the classical integral. So, this method determines the center of the area below the combined membership function. This defuzzification method is computationally rather complex and therefore results in quite slow inference cycles.

2.2 Center of Sums Defuzzification (COS)

This type of defuzzification method is similar to COG but computationally faster. The idea is to consider the contribution of the area of each $\overline{CLU}^{(k)}$ individually. The motivation for using this method is to avoid the computation of $\tilde{U}$. Mathematically, the COG method builds $\tilde{U}$ by taking the union of all $\overline{CLU}^{(k)}$. Center of Sums, however, takes the sum of the $\overline{CLU}^{(k)}$. Thus, overlapping areas, where exists, are considered more than once. This method of defuzzification requires only one multiplication as compared to n-multiplication in COG. Hence, this method is faster than COG. Center of Sums is formally defined in (2) for discrete and continuous cases respectively,

$$u^{\star}=\frac{\sum_{i=1}^{l} \mu_i \cdot \sum_{k=1}^{l} \mu_{CLU^{(k)}}(u_i)}{\sum_{i=1}^{l} \cdot \sum_{k=1}^{l} \mu_{CLU^{(k)}}(u_i).} \text{ and } u^{\star}=\frac{\int_U u \cdot \sum_{k=1}^{n} \mu_{CLU^{(k)}(u)du}}{\int_U \sum_{k=1}^{n} \mu_{CLU^{(k)}(u)du\cdot}} \tag{2}$$

2.3 Height Defuzzification Method (HM)

This defuzzification method uses the individual clipped outputs instead of $\tilde{U}$. Accordingly, this method takes the peak value of each clipped output fuzzy sets and builds the weighted sum of these peak values. Let $c^{(k)}$ be the peak value and f_k be the height of kth fuzzy set ($k = 1, 2 \ldots, m$). Then, the height defuzzification method for an output from a system of m rules is formally given by

$$u^{\star} = \frac{\sum_{k=1}^{n} c^{(k)} \cdot f_k}{\sum_{k=1}^{n} \cdot f_k}. \tag{3}$$

2.4 Center of Largest Area (COLA)

The Center of Largest Area is used in the case where the fuzzy set is nonconvex, i.e., it consists of at least two convex fuzzy subsets [1]. The method determines the convex fuzzy subset with the largest area and defines the crisp output value u^* to be the Center of Area of this particular fuzzy subset. It is difficult to present this defuzzification method formally, because it involves finding the convex fuzzy subsets, their areas, etc.

2.5 First of Maxima (FOM)

First of Maxima uses the fuzzy set $\tilde{U}$ of U and takes the smallest value from the Universe U with maximal membership degree in $\tilde{U}$. Mathematically, it is understood by the following steps. Let

$$hgt(U) = sup_{u \in U} \mu_U(u) \tag{4}$$

be the highest membership degree of the elements in $\tilde{U}$, and let

$$\{u \in U | \mu_U(u) = hgt(U)\} \tag{5}$$

be the set of elements with degree of membership equal to $hgt(U)$. Then $u^{\star}$is given by

$$u^{\star} = inf_{u \in U}\{u \in U | \mu_U(u) = hgt(U)\}. \tag{6}$$

An alternative method is called the Last of Maxima and is given by

$$u^{\star} = sup_{u \in U}\{u \in U | \mu_U(u) = hgt(U)\}. \tag{7}$$

2.6 Middle of Maxima (MOM)

Middle of maxima is almost equal to First of Maxima or Last of Maxima. Instead of considering u^* to be the first or last from all values U having maximal membership degree, this method takes the average of these two values. Formally,

$$u^{\star} = \frac{inf_{u \in U}\{u \in X | \mu_U(u) = hgt(U)\} + sup_{u \in U}\{u \in U | \mu_U(u) = hgt(U)\}}{2} \tag{8}$$

3 Specificity Measure—Defuzzification

The result of rule firing, for a typical observation, is a fuzzy set. This is interpreted, at the semantic level, as the desired output. Often, we need to determine a precise action as output. The purpose of defuzzification is to obtain a scalar value $u \in U$, from the said output fuzzy set, as the action. Then, if necessary, denormalization is performed on the output so as to obtain the corresponding action on its physical domain. Specificity measure of a fuzzy set estimates the precision of an information when represented by it. In order to provide a definition for a specificity measure of a fuzzy set, a number of factors must be considered. A fuzzy set with maximum specificity value corresponds to a precise assessment of the values of a variable. According to Dubois and Prade, a specificity measure $Sp(A)$ [2] should satisfy the following properties. Let X be a linguistic variable defined on a universe of discourse U. A and B are normalized fuzzy subsets of U.

A1. $Sp(A) \in [0, 1]$.
A2. $Sp(A) = 1$ if and only if A is a singleton of U.
A3. If $A \subseteq B$ then $Sp(A) \geq Sp(B)$.

Yager [3] introduced one such measure of specificity that satisfies the above properties. When U is finite, Yager proposed an expression for defining the specificity. Let us assume that A be a fuzzy set defined over the universal set U and A_α be the α-level set of A. The specificity associated with A is denoted as Sp(A) and is defined as

$$\text{Sp}(A) = \int_0^{\alpha_{max}} \frac{1}{Card A_\alpha} d\alpha \tag{9}$$

where $\alpha_{max} = sup_{u \in U} \mu_A(u)$. Let us now list some properties [3] associated with the above definition.

P1. For all A, $Sp(A)$ assumes its maximum value 1, when $A = \{1/u\}$, i.e., fuzzy subset A consists of a singleton $u \in U$ having membership value 1.
P2. For all A, $Sp(A) \in [0, 1]$ and it assumes its minimum value 0, when $A = \Phi$.
P3. If $\mu_A(u) = k$ for all $u \in U$ then $Sp(A) = \frac{k}{n}$ where n is the cardinality of the ordinary set U.

Defuzzification is a procedure applied to reduce the anxiety in a decision. Specificity estimates the precision of an information on the values of a linguistic variable restricted by a fuzzy set. As suggested by the axioms and further properties of specificity, a crisp set can be less specific than a fuzzy set for restricting the possible values of a variable. Accordingly, we propose a new technique for defuzzification based on measure of precision. Let there be m clipped fuzzy sets $\{A^{(k)}; k = 1, 2, \dots, m\}$ and let $\{s_{(k)}, p^{(k)}; k = 1, 2, \dots, m\}$ be the specificity associated with $A^{(k)}$ as well as, the peak point of the consequent fuzzy set of the kth-rule. Let $p^{(k)}$ be the peak value of $A^{(k)}$ and $h^{(k)}$ be the corresponding height of the clipped version of $A^{(k)}$. Then, using height method the defuzzified value will be given by

$$u^* = \frac{\sum\limits_{k=1}^{m} p^{(k)}.h^{(k)}}{\sum\limits_{k=1}^{m} h^{(k)}}. \tag{10}$$

whereas, the specificity-based defuzzified value u^* will be given by

$$u^* = \frac{\sum_{k=1}^{m} p^{(k)}.s_{(k)}}{\sum_{k=1}^{m} s_{(k)}} \tag{11}$$

This definition is similar to the **Height** method of defuzzification which demands strictly convex fuzzy sets whereas the proposed one is applicable to all. The individual peak values of consequent fuzzy sets of the fired rules are used to generate the weighted average of these peak values. It is a simple method and works faster than the Center-of-sums method. Again, we can use the specificity, height, and peak values simultaneously to compute a modified defuzzified value as in the following:

$$u^* = \frac{\sum\limits_{k=1}^{m} p^{(k)}.h^{(k)}.s_{(k)}}{\sum\limits_{k=1}^{m} h^{(k)}.s_{(k)}}. \tag{12}$$

3.1 Approximate Reasoning with Specificity Based Defuzzification

In this section, we would like to validate our proposal considering defuzzification as an important integral part of approximate reasoning methodology [4].

Let there be m rules for a linguistic variable X taking values from the universe of discourse U. For each value of $j = \{1, 2, \dots, m\}$, A_j is a fuzzy subset of U and B_j

of V. In this case, the consequence can be obtained in two different ways. One can first compute the conclusion from each rule consistent with the premise q and then compose them by some rule of composition. The other approach is to first compute a fuzzy relation from the combination of all the rules (each of them is a fuzzy relation) conformal for firing on the product space and then use q to obtain the desired result. Both approaches are frequently used in many rule-based systems viz., fuzzy control. In finding a conclusion B' using the max-min compositional rule, the premise p is translated first into a fuzzy relation between the inputs (rule-antecedents) and the output (the consequent). There are different ways to obtain such a relation. The premise q is translated into another fuzzy relation between the input variables and then cylindrically extended over the product space of the input and output variables. These two relations are composed together using the (min)conjunction principle and finally, (max) projected over the universe (V) of the output variable, for the desired conclusion (Table 1).

We propose a new scheme, for computing the final conclusion, based on a measure of similarity. Our method is based on first rule-selection and then rule-execution. In both cases, we use the concept of similarity between fuzzy sets as a basis of the task. For that, first of all, we compute $s_i = S(A, A_i); i = 1, 2, \ldots, m$. From among the m distinct rules, we choose those rules for which $s_i > \epsilon$. This ϵ can be interpreted as a threshold in our case. We then apply algorithm **APPR** to generate a conclusion from each rule conformal for firing. The output can be generated using the intersection of the output fuzzy sets. It is important to note that the intersection operation is chosen in order to justify the rule-selection procedure. Here, fewer rules are fired and the output of each rule is significant.

Algorithm APPR:
Step 1. Compute s_i, the similarity of the input fuzzy set from the ith-rule for $i = 1, 2, \ldots, m$;.
Step 2. Define ϵ and find the rules conformal for firing.
Step 3. Translate the ith-rule, provided $s_i > \epsilon$ and compute the relation R_i using any suitable translating rule possibly, a T-norm operator.
Step 4. Modify R_i with s_i to obtain the modified conditional relation R_i'.
Step 5. Use sup-projection operation on R_i' to obtain B_i' as given in Eq. (13)

Table 1 Rule-based approximate reasoning

p1 :	if X is A_1 then Y is B_1
p2 :	if X is A_2 then Y is B_2
	$\vdots$ $\vdots$
pm :	if X is A_m then Y is B_m
q :	X is A
Conclusion	Y is B

$$\mu_{B_i'}(v) = \sup_{u_1,u_2,\ldots,u_k} \mu_{R(A_1|A_{i1},A_2|A_{i2},\ldots,A_k'|A_{ik},B)}(u_1, u_2, \ldots, u_k, v). \quad (13)$$

Step 6. Compute sp_i, the specificity of the output B_i'. Choose the most specific among the output fuzzy sets and apply the specificity-based defuzzification as defined earlier.

Example 1 : Let us consider for an input fuzzy set A, four rules are fired by the above algorithm and we get four output fuzzy sets such as $B_{i1}, B_{i2}, B_{i3}, B_{i4}$. over $V = \{400, 410, \ldots, 800\}$. where

$B_{i1} = 0.10/460 + 0.10/470 + 0.10/480 + 0.10/490 + 0.10/500 + 0.10/510 + 0.10/520 + 0.10/530 + 0.10/540$;

$B_{i2} = 0.10/510 + 0.20/520 + 0.20/530 + 0.20/540 + 0.20/550 + 0.20/560 + 0.20/570 + 0.20/580 + 0.10/590$;

$B_{i3} = 0.20/560 + 0.40/570 + 0.60/580 + 0.60/590 + 0.60/600 + 0.60/610 + 0.60/620 + 0.60/630 + 0.20/640$;

$B_{i4} = 0.20/610 + 0.40/620 + 0.60/630 + 0.80/640 + 0.80/650 + 0.80/660 + 0.60/670 + 0.40/680 + 0.20/690$.

Now, $Sp(B_{i1}) = 0.010$, $Sp(B_{i2}) = 0.014$, $Sp(B_{i3}) = 0.091$, $Sp(B_{i4}) = 0.157$.

U^* (Specificity based defuzzification) $= 630$

U^* (Centroid based defuzzification) $= 610$

U^* (Centroid based defuzzification of fuzzy sets where sp ≥ 0.05) $= 628.330$

Observation: The above-mentioned result shows that the specificity-based defuzzification technique works for fuzzy systems. It is intuitively plausible, physically significant, and computationally efficient. In taking a decision under imprecise environment, specificity will play an important role, particularly in defuzzification.

4 A Case Study on DC Motor

In this section, let us consider the DC Motor as in [5].

The human expert observed the behavior of the DC Motor and described the relation between current and speed in the form of fuzzy conditional statement as in the following (Fig. 2).

The data for the fuzzy model is given in Table 2. For a particular observed value of current, expressed in natural language, we first translate the inexact concepts into fuzzy sets(the simple observation) or fuzzy relations(the complex rule) over the specified domain using triangular membership functions. We then perform approximate reasoning to obtain the corresponding speed of the DC Motor using algorithm APPR. The defuzzified input/output are plotted for a comparative assessment of the utility of the proposed similarity-based approximate reasoning methodology. The simulation results are presented in the following self-explanatory diagrams.

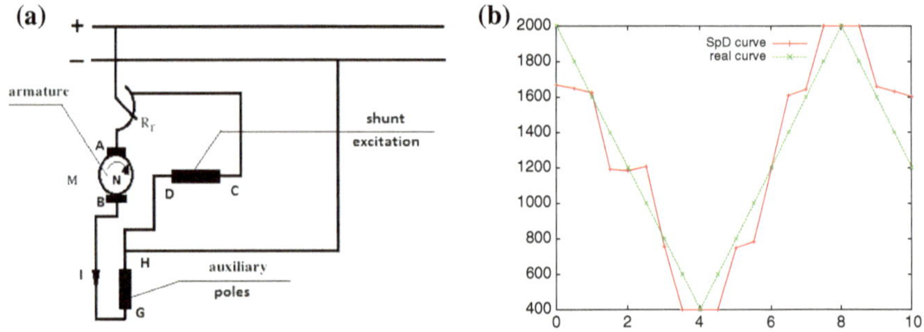

If I=$null$ $then$ N=$verylarge$ also
If I=$zero$ $then$ N=$medium$ also
If I=$small$ $then$ N=$zero$ also
If I=$medium$ $then$ N=$medium$ also
If I=$large$ $then$ N=$verylarge$ also
If I=$verylarge$ $then$ N=$medium$

Fig. 2 Rule base of DC Motor. **a** Diagram of DC Motor. **b** A Comparison of real and proposed inference result

Table 2 Real data of a DC Motor

I	0.0	0.5	1.0	1.5	2.0	2.5	3.0	3.5	4.0	4.5	5.0
N	2000	1800	1600	1400	1200	1000	800	600	400	600	800
I	5.5	6.0	6.5	7.0	7.5	8.0	8.5	9.0	9.5	10.0	
N	1000	1200	1400	1600	1800	2000	1800	1600	1400	1200	

5 Conclusion

Development of suitable mathematics for the realization of intelligent systems becomes necessary to handle modern computer-based technologies that manage different kinds of information and knowledge. This paper discusses one such tool required to help in the design of solutions to difficult problems in the construction of intelligent systems where the available information is supplied by human experts the later at times are found to be incomplete, imprecise, or even uncertain in nature and therefore, inherently ambiguous. It requires a logical framework which will enable one to reason and make decisions in an environment of imprecision, uncertainty, incompleteness of information and partiality of truth.

An imprecise/incomplete description of the input-output behavior of a system, as obtained from human experts, containing vague concepts is represented as fuzzy if-then rules—transforming the system into a simple fuzzy rule-based one. Approximate reasoning methodology has been used to predict the possible behavior of the system.

Defuzzification process allows us to select a member from the universal set as a representative given the fuzzy output inferred from the algorithm. The use of fuzzy logic allows us to use different interpretation of the logical operators for flexibility.

The achievement of human-level machine intelligence will have a profound impact on the contemporary society. It is hoped that upgradation of existing methodologies through addition of concepts and techniques drawn from fuzzy set theory will open possibilities for a substantial enhancement of our ability to model reality. Further research on the use of similarity and approximate reasoning is necessary for better understanding of the effect of the same on the cognitive process involved in the modeling and simulation of fuzzy systems. We have suggested relevant issues involved in the design of fuzzy systems—introduced similarity in reasoning and the concept of specificity measure in defuzzification.

References

1. Drainkov, D., Hellendoorn, H., Reinfrank, M.: An Introduction to Fuzzy Control. Narosa Publishing House, New Delhi (1996)
2. Dubois, D., Prades, H.: Fuzzy sets in approximate reasoning, part I: inference with possibility distribution. Fuzzy Sets Syst. **40**, 143–202 (1991)
3. Yager, R.R.: Measuring tranquility and anxiety in decision making: an application of fuzzy sets. Int. J. Gen. Syst. **8**, 139–146 (1982)
4. Mondal, B., Mazumdar, D., Raha, S.: Similarity in approximate reasoning. Int. J. Comput. Cogn. **4**(3), 46–56 (2006)
5. Goralczany, M.B., Kiszka, J., Stachowicz, M.: Some problems of studying adequacy of fuzzy models. In: Yager, R.R., (ed.) Fuzzy Set and Possibility Theory. pp. 14–31, Pergamon Press, USA (1982)

Proto-Fuzzy Concepts Generation Technique Using Fuzzy Graph

Partha Ghosh and Krishna Kundu

Abstract Since one major disadvantage of application of fuzzy formal concept analysis is that large numbers of fuzzy concepts are generated from fuzzy context, it is practically impossible to analyze such a large amount of concepts. Often it may be required to consider some particular concepts. For example, one might be interested to find out the fuzzy concepts containing all those objects which share some specific property with a specific/required degree from a given fuzzy context. Given such a situation, proto-fuzzy concepts may play a very useful role. This paper proposes a proto-fuzzy concept generation technique using fuzzy graph on uncertainty data. In this paper, we begin with defining a fuzzy graph corresponding to the **L**-context (fuzzy context). We then go on to demonstrate that t-concepts can be found to correspond with each maximal cliques of t-level graph of the defined fuzzy graph. After that, we determine all those cliques which corresponds to the proto-fuzzy concepts of degree t. Finally, a demonstration has been made using an example with the proposed technique.

Keywords Fuzzy sets · Fuzzy graph · Formal concept analysis (FCA) · Fuzzy concepts · Proto-fuzzy concepts

P. Ghosh (✉)
Department of Mathematics, Haldia Government College, P.O. Debhog,
Purba Medinipur District 721657, India
e-mail: partha_0377@rediffmail.com

K. Kundu
Department of Applied Mathematics, University of Calcutta, 92 A.P.C. Road,
Kolkata 700009, India

M.K. Chakraborty et al. (eds.), *Facets of Uncertainties and Applications*,
Springer Proceedings in Mathematics & Statistics 125,
DOI 10.1007/978-81-322-2301-6_5

1 Introduction

The work on Formal Concept Analysis (FCA) by Wille and Ganter [15, 16, 31] has introduced a new perspective and opened up new areas of its application. In present times, FCA has been extensively applied as an effective tool for data analysis to fields such as decision making, information retrieval, data mining, knowledge discovery, etc. The basic setting of FCA is based on bivalent logic. However, bivalent logic cannot be of much use as the bulk of information that encounters us is usually fuzzy and imprecise. As far as our knowledge is concerned, Burusco and Fuentes-Ganzález [9, 10] first introduced the theory of FCA in fuzzy setting. Later, Pollandt [27], and independently Bělohlávek [1] proposed the L-fuzzy context to combine fuzzy logic with FCA.

Nevertheless, the downside of FCA has been the existence of a large number of clusters [2] combined with the fact that many of the existing approaches require computation of a whole fuzzy concept lattice which, often, is too large. Handling such large amount of clusters become an unwieldy task and usually impossible. To cope with this situation, different techniques has been proposed [3, 4, 24]. However, as the size of data sets grows, the crisply generated fuzzy concept lattice [3] or, one-sided fuzzy concept lattice [24] continues to grow inexorably in size. Recently, another work has been presented by Krídlo and Krajči [25, 26]. The concepts, introduced in their paper, are known as proto-fuzzy concepts.

In this paper, we propose a graph based technique for computing proto-fuzzy concepts from a fuzzy context. The use of graph in FCA is not new. In [5], Berry et al. presented a graph-theoretic approach for generating formal concepts. They showed that a particular graph can be derived from a given formal context in a manner that there is a one-to-one correspondence between the formal concepts and the minimal separators of the graph. They also describe a process to determine sub-lattices of a whole lattice by saturating the minimal separators. This process could be proved more useful than those which consider whole lattices in applications. One more work has been presented in [17]. In this work, the authors have shown a relation between a fuzzy concept lattice and a fuzzy graph defined for a given fuzzy context.

Our present work presents another fuzzy graph theoretic approach. At the beginning, by defining a fuzzy graph for a given fuzzy context, we show a one-to-one correspondence between the t-concepts and cliques of t-level fuzzy graph. Then, we determine all those cliques which corresponds to the proto-fuzzy concepts of degree $t \in L$.

This paper is organized as follows. In Sect. 2 of this paper, we briefly discuss about fuzzy sets, fuzzy graph, fuzzy contexts, fuzzy concepts and proto-fuzzy concepts. In the Sect 3.1, for a given fuzzy context, we have defined a fuzzy graph. Then, in Sect. 3.2, we have proved that t-concepts can be generated from the cliques of t-level graph of fuzzy graph. After that, in Sect. 3.3, we determine all those cliques which corresponds to the proto-fuzzy concepts of degree t. Finally, we round up and demonstrate all these procedures using a sample example.

2 Mathematical Background—Explanations on the Fundamentals Applied

2.1 Basics of Fuzzy Logic and Fuzzy Graph

In this sub-section we first recall the basics of fuzzy logic [14, 19, 22, 33] and fuzzy graph [6, 12, 21, 28, 29, 32].

Since fuzzy logic are developed using general structure of truth degree, in this paper we use a complete residuated lattice $\mathbf{L} = \langle L, \wedge, \vee, \bigotimes, \rightarrow, 0, 1\rangle$ such (see in [19]) as a basic structure of truth degree. Operations $\otimes$ and $\rightarrow$ are known as "fuzzy conjunction" and "fuzzy implication". All elements a of L are called truth degrees. Usually, the common choice of $\mathbf{L}$ is a structure with $L = [0, 1]$, with $\vee$ and $\wedge$ being maximum and minimum, respectively, $\otimes$ being a left-continuous t–norm with the corresponding $\rightarrow$. Now considering $\mathbf{L}$ as a structure of truth degree, we present the notions of $\mathbf{L}$-set (fuzzy set) and $\mathbf{L}$-relation (fuzzy relation). The concept of $\mathbf{L}$-set was introduced by Goguen [18]. An $\mathbf{L}$-set μ in a universe set X is a mapping $\mu : X \rightarrow L$. $\mu(x)$ is called the truth value (or membership value) of x in μ which maps X to the membership space L. Similarly, an $\mathbf{L}$-relation I is a mapping $I : X \times Y \rightarrow L$ assigning to any $x \in X$ and $y \in Y$ a truth value $I(x, y)$ to which x and y is related under I. The collection of all $\mathbf{L}$-sets in X is denoted by the set L^X. For every $t \in L$, $\mu^t = \{x \in X \mid \mu(x) \geq t\}$ are called level sets or t-cut of μ. We let $supp(\mu) = \{x \in X \mid \mu(x) > 0\}$. We call $supp(\mu)$ the *support* of μ. An $\mathbf{L}$-set μ is nontrivial if $supp(\mu) \neq \phi$. In this paper, we use the notation $\vee$ for supremum and $\wedge$ for infimum. Let h be the function of $\wp(X)$ into $[0, 1]$ defined by $h(\mu) = \vee\{\mu(x) \mid x \in X\}$ for all $\mu \in \wp(X)$. Then $h(\mu)$ is called the height of μ. Let μ, ν be any two fuzzy subsets of X then $\mu \subseteq \nu$ if $\mu(x) \leq \nu(x)$ for all $x \in X$. The union $\mu \cup \nu$ of μ, ν is a subset of X defined by $(\mu \cup \nu)(x) = \mu(x) \vee \nu(x)$ for all $x \in X$ and intersection $\mu \cap \nu$ of μ, ν is also a subset of X defined by $(\mu \cap \nu)(x) = \mu(x) \wedge \nu(x)$ for all $x \in X$.

The fuzzy graphs used in this work are finite and undirected. As far as our knowledge is concerned, fuzzy graph was first proposed by Rosenfield [28]. A fuzzy graph is practically a fuzzy relation and fuzzy sets defined long ago by Chakraborty and Das [11], i.e., a fuzzy graph $G = (V, \mu, \rho)$ is a non empty set V together with a pair of functions $\mu : V \rightarrow [0, 1]$ and $\rho : V \times V \rightarrow [0, 1]$ such that for all x, y in V, $\rho(x, y) \leq \mu(x) \wedge \mu(y)$. μ is said to be the fuzzy vertex set of G and ρ the fuzzy edge set of G, respectively. For $P \subseteq V$, $H = (P, \nu, \tau)$ is called a fuzzy subgraph of $G = (V, \mu, \rho)$ induced by P if $\mu(x) = \nu(x)$ for all $x \in P$ and $\tau(x, y) = \rho(x, y)$ for all $x, y \in P$. For the sake of simplicity, we sometimes call H a fuzzy subgraph of G. Similarly $H = (P, \nu, \tau)$ is said to be partial fuzzy subgraph of $G= (V, \mu, \rho)$ if $\nu \subseteq \mu$ and $\tau \subseteq \rho$. Let $G = (V, \mu, \rho)$ be a fuzzy graph. For any threshold $t \in [0, 1]$, $\mu^t = \{x \in V|\mu(x) \geq t\}$ and $\rho^t = \{(x, y) \in V \times V|\rho(x, y) \geq t\}$. If $\mu^t \neq \phi$, then the crisp graph $G^t = (\mu^t, \rho^t)$ is said to be t-level graph of $G = (V, \mu, \rho)$.

2.2 Fuzzy Contexts and Fuzzy Concepts

The theory of concept lattices has been generalized from the point of view of fuzzy logic in [1, 9, 24, 27]. In this sub-section, first we discuss the approach made by Bělohlávek. The next sub-section gives the basics of proto-fuzzy concepts.

2.2.1 Fuzzy Concept Lattice Introduced by Bělohlávek

We start with a set X of objects, a set Y of attributes, a complete residuated lattice $\mathbf{L}$ and a fuzzy relation I between X and Y. The key idea of a fuzzy context ($\mathbf{L}$-context) is as follows: it is a triplet $\langle X, Y, I\rangle$, where $I(x, y) \in L$ (the set of truth values of complete residuated lattice L) is interpreted as the truth value of the fact, "the object $x \in X$ has the attribute $y \in Y$". For fuzzy sets $A \in L^X$ and $B \in L^Y$, Bělohlávek [1] and, independently, Pollandt [27] defined the fuzzy sets $A^{\uparrow} \in L^Y$ and $B^{\downarrow} \in L^X$ according to the formulas

$$A^{\uparrow}(y) = \bigwedge_{x\in X}\{A(x) \to I(x, y)\}$$
$$B^{\downarrow}(x) = \bigwedge_{y\in Y}\{B(y) \to I(x, y)\}$$

One can easily interpret the element $A^{\uparrow}(y) \in A^{\uparrow}$ as the truth degree of "y is shared by all objects from A" and $B^{\downarrow}(x) \in B^{\downarrow}$ as the truth degree of "x has all attributes from B".

A fuzzy concept $\langle A, B\rangle$ consists of a fuzzy set A of objects (the extent of the concept) and a fuzzy set B of attributes (the intent of the concept) such that $A^{\uparrow} = B$ and $B^{\downarrow} = A$. As the size of dataset grows, the fuzzy concepts generated from fuzzy context become larger in number. Since it is very hard to deal with a large number of fuzzy concepts, Bělohlávek et al. [3] introduced the notion of crisply generated fuzzy concepts. The fuzzy concept $\langle A, B\rangle$ is called crisply generated if there is a crisp set $B_c \subseteq Y$ such that $A = B_c^{\downarrow}$ (and thus $B = B_c^{\downarrow\uparrow}$). If $B\langle X, Y, I\rangle = \{\langle A, B\rangle \mid A^{\uparrow} = B, B^{\downarrow} = A\}$ denotes the set of all fuzzy concepts of the fuzzy context $\langle X, Y, I\rangle$, then the set $B\langle X, Y, I\rangle$ with the order relation:

$\langle A_1, B_1\rangle \leq \langle A_2, B_2\rangle$ if and only if $A_1 \subseteq A_2$ (or, equivalently $B_1 \supseteq B_2$) is a complete lattice. The lattice $(B\langle X, Y, I\rangle, \leq)$ is called a fuzzy concept lattice.

2.2.2 Proto-Fuzzy Concepts Introduced by Krídlo and Krajči [25, 26]

Let $\langle X, Y, I\rangle$ be an fuzzy context, where X and Y are set of objects (X) and set of attributes (Y), respectively and I is a fuzzy relation between X and Y. Since the value $I(x, y)$ express the degree to which the object x carries the attribute y. If we set a threshold value $t \in L$ to eliminate the lower degree membership value from fuzzy relation then the resulting relation is called t-cut of fuzzy context which is basically a binary relation between X and Y and is denoted by I_t. For every confidence threshold $t \in L$, consider two sets: $A' = \{y \in Y \mid \forall x \in A : I(x, y) \geq t\}$ for $A \subseteq X$, i.e., the set of all attributes from Y shared by all objects of A at least with the degree t and

$B' = \{x \in X \mid \forall y \in B : I(x, y) \geq t\}$ for $B \subseteq Y$, i.e., the set of all objects from X sharing attributes from B at least to the degree t. The pair $\langle A, B\rangle \in 2^X \times 2^Y$ is called t-concept iff $A' = B$, $B' = A$. The set of all t-concept in the t-cut is denoted by C_t.

The Triples $\langle A, B, t\rangle \in 2^X \times 2^Y \times L$ such that $\langle A, B\rangle \in \bigcup_{k \in L} C_k$ and $t = sup\{k \in L : \langle A, B\rangle \in C_k\}$ are called proto-fuzzy concepts. i.e., the proto-fuzzy concept is triple of a subset of objects, a subsets of attributes and a value as a best common degree of membership of all pairs of objects and attributes from the above-mentioned sets to the fuzzy context. The set of proto-fuzzy concepts denoted by C^P.

3 Proposed Fuzzy Graph Based Proto-Fuzzy Concepts Generation Technique

Since one major disadvantage of application of fuzzy formal concept analysis is that large numbers of fuzzy concepts are generated from fuzzy context, it is practically impossible to analyze such a large amount of concepts. Given such a situation, without generating a whole set of fuzzy concepts, it would be of use if one could develop an useful technique to determine proto-fuzzy concepts for each $t \in L$ as per users requirement.

In this section, for a given fuzzy context $\langle X, Y, I\rangle$, we define a fuzzy graph and we show that t-concepts can be found corresponding to each maximal cliques of t-level graph of the defined fuzzy graph. After that, we determine all those cliques which corresponds to the proto-fuzzy concepts of degree t. Now we present our approach step-by-step.

1. For a given fuzzy context a fuzzy graph has been defined.
2. Using this fuzzy graph, proto-fuzzy concepts have been derived. Detailed steps are given below.
   (a) In Theorem 1, it has been derived that each t-concept correspond to a maximal clique of t-level graph of fuzzy graph.
   (b) In Theorem 2, it has been proved that the proto-fuzzy concepts can also be generated as maximal cliques of t-level graph of fuzzy graph.

3.1 Fuzzy Graph Defined for a Formal Context

In this sub-section, we define an underlying fuzzy graph from a given fuzzy context in the following way:

Let $K = \langle X, Y, I\rangle$ be a fuzzy context where X is the set of objects and Y is the set of properties. Let $\{\alpha_{o_i} | o_i \in X\}$ and $\{\beta_{p_j} | p_j \in Y\}$ be two family of fuzzy subsets of Y and X , respectively, where for each $o_i \in X$, $\alpha_{o_i}(p_j) = I(o_i, p_j)$ for all $p_j \in Y$ and for each $p_j \in Y$, $\beta_{p_j}(o_i) = I(o_i, p_j)$ for all $o_i \in X$. We construct the fuzzy graph $G_I = (\mu, \rho)$, where $\mu : X \bigcup Y \rightarrow [0, 1]$ is defined by

$\mu(o_i) = h(\alpha_{o_i})$ for all $o_i \in X$ and $\mu(p_j) = h(\beta_{p_j})$ for all $p_j \in Y$ and $\rho : (X \bigcup Y) \times (X \bigcup Y) \rightarrow [0, 1]$ is defined by

$$\rho(o_i, o_j) = \begin{cases} h(\alpha_{o_i} \bigcap \alpha_{o_j}) \; if \; o_i, o_j \in X, i \neq j \\ 0 \; if \; i = j \end{cases}$$
$$\rho(p_i, p_j) = \begin{cases} h(\beta_{p_i} \bigcap \beta_{p_j}) \; if \; p_i, p_j \in Y, i \neq j \\ 0 \; if \; i = j \end{cases}$$
$$\rho(o_i, p_j) = I(o_i, p_j) \; if \; o_i \in X \; and \; p_j \in Y$$

It is to be noticed that the computations of proto-fuzzy concepts of a specific degree t or, computing lattice **L** involved in proto-fuzzy concepts of a specific degree t may be useful in different applications. But the major disadvantage of computing such concepts or, lattice is that we need to compute all t-concepts from t-cut of the fuzzy context for each values of $t \in L$. In the next sub-sections, we show that how the above defined fuzzy graph is used to compute proto-fuzzy concepts of a specific degree t or, computing lattice **L** involved in proto-fuzzy concepts of a specific degree t without generating all t-concepts from t-cut of the fuzzy context for each values of $t \in L$.

3.2 Generation of t-Concepts

The following theorem shows how the above fuzzy graph could be used to generate all t-concepts for the fuzzy fuzzy context $K = \langle X, Y, I \rangle$.

Theorem 1 *Let $K = \langle X, Y, I \rangle$ be a fuzzy context and G_I be the corresponding fuzzy graph. If $t \in [0, 1]$, then for each clique C of t-level graph, G_I^t there always is a unique t-concept of the fuzzy context $K = \langle X, Y, I \rangle$. Conversely, a unique clique of t-level graph G_I^t exists for each t-concept of $K = \langle X, Y, I \rangle$.*

Proof For each $t \in [0, 1]$, let C be any maximal clique of the t-level graph G_I^t. The vertex set C of any clique can be partitioned into two sets $A = \{o \mid o \in X\}$ and $B = \{p \mid p \in Y\}$. We claim that $\langle A, B \rangle$ is a t-concept with extent A and intent B. Since C is a maximal clique of G_I^t. Therefore, no other vertices in $G_I^t - C$ could be connected to all vertices in C, and also, $A = \{o \in X \mid$ for all $p \in B : I(o, p) \geq t\} = B'$ and $B = \{p \in Y | I(o, p) \geq t$ for all $o \in A\} = A'$.

Conversely, let G_I be the graph for the fuzzy context $K = \langle X, Y, I \rangle$ and $\langle A, B \rangle$ be a t-concept of $K = \langle X, Y, I \rangle$. By the definition of t-concept, $I(o, p) \geq t$ for all $o \in A$, $p \in B$, i.e., $\rho(o, p) \geq t$ for all $o \in A$, $p \in B$ in G_I. This implies, $\rho(o, o') = h(\alpha_o \bigcap \alpha_{o'}) \geq t$ for $o, o' \in A$ and $\rho(p, p') = h(\alpha_p \bigcap \alpha_{p'}) \geq t$ for $p, p' \in B$. Therefore, $C = A \bigcup B$ is a maximal clique t-level graph G_I^t.

3.3 Proto-Fuzzy Concepts Determination from the Graph of Fuzzy Context

In this next section, we show the graph can also be used to generate proto-fuzzy concepts.

Theorem 2 *Let $K = \langle X, Y, I\rangle$ be a fuzzy context and G_I be the corresponding underlying fuzzy graph. If G_I^t is a t-level graph of G_I, then corresponding to each clique of G_I^t, containing the vertices or the edges of degree t in G_I, there exists unique proto-fuzzy concepts of degree t.*

Proof For $t \in [0, 1]$, let $C = A \bigcup B$ be any maximal clique of the t-level graph G_I^t, where $A = \{o_i | o_i \in X\} \subseteq X$ and $B = \{p_j | p_j \in Y\} \subseteq Y$. Also let, C contains some vertices or some edges of which the degree in G_I are t. From Theorem 1, the t-concept of t-cut of fuzzy context corresponding to the clique C is $\langle A, B\rangle$. Since there is one-to-one correspondence between t-concepts and cliques of G_I^t, the t-level graph of G_I. Therefore for each $t' \in [0, 1]$ and $t' > t$, C can not be a clique of the t'-level graphs of G_I, i.e., $\langle A, B\rangle$ can not be a t'-concept of t'-cut of fuzzy context. Hence $\langle A, B, t\rangle$ is a proto-fuzzy concept of degree t.

Generating maximal cliques or maximal independent sets of a given graph is one of the fundamental problems in the theory of graphs. There are several number of algorithm exists for this problem [7, 8, 13, 23, 30], and time complexity of most of these algorithms depends on the number of vertices and number of cliques of G. In this article,our proposed methodology for computing t-concepts or proto-fuzzy concepts involves two steps: (i) computation of a fuzzy graph from the fuzzy context, and (ii) enumeration of all maximal cliques in t-level graph of the fuzzy graph. For computing the fuzzy graph we need to compute $\frac{m(m-1)}{2} + \frac{n(n-1)}{2}$ number of fuzzy sets by performing fuzzy intersection. Therefore The computational work to construct fuzzy graph can be done in $\mathbf{O}((m+n)^2)$ time. Also all maximal cliques can be generated in $\mathbf{O}((m+n)^3)$ time delay (see in [20]).

Example 3 Consider the fuzzy context given in Table 1. The context Table 1 presents the marks of an interview obtained by five students, namely, Adi(o_1), Bimal(o_2), Chaki(o_3), Dipak(o_4) Eva(o_5) in five areas: Physics(p_1), Mathematics(p_2), Chemistry(p_3), English(p_4), Statistics(p_5).

Table 1 Fuzzy context of the given example

	p_1	p_2	p_3	p_4	p_5
o_1	0.9	0.7	0.2	0.4	1
o_2	0.8	1	0.3	0.7	0.9
o_3	0.2	0.2	0.2	0.1	0.3
o_4	0.3	0.6	0.3	0.2	0.2
o_5	0.5	0.8	0.4	0.3	0.4

The calculation of truth values of a fuzzy context, i.e., generation of fuzzy context from raw data depends on the nature of the real-life problem. In a real-life applications, to compute the truth values for above type of problem, one may follow the rule introduced by Zaman et al. [34]. The rule introduced in [34] has also been executed in matlab interface [35] on large data sets.

Now we construct the fuzzy graph $G_I = (\mu, \rho)$ for the above fuzzy context as defined in Sect. 3.1. The fuzzy graph G_I is given by the incidence matrix G_I below, where for $i = 1, 2, \ldots 5$, $\mu(o_i)$ are 1.0, 1.0, 0.3, 0.6, 0.8, respectively, and for $j = 1, 2, \ldots 5$, $\mu(p_j)$ are 0.9, 1.0, 0.4, 0.7, 1.0, respectively. In Fig. 1, (a) represents $G_I^{0.1}$ (b) represents $G_I^{0.2}$ (c) represents $G_I^{0.3}$ (d) represents $G_I^{0.4}$ (e) represents $G_I^{0.5}$

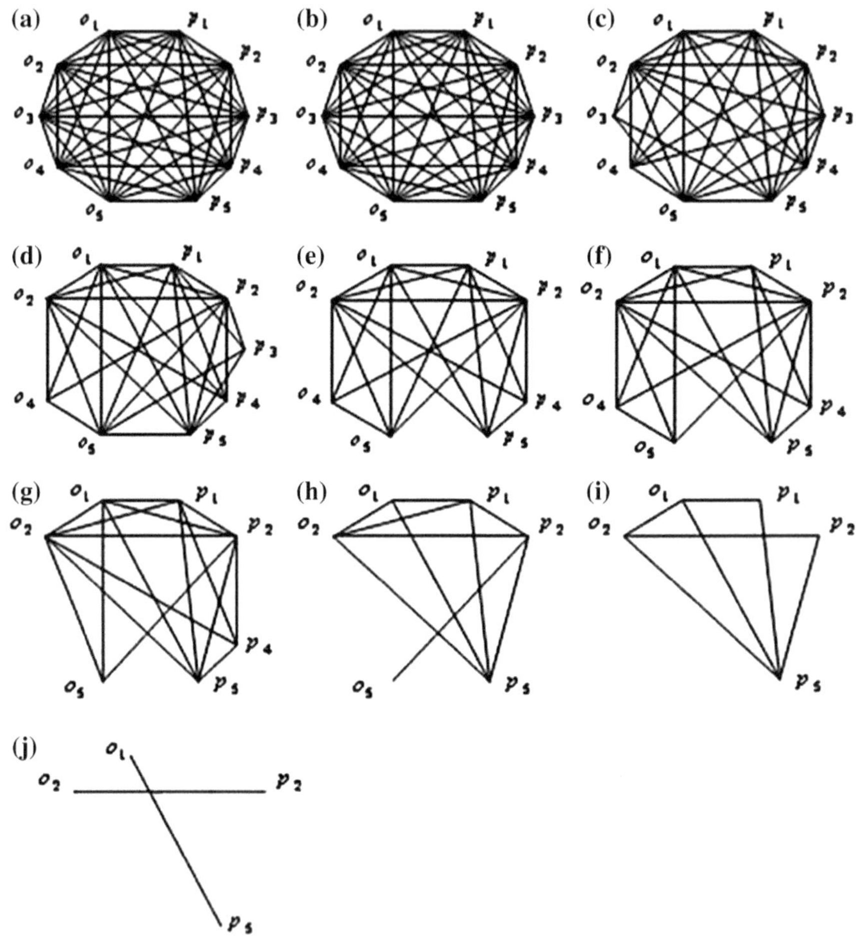

Fig. 1 Cut graph of the fuzzy graph G_I

(f) represents $G_I^{0.6}$ (g) represents $G_I^{0.7}$ (h) represents $G_I^{0.8}$ (i) represents $G_I^{0.9}$ (j) representsG_I^1.

$$G_I = \begin{array}{c} \\ o_1 \\ o_2 \\ o_3 \\ o_4 \\ o_5 \\ p_1 \\ p_2 \\ p_3 \\ p_4 \\ p_5 \end{array} \begin{array}{c} \begin{array}{cccccccccc} o_1 & o_2 & o_3 & o_4 & o_5 & p_1 & p_2 & p_3 & p_4 & p_5 \end{array} \\ \begin{pmatrix} 0.0 & 0.9 & 0.3 & 0.6 & 0.7 & 0.9 & 0.7 & 0.2 & 0.4 & 1.0 \\ 0.9 & 0.0 & 0.3 & 0.6 & 0.8 & 0.8 & 1.0 & 0.3 & 0.7 & 0.9 \\ 0.3 & 0.3 & 0.0 & 0.2 & 0.3 & 0.2 & 0.2 & 0.2 & 0.1 & 0.3 \\ 0.6 & 0.6 & 0.3 & 0.0 & 0.6 & 0.3 & 0.6 & 0.3 & 0.2 & 0.2 \\ 0.7 & 0.8 & 0.3 & 0.6 & 0.0 & 0.5 & 0.8 & 0.4 & 0.3 & 0.4 \\ 0.9 & 0.8 & 0.2 & 0.3 & 0.5 & 0.0 & 0.8 & 0.4 & 0.7 & 0.9 \\ 0.7 & 1.0 & 0.2 & 0.6 & 0.8 & 0.8 & 0.0 & 0.4 & 0.7 & 0.9 \\ 0.2 & 0.3 & 0.2 & 0.3 & 0.4 & 0.4 & 0.4 & 0.0 & 0.3 & 0.4 \\ 0.4 & 0.7 & 0.1 & 0.2 & 0.3 & 0.7 & 0.7 & 0.3 & 0.0 & 0.7 \\ 1.0 & 0.9 & 0.3 & 0.2 & 0.4 & 0.9 & 0.9 & 0.4 & 0.7 & 0.0 \end{pmatrix} \end{array}$$

Now, using Theorem 2, we generate proto-fuzzy concepts of degree $t \in L$ corresponding to each maximal cliques, containing the vertices or the edges of degree t in G_I, of the above t-level graphs. In order to consider only objects related to attributes with relevant grades of membership, a threshold is fixed such that the pairs with membership values less than the threshold are ignored. In real-life applications threshold t can be set by users, or would be selected automatically based on user's query according to the application or the domain knowledge. For instance, assume that a threshold value is fixed equal to 0.3 by user for this problem. Then $\langle\{o_1, o_2, o_5\}, \{p_1, p_2, p_4, p_5\}, 0.3\rangle$ is a proto-fuzzy concept corresponding to the graph $G_I^{0.3}$. The objects o_1, o_2, and o_5 sharing the attributes p_1, p_2, p_4, p_5, and, vice versa, these four attributes are shared by the objects o_1, o_2, and o_5 with best common membership value 0.3 which is not greater than threshold value.

In Table 2, t-level graphs, each maximal clique and its corresponding proto-fuzzy concepts of degree $t \in L$ are shown.

Scalability is a real issue for FCA with fuzzy attributes, since fuzzy concepts can be large in number in the context. The generation of proto-fuzzy concepts [25] with the help of cuts and projections to the objects-values or attribute-values plains of a complex data sets is also an unwieldy task. In this paper, we emphasize on the generation of all proto-fuzzy concepts by segregating them according to their degree. In fact the segregation of proto-fuzzy concepts according to their degree may somehow become useful in different field of applications such as information retrieval, data mining, knowledge discovery etc. But generation of proto-fuzzy concepts by using our proposed method is also a time consuming task for complex data set, since maximal clique problem of a general graph is an NP-hard problem. In our future work, our aim is to determine the rules by which we can generate all proto-fuzzy concepts directly from fuzzy context.

Table 2 Proto-fuzzy concepts of the fuzzy context given in Table 1

t-level graph	Maximal cliques	Proto-fuzzy concepts of degree t
$G_I^{0.1}$	$\{o_1, o_2, o_3, o_4, o_5, p_1, p_2, p_3, p_4, p_5\}$	$\langle\{o_1, o_2, o_3, o_4, o_5\}, \{p_1, p_2, p_3, p_4, p_5\}, 0.1\rangle$
$G_I^{0.2}$	$\{o_1, o_2, o_4, o_5, p_1, p_2, p_3, p_4, p_5\}$	$\langle\{o_1, o_2, o_4, o_5\}, \{p_1, p_2, p_3, p_4, p_5\}, 0.2\rangle$
	$\{o_1, o_2, o_3, o_4, o_5, p_1, p_2, p_3, p_5\}$	$\langle\{o_1, o_2, o_3, o_4, o_5\}, \{p_1, p_2, p_3, p_5\}, 0.2\rangle$
$G_I^{0.3}$	$\{o_1, o_2, o_5, p_1, p_2, p_4, p_5\}$	$\langle\{o_1, o_2, o_5\}, \{p_1, p_2, p_4, p_5\}, 0.3\rangle$
	$\{o_2, o_5, p_1, p_2, p_3, p_4, p_5\}$	$\langle\{o_2, o_5\}, \{p_1, p_2, p_3, p_4, p_5\}, 0.3\rangle$
	$\{o_1, o_2, o_3, o_5, p_5\}$	$\langle\{o_1, o_2, o_3, o_5\}, \{p_5\}, 0.3\rangle$
	$\{o_2, o_4, o_5, p_1, p_2, p_3\}$	$\langle\{o_2, o_4, o_5\}, \{p_1, p_2, p_3\}, 0.3\rangle$
	$\{o_1, o_2, o_4, o_5, p_1, p_2\}$	$\langle\{o_1, o_2, o_4, o_5\}, \{p_1, p_2\}, 0.3\rangle$
$G_I^{0.4}$	$\{o_1, o_2, p_1, p_2, p_4, p_5\}$	$\langle\{o_1, o_2\}, \{p_1, p_2, p_4, p_5\}, 0.4\rangle$
	$\{o_5, p_1, p_2, p_3, p_5\}$	$\langle\{o_5\}, \{p_1, p_2, p_3, p_5\}, 0.4\rangle$
	$\{o_1, o_2, o_5, p_1, p_2, p_5\}$	$\langle\{o_1, o_2, o_5\}, \{p_1, p_2, p_5\}, 0.4\rangle$
$G_I^{0.5}$	$\{o_1, o_2, o_5, p_1, p_2\}$	$\langle\{o_1, o_2, o_5\}, \{p_1, p_2\}, 0.5\rangle$
$G_I^{0.6}$	$\{o_1, o_2, o_4, o_5, p_2\}$	$\langle\{o_1, o_2, o_4, o_5\}, \{p_2\}, 0.6\rangle$
$G_I^{0.7}$	$\{o_1, o_2, p_1, p_2, p_5\}$	$\langle\{o_1, o_2\}, \{p_1, p_2, p_5\}, 0.7\rangle$
	$\{o_2, p_1, p_2, p_4, p_5\}$	$\langle\{o_2\}, \{p_1, p_2, p_4, p_5\}, 0.7\rangle$
	$\{o_1, o_2, o_5, p_2\}$	$\langle\{o_1, o_2, o_5\}, \{p_2\}, 0.7\rangle$
$G_I^{0.8}$	$\{o_1, o_2, p_1, p_5\}$	$\langle\{o_1, o_2\}, \{p_1, p_5\}, 0.8\rangle$
	$\{o_2, p_1, p_2, p_5\}$	$\langle\{o_2\}, \{p_1, p_2, p_5\}, 0.8\rangle$
	$\{o_2, o_5, p_2\}$	$\langle\{o_2, o_5\}, \{p_2\}, 0.8\rangle$
$G_I^{0.9}$	$\{o_1, p_1, p_5\}$	$\langle\{o_1\}, \{p_1, p_5\}, 0.9\rangle$
	$\{o_2, p_2, p_5\}$	$\langle\{o_2\}, \{p_2, p_5\}, 0.9\rangle$
	$\{o_1, o_2, p_5\}$	$\langle\{o_1, o_2\}, \{p_5\}, 0.9\rangle$
G_I^{1}	$\{o_1, p_5\}$	$\langle\{o_1\}, \{p_5\}, 1\rangle$
	$\{o_2, p_2\}$	$\langle\{o_2\}, \{p_2\}, 1\rangle$

4 Conclusion

In this paper, we begin with defining a fuzzy graph for a given fuzzy context. We then show that t-concepts as well as proto-fuzzy concepts correspond to each maximal cliques of t-level graphs of the defined fuzzy graph. As far as our knowledge is concerned, there is no general method to generate proto-fuzzy concepts. Therefore, presentation of fuzzy context by a fuzzy graph and generating all proto-fuzzy concepts using the fuzzy graph creates an important relationship between the two fields of fuzzy concept lattice theory and fuzzy graph theory. Also the apparent advantage of our approaches is that one can keep control over the values of the lattice **L** involved in proto-fuzzy concepts.

References

1. Bělohlávek, R.: Lattices generated by binary fuzzy relations (extended abstract). In: Abstract of the 4th International Conference on Fuzzy Sets Theory and Its Applications, Liptovsky Jan', Slovakia, p. 11 (1998)
2. Bělohlávek, R.: Fuzzy Relational Systems: Foundations and Principles. Kluwer Academic/Plenum Publishers, New York (2002)
3. Bělohlávek, R., Sklenar, V., Zacpal, J.: Crisply generated fuzzy concepts. In: Ganter, B., Godin, R. (eds.) ICFCA 2005. Lecturer Notes in Computer Science, vol. 3403, pp. 268–283. Springer, Berlin (2005)
4. Bělohlávek, R., Vychodil, V.: Reducing the size of fuzzy concept lattices by hedges. In: The IEEE International Conference on Fuzzy Systems, FUZZ-IEEE 2005 (Proceedings on CD), pp. 663–668, Reno, Nevada, USA, 22–25 May 2005, Abstract in Printed Proceedings, p. 44, ISBN: 0-7803-9158-6
5. Berry, A., Sigayret, A.: Representation a concept lattice by a graph. Discret. Appl. Math. **144**, 27–42 (2004)
6. Bhattacharya, P.: Some remarks on fuzzy graphs. Pattern Recognit. Lett. **6**, 297–302 (1987)
7. Bomze, I.M., Budinich, M., Pardalos, P.M., Pelillo, M.: The maximum clique problem. In: Du, D.-Z., Pardalos, P.M. (eds.) Handbook of Combinatorial Optimization, Supplement vol. A, pp. 1–74. Kluwer, Dordrecht (1999)
8. Bron, C., Kerbosch, J.: Algorithm 457, finding all cliques of an undirected graph. Commun. ACM **16**, 575–577 (1973)
9. Burusco, A., Fuentes-Gonzalez, R.: The study of the L-fuzzy concept lattices. Mathw. Soft Comput. **3**, 209–218 (1994)
10. Burusco, A., Fuentes-Gonzalez, R.: Construction of the L-fuzzy concept lattice. Fuzzy Sets Syst. **97**, 109–114 (1998)
11. Chakroborthy, M.K., Das, M.: Study in fuzzy relations over fuzzy subsets. Fuzzy Sets Syst. **9**, 79–89 (1983)
12. Chakroborthy, M.K., Das, M.: Fuzzy hypergraphs. In: Proceedings of International Conference on Logic Programming Born (1997)
13. Chiba, N., Nishizeki, T.: Arboricity and subgraph listing algorithms. SIAM J. Comput. **14**, 210–223 (1985)
14. Dubois, D., Prade, H.: Fuzzy Sets and Systems, Theory and Applications. Academic Press, New York (1980)
15. Ganter, B.: Two Basic Algorithms in Concept Analysis, Preprint 831. Technische Hochschule Darmstadt, Darmstadt (1984)
16. Ganter, B., Wille, R.: Formal Concept Analysis, Mathematical Foundation. Springer, Berlin (1999)
17. Ghosh, P., Kundu, K., Sarkar, D.: Fuzzy graph representation of a fuzzy concept lattice. Fuzzy Sets Syst. **161**, 1669–1675 (2010)
18. Goguen, J.A.: L-fuzzy set. J. Math. Anal. Appl. **18**(1), 145–174 (1967)
19. HöHle, U.: On the fundamentals of fuzzy set theory. J. Math Anal. Appl. **201**, 786–826 (1996)
20. Jhonson, D.S., Yannakakis, M., Papadimitriou, C.H.: On generating all maximal independent sets. Inf. Process. Lett. **27**, 119–123 (1988)
21. John, N.M., Premchand, S.N.: Fuzzy Graphs Fuzzy Hyper Graph. Springer, Berlin (2000)
22. Kaufmann, A.: Introduction a lathèorielath des sous-ensembles flous. Masson, Paris (1973)
23. Koch, I.: Enumerating all connected maximal common subgraphs in two graphs. Theor. Comput. Sci. **250**, 1–30 (2001)
24. Krajči, S.: Cluster based efficient generation of fuzzy concepts. Neural Netw. World **5**, 521–530 (2003)
25. Krídlo, O., Krajči, S., Proto-fuzzy concepts, their retrieval and usage. In: Proceedings of Sixth International Conference on Concept Lattices and their Applications (CLA 2008), pp. 83–96
26. Krídlo, O., Krajči, S., Fuzzy concept lattice is made by proto-fuzzy concepts. IFSA-EUSFLAT (2009)

27. Pollandt, S.: Fuzzy Begriffe. Springer, Berlin (1997)
28. Rosenfield, A.: Fuzzy graphs. In: Zadeh, L.A., Fu, K.-S., Tanaka, K., Shimura, M. (eds.) Fuzzy Sets and their Applications to Cognitive and Decision Process, pp. 77–96. Academic Press, New York (1975)
29. Takeda, E.: Connectvity in fuzzy graphs. Tech. Rep. Osaka Univ. **23**, 343–352 (1973)
30. Tomitaa, E., Tanakaa, A., Takahashia, H.: The worst-case time complexity for generating all maximal cliques and computational experiments. Theor. Comput. Sci. **363**, 28–42 (2006)
31. Wille, R.: Restructuring lattice theory: an approach based on hierarchies of concepts. In: Rival, I. (ed.) Ordered Sets, pp. 445–470. Reidel, Dordrecht, Boston (1982)
32. Yeh, R.T., Bang, S.Y.: Fuzzy relations, fuzzy graphs, and their applications to clustering analysis. In: Zadeh, L.A., Fu, K.S., Shimura, M. (eds.) Fuzzy Sets and their Applications, pp. 125–149. Academic Press, New York (1975)
33. Zadeh, L.A.: Fuzzy sets. Inf. Control **8**, 338–353 (1965)
34. Zaman, W., Ghosh, P., Dutta, K.K., Basu, P.N.: A framework to incorporate domain knowledge modeling. Math. Sci. Int. Res. J. **2**, 474–478 (2013)
35. Zaman, W., Ghosh, P., Basu, P. N.: A Prototype to incorporate domain knowledge computation in an education system. Br. J. Educ. Technol. (communicated)

Part II
Logic of Uncertainty

Open World Models: A View from Rough Set Theory

Mohua Banerjee, Shier Ju, Md. Aquil Khan and Liping Tang

Abstract In rough set theory, *information systems* are used to represent knowledge-bases, particularly when practical applications are involved. We explore one such use of information systems, for representing the *open world model*. Open world information systems are defined, and a temporal logic, including descriptors and the global modality, is proposed as a formal reasoning framework for these structures.

Keywords Open world model · Rough sets · Information systems · Temporal logics

1 Introduction

Ignorance and uncertainty are part of the knowledge[1] system of any community. An 'open world model' is based on this premise, and thus embodies a variety of situations. For instance, categories of concepts that applied to some members of a community at some time point in a world model, may not apply to them any further with the arrival of new information. Communities constituting the world may also change, with the arrival of new members or departure of old ones. In this sense, the world is 'open', subject to change.our study on this article on a specific scenario

[1] In this discourse, we identify 'knowledge' with 'information'.

The author acknowledges support of the Institute of Logic and Cognition, Sun Yat-Sen University, Guangzhou, during a visit from which the work was initiated.

M. Banerjee (✉)
Department of Mathematics and Statistics, Indian Institute of Technology, Kanpur, India
e-mail: mohua@iitk.ac.in

S. Ju · L. Tang
Institute of Logic and Cognition, Sun Yat-Sen University, Guangzhou, China
e-mail: hssjse@mail.sysu.edu.cn

Md.A. Khan
Discipline of Mathematics, Indian Institute of Technology, Indore, India
e-mail: aquilk@iiti.ac.in

M.K. Chakraborty et al. (eds.), *Facets of Uncertainties and Applications*,
Springer Proceedings in Mathematics & Statistics 125,
DOI 10.1007/978-81-322-2301-6_6

Fig. 1 An open world model

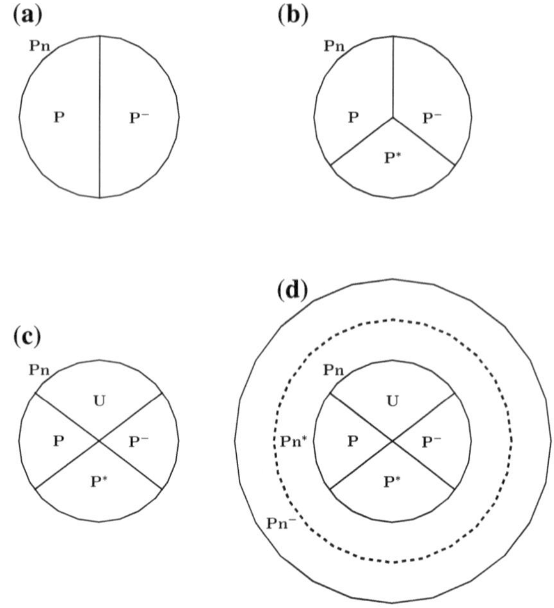

Pn: pneumonia **P**: typical pneumonia **P**$^-$: non-typical pneumonia
P*: SARS **U**: unknown pneumonia **Pn***: category unknown
Pn$^-$: non-pneumonia

illustrating an open world model. However, as we point out at the end, the framework may easily be generalized.

Acute pneumonia, till late 2002, was clinically classified into the 'typical' and 'non-typical' types. One may thus begin with a world model (Fig. 1a) of the 'category' pneumonia that consists of patients classified into the P (typical) and P^- (non-typical) regions. In 2002–2003, the disease SARS (Severe Acute Respiratory Syndrome) was encountered. The pathogeny of this disease could not be totally figured out, although it was indicative of a kind of pneumonia that was different from the typical and non-typical types. The previous world model thus 'opens' into one given in Fig. 1b, which accommodates patients afflicted by SARS in the 'ignorance' region P^*.

In fact, considering that there may well be another kind of pneumonia currently not classifiable into any of these three regions, the second model opens into the one in Fig. 1c, having the added 'uncertainty' region U.

One may further enhance the world model by considering the categorial picture in entirety, i.e. by including the region Pn^- of non-pneumonia, and the region Pn^* consisting of patients currently not classifiable into either Pn or Pn^-. Thus we get the open world model in Fig. 1d.

With time then, this world model evolves. The classifications may get modified with new information. New members may also enter the picture, altering some classification regions and expanding the world model at large.

Of course, a more precise model of the world at any time would include, for instance, (i) several categories (we have only Pn here), (ii) other distinct classifications inside a category (only P and P^- here), or (iii) more than one ignorance region (only P^* here). In this article, we address the above simple model and we shall see that these generalizations of the model would not be hard to obtain.

In rough set theory, information about objects comes in terms of attributes and attribute-values—the basic entities in an information system. In the following section, we define an *open world information system* (OWIS) to represent the open world model, and observe some of its features. These structures incorporate an element of time, in order to capture the world changing with time as described earlier. Logics of open world have been studied in the literature. In [2], a three-valued logic of open world is defined, where the behaviour of open world is characterized by a special Boolean negation. For formal reasoning with OWIS, we propose a temporal logic $\mathcal{L}_{OW}$ here (cf. Sect. 3). The logic contains, apart from temporal modalities, 'descriptors' as part of the collection of atomic formulae, and the global modality. Descriptors represent sets of objects that take a particular value for a particular attribute. These were first used in Pawlak's *decision logic* [8], bringing 'information' explicitly into the syntax of a formal reasoning framework. We follow this line to formulate $\mathcal{L}_{OW}$. Descriptors have been included in the language to obtain logics of different kinds of information systems, e.g. in [4, 5], but none of these contain temporal modalities. On the other hand, temporal logics for rough sets have been studied in [1, 3], but these do not contain descriptors. The logic $\mathcal{L}_{OW}$ proposed here introduces in its syntax, a combination of (i) information as dealt with in rough set theory (through descriptors) and (ii) time. This amalgamation appears just right for reasoning with open world models. Section 4 concludes the article, indicating future lines of work.

2 Open World Information Systems

Let us recall the definition of an information system [8].

Definition 1 An *information system* $\mathcal{S} := (W, \mathcal{A}, \{V_a\}_{a \in A}, f)$, comprises a non-empty set W of objects, a non-empty set $\mathcal{A}$ of attributes, for each $a \in \mathcal{A}$, a non-empty set V_a of attribute-values of the attribute a, and an assignment $f : W \times \mathcal{A} \to \bigcup_{a \in \mathcal{A}} V_a$ such that $f(x, a) \in V_a$, for any $x \in W, a \in \mathcal{A}$.

In [7], the notion was extended by adding the concept of time. A set T of 'time points' and a linear order $<$ on T were included to define a *dynamic information system* $\mathcal{DS} := (W, \mathcal{A}, \{V_a\}_{a \in \mathcal{A}}, T, <, \{f_t\}_{t \in T})$, where for each $t \in T$, $(W, \mathcal{A}, \{V_a\}_{a \in \mathcal{A}}, f_t)$ is an information system.

In a dynamic information system, the domain of discourse is fixed and does not change with time. In order to get to the open world model, we consider the case where the domain may expand with time. We make a restriction, however: objects may be added, but not discarded. The time line is taken to be an initial segment of $\mathbb{N}$, the set of natural numbers, together with the natural ordering on it. We have the following structure:

Definition 2 A *temporal information system* is a tuple

$$\mathcal{K} := (W, N, \mathcal{A}, \{V_a\}_{a\in\mathcal{A}}, \{W_t\}_{t\in N}, \{f_t\}_{t\in N}),$$

where

- N is an initial segment of $\mathbb{N}$,
- for each $t \in N$, $W_t \subseteq W$ such that $W_t \subseteq W_{t+1}$ for $t, t+1 \in N$,
- for each $t \in N$, $(W_t, \mathcal{A}, \{V_a\}_{a\in\mathcal{A}}, f_t)$ is an information system.

It is to be noted that this structure is not the one discussed in temporal databases, and is also different from the one considered in [9].

We now define an open world information system to be an extension of the temporal information system, where we provide sets of distinguished attributes and attribute-values.

Definition 3 An *open world information system* (OWIS) is a tuple

$$\mathcal{S} := (\mathcal{K}, B, C_B, \{D_B^t\}_{t\in N}, \{v_b\}_{b\in B}, \{v'_c, v''_c\}_{c\in C_B}, \{v_d\}_{d\in D_B^t}),$$

where

- $\mathcal{K} := (W, N, \mathcal{A}, \{V_a\}_{a\in\mathcal{A}}, \{W_t\}_{t\in N}, \{f_t\}_{t\in N})$ is a temporal information system,
- $B, C_B \subseteq \mathcal{A}$ are non-empty sets of *special attributes*,
- for each $t \in N$, $D_B^t \subseteq \mathcal{A}$ is a set of *special attributes*,
- for each $b \in B$, v_b is a distinguished value from V_b,
- for each $c \in C_B$, v'_c, v''_c are distinct distinguished values from V_c,
- for each $d \in D_B^t$, v_d is a distinguished value from V_d,
- for each $b \in B$, $f_t(x, b) = v_b$ implies $f_{t'}(x, b) = v_b$, where $t \leq t'$,
- for each $c \in C_B$, $f_t(x, c) = v'_c$ implies $f_{t'}(x, c) = v'_c$, and $f_t(x, c) = v''_c$ implies $f_{t'}(x, c) = v''_c$, where $t \leq t'$.

So in an OWIS, the domain of objects also evolves with time with new objects being added. Moreover, objects which take the values v_b and v'_c, v''_c for the attributes $b \in B$ and $c \in C_B$, respectively, continue to do so at all future time points.

Given an OWIS $\mathcal{S}$ as above, for each $t \in N$, we obtain the following subsets of W.

- $P_t := \{x \in W_t : f_t(x, b) = v_b$, for each $b \in B$, $f_t(x, c) = v'_c$, for each $c \in C_B\}$,
- $P_t^- := \{x \in W_t : f_t(x, b) = v_b$, for each $b \in B$, $f_t(x, c) = v''_c$, for each $c \in C_B\}$,

- $P_t^* := \{x \in W_t : f_t(x, b) = v_b$, for each $b \in B$, $f_t(x, d) = v_d$, for each $d \in D_B^t$, $f_t(x, c_1) \neq v'_{c_1}$ and $f_t(x, c_2) \neq v''_{c_2}$, for some $c_1, c_2 \in C_B\}$,
- $U_t := \{x \in W_t : f_t(x, b) = v_b$, for each $b \in B$, $f_t(x, d) \neq v_d$, for some $d \in D_B^t$, $f_t(x, c_1) \neq v'_{c_1}$ and $f_t(x, c_2) \neq v''_{c_2}$, for some $c_1, c_2 \in C_B\}$,
- $Pn_t^- := \{x \in W_t : f_t(x, b) \neq v_b$, for each $b \in B\}$,
- $Pn_t^* := \{x \in W_t : f_t(x, b) \neq v_b$, for some $b \in B$, and $f_t(x, b') = v_{b'}$, for some $b' \in B\}$.

Proposition 1 *For each $t, t_1, t_2 \in N$, we have the following.*

1. *The sets P_t, P_t^-, P_t^*, U_t, Pn_t^-, Pn_t^* form a partition of W_t.*
2. *$P_{t_1} \subseteq P_{t_2}$ for $t_1 \leq t_2$.*
3. *$P_{t_1}^- \subseteq P_{t_2}^-$ for $t_1 \leq t_2$.*

We can take the pneumonia case to illustrate again. The special attributes in B are used to distinguish pneumonia and non-pneumonia, while C_B is used to distinguish typical pneumonia and non-typical pneumonia. D_B^t is used to classify new types of pneumonia such as SARS, which may have come into the picture at time point t. Therefore, in the partition $\{P_t, P_t^-, P_t^*, U_t, Pn_t^-, Pn_t^*\}$, we may interpret P_t, P_t^- and P_t^* to be, respectively, the classes of objects (patients) with typical pneumonia, non-typical pneumonia and a third type of pneumonia. U_t represents the class of patients who have pneumonia, but of a kind that is currently (at time point t) not classifiable into any of these three types. Pn_t^- represents the class of patients who do not have pneumonia, and Pn_t^* those for whom it is not known whether they have pneumonia or not.

Note that D_B^t may be empty at some time point t, making U^t empty. The information system at t would then, for example, represent the scenario in the second world model (cf. Fig. 1b, Sect. 1). On the other hand, it is possible that we get progressively better characterization of the third type of pneumonia over time, and at some t_0, the set $D_B^{t_0}$ of attributes fully characterizes it. In that case, one may wish to stipulate that, for all $t' \geq t_0$, $D_B^{t_0} = D_B^{t'}$ and also that, for each $d \in D_B^{t_0}$, $f_{t_0}(x, d) = v_d$ implies $f_{t'}(x, d) = v_d$. Another possibility is that the assignments f_t of values v_d alter so that an object currently in region U^t is able to enter $P_{t'}^*$ at some later time point t'. Similarly, an object currently in Pn_t^* may enter one of $P_{t'}$, $P_{t'}^-$, $P_{t'}^*$, $U_{t'}$, or even $Pn_{t'}^-$, at a later t'.

Thus we get information systems which formalize the open world model described in Sect. 1. The temporal logic $\mathcal{L}_{OW}$ for open world information systems is proposed in this backdrop.

3 A Logic for Open World Information Systems

The syntax and semantics of the logic $\mathcal{L}_{OW}$ are given as follows.

3.1 Syntax

The language of $\mathcal{L}_{OW}$ consists of

(i) a non-empty finite set $\mathcal{A}$ of attribute constants,
(ii) for each $a \in \mathcal{A}$, a non-empty finite set V_a of attribute-value constants,
(iii) non-empty subsets B, C_B of $\mathcal{A}$ of distinguished attribute constants,
(iv) for each $t \in \mathbb{N}$, subset D^t_B of $\mathcal{A}$ of distinguished attribute constants,
(v) corresponding to each $b \in B$, a distinguished attribute-value constant $v_b \in V_b$,
(vi) corresponding to each $c \in C_B$, two distinct distinguished attribute-value constants $v'_c, v''_c \in V_c$,
(vii) corresponding to each $d \in D^t_B$, a distinguished attribute-value constant $v_d \in V_d$,
(viii) a non-empty countable set PV of propositional variables, and
(ix) constants $\top$ and $\bot$.

Atomic well-formed formulae (wffs) are the constants $\top$ and $\bot$, propositional variables p from PV, and *descriptors*, i.e. pairs (a, v) for each $a \in \mathcal{A},\ v \in V_a$. The set of all descriptors is denoted as $\mathcal{D}$.
Using the Boolean logical connectives $\neg$ (negation) and $\wedge$ (conjunction), temporal operators $\mathcal{U}$ (until), $\mathcal{S}$ (since), $\oplus$ (next), $\ominus$ (previous), global modal operator A, wffs of $\mathcal{L}_{OW}$ are then defined recursively as:

$$(a, v) \mid p \mid \neg\alpha \mid \alpha \wedge \beta \mid A\alpha \mid \alpha\mathcal{U}\beta \mid \alpha\mathcal{S}\beta \mid \oplus\alpha \mid \ominus\alpha.$$

Apart from the usual derived connectives $\wedge, \longrightarrow, \longleftrightarrow$, there are the following:
$E\alpha := \neg A\neg\alpha$;
$F\alpha := \top\mathcal{U}\alpha$ (*some time in the future*);
$G\alpha := \neg F\neg\alpha$ (*always in the future*);
$P\alpha := \top\mathcal{S}\alpha$ (*some time in the past*);
$H\alpha := \neg P\neg\alpha$ (*always in the past*).
We use the symbol $\mathcal{L}_{OW}$ to denote the set of all wffs as well.

3.2 Semantics

The semantics of $\mathcal{L}_{OW}$ is based on the temporal information systems. Formally, we have the following.

Definition 4 A *model*, is a tuple $\mathfrak{M} := (\mathcal{K}, V)$, where

- $\mathcal{K} := (W, N, \mathcal{A}, \{V_a\}_{a\in\mathcal{A}}, \{W_t\}_{t\in N}, \{f_t\}_{t\in N})$ is a temporal information system.
- $V : PV \to 2^{W\times N}$.

Satisfiability of a wff α in a model $\mathfrak{M} := (\mathcal{K}, V)$ as above at time point $t \in N$, at an object w of the domain W at time t, i.e. $w \in W_t$, denoted as $\mathfrak{M}, t, w \models \alpha$, is given as follows.

Definition 5

- $\mathfrak{M}, t, w \models \top$, and $\mathfrak{M}, t, w \not\models \bot$, for each $w \in W_t$.
- $\mathfrak{M}, t, w \models p$, if and only if $(w, t) \in V(p)$, for $p \in PV$.
- $\mathfrak{M}, t, w \models (a, v)$ if and only if $f_t(w, a) = v$, for any descriptor (a, v).
- $\mathfrak{M}, t, w \models \neg\alpha$, if and only if $\mathfrak{M}, t, w \not\models \alpha$.
- $\mathfrak{M}, t, w \models \alpha \wedge \beta$, if and only if $\mathfrak{M}, t, w \models \alpha$ and $\mathfrak{M}, t, w \models \beta$.
- $\mathfrak{M}, t, w \models A\alpha$, if and only if $\mathfrak{M}, t, u \models \alpha$ and for all $u \in W_t$.
- $\mathfrak{M}, t, w \models \oplus\alpha$, if and only if $t < |N|$, and $\mathfrak{M}, t+1, w \models \alpha$.
- $\mathfrak{M}, t, w \models \ominus\alpha$, if and only if $t > 1$, and $\mathfrak{M}, t-1, w \models \alpha$.
- $\mathfrak{M}, t, w \models \alpha\mathcal{U}\beta$, if and only if there exists j with $t \leq j \leq |N|$ such that $\mathfrak{M}, j, w \models \beta$, and for all k such that $t \leq k < j$, $\mathfrak{M}, k, w \models \alpha$;
- $\mathfrak{M}, t, w \models \alpha\mathcal{S}\beta$, if and only if there exists j with $1 \leq j \leq t$ such that $\mathfrak{M}, j, w \models \beta$, and for all k such that $j < k \leq t$, $\mathfrak{M}, k, w \models \alpha$;

Conditions of satisfiability of the derived connectives E, F and G are then obtained as follows:

- $\mathfrak{M}, t, w \models E\alpha$, if and only if $\mathfrak{M}, t, u \models \alpha$ and for some $u \in W_t$.
- $\mathfrak{M}, t, w \models F\alpha$ if and only if there exists a j with $t \leq j \leq |N|$ such that $\mathfrak{M}, j, w \models \alpha$;
- $\mathfrak{M}, t, w \models G\alpha$ if and only if for all j with $t \leq j \leq |N|$, $\mathfrak{M}, j, w \models \alpha$;

Satisfiability of the connectives P and H can be obtained similarly.

The extension of a wff α relative to a model $\mathfrak{M} := (\mathcal{K}, V)$ and the time point t, denoted as $[\![\alpha]\!]_{\mathfrak{M},t}$, is given by the set $\{w \in W_t : \mathfrak{M}, t, w \models \alpha\}$. A wff α is said to be *valid in a model* $\mathfrak{M}$, if $[\![\alpha]\!]_{\mathfrak{M},t} = W_t$ for all t. A wff is said to be *1-valid in* $\mathfrak{M}$, if $[\![\alpha]\!]_{\mathfrak{M},1} = W_1$. A wff α is said to be *valid* and *1-valid* according as α is valid and 1-valid in all models respectively. We shall use $\models \alpha$ to denote that the wff α is a valid wff.

We base the semantics of $\mathcal{L}_{OW}$ on temporal information systems (TIS) instead of OWIS, for a neater presentation. This can be done without any loss, as the attribute and attribute-value sets which generate an OWIS from a TIS are already embedded in the language of $\mathcal{L}_{OW}$. In other words, given any TIS $\mathcal{K} := (W, N, \mathcal{A}, \{V_a\}_{a \in \mathcal{A}}, \{W_t\}_{t \in N}, \{f_t\}_{t \in N})$, the attribute and attribute-value constants in the language of $\mathcal{L}_{OW}$ determine an OWIS $\mathcal{S} := (\mathcal{K}, B, C_B, \{D_B^t\}_{t \in N}, \{v_b\}_{b \in B}, \{v'_c, v''_c\}_{c \in C_B}, \{v_d\}_{d \in D_B^t})$. Now consider the following wffs in $\mathcal{L}_{OW}$.

- $\alpha_P := \bigwedge_{b \in B}(b, v_b) \wedge \bigwedge_{c \in C_B}(c, v'_c)$.
- $\alpha_{P^-} := \bigwedge_{b \in B}(b, v_b) \wedge \bigwedge_{c \in C_B}(c, v''_c)$.
- $\alpha_{P^*} := \bigwedge_{b \in B}(b, v_b) \wedge \bigwedge_{d \in D_B}(d, v_d) \wedge \left(\bigvee_{c \in C_B} \neg(c, v'_c)\right) \wedge \left(\bigvee_{c \in C_B} \neg(c, v''_c)\right)$.
- $\alpha_U := \bigwedge_{b \in B}(b, v_b) \wedge \bigvee_{d \in D_B} \neg(d, v_d) \wedge \left(\bigvee_{c \in C_B} \neg(c, v'_c)\right) \wedge \left(\bigvee_{c \in C_B} \neg(c, v''_c)\right)$.
- $\alpha_{Pn^-} := \bigwedge_{b \in B} \neg(b, v_b)$.
- $\alpha_{Pn^*} := \bigvee_{b \in B}(b, v_b) \wedge \neg \bigwedge_{b \in B}(b, v_b)$.

For the sets $P_t, P_t^-, P_t^*, U_t, Pn_t^-, Pn_t^*$ of the OWIS $\mathcal{S}$, we have

Proposition 2

- $P_t = [\![\alpha_P]\!]_{\mathfrak{M},t} = [\![\oplus^{t-1}\alpha_P]\!]_{\mathfrak{M},1}$.
- $P_t^- = [\![\alpha_{P^-}]\!]_{\mathfrak{M},t} = [\![\oplus^{t-1}\alpha_{P^-}]\!]_{\mathfrak{M},1}$.
- $P_t^* = [\![\alpha_{P^*}]\!]_{\mathfrak{M},t} = [\![\oplus^{t-1}\alpha_{P^*}]\!]_{\mathfrak{M},1}$.
- $U_t = [\![\alpha_U]\!]_{\mathfrak{M},t} = [\![\oplus^{t-1}\alpha_U]\!]_{\mathfrak{M},1}$.
- $Pn_t^- = [\![\alpha_{Pn^-}]\!]_{\mathfrak{M},t} = [\![\oplus^{t-1}\alpha_{Pn^-}]\!]_{\mathfrak{M},1}$.
- $Pn_t^* = [\![\alpha_{Pn^*}]\!]_{\mathfrak{M},t} = [\![\oplus^{t-1}\alpha_{Pn^*}]\!]_{\mathfrak{M},1}$.

The following proposition lists a few valid wffs.

Proposition 3

1. $\models G(\alpha_P \vee \alpha_{P^-} \vee \alpha_{P^*} \vee \alpha_U \vee \alpha_{Pn^-} \vee \alpha_{Pn^*})$.
2. $\models G\neg(\beta_1 \wedge \beta_2)$,
   where β_1, β_2 are distinct wffs from the set $\{\alpha_P, \alpha_{P^-}, \alpha_{P^*}, \alpha_U, \alpha_{Pn^-}, \alpha_{Pn^*}\}$.
3. $\models (b, v_b) \rightarrow G(b, v_b)$, $b \in B$.
4. $\models (c, v_c') \rightarrow G(c, v_c')$, $c \in C_B$.
5. $\models (c, v_c'') \rightarrow G(c, v_c'')$, $c \in C_B$.
6. $\models G(\alpha_P \rightarrow G\alpha_P)$.
7. $\models G(\alpha_{P^-} \rightarrow G\alpha_{P^-})$.

The validity of the wffs in items 1 and 2 captures the fact that the sets P_t, P_t^-, P_t^*, U_t, Pn_t^-, Pn_t^* form a partition of the domain W_t at time point t. Item 3 shows that if an object takes the attribute-value v_b for an attribute $b \in B$, then it takes the same value at all future time points. Items 6 and 7 indicate that the sets P_t and P_t^- cannot shrink over time.

Observation 1 Here are some examples of what we can express through the language of $\mathcal{L}_{OW}$, as satisfiable wffs.

- $F(A(\alpha_P \vee \alpha_{P^-}) \rightarrow PA(\alpha_P \vee \alpha_{P^-}))$.
  This wff can be interpreted to represent the fact that if only two types of pneumonia, typical and non-typical, are known (i.e. there is no instance of a third type of pneumonia) at a time point, then this is also true for all the past time points as well.
- $P(\neg \oplus \top \rightarrow (E\alpha_P \wedge E\alpha_{P^-}))$.
  This wff says that at the very beginning (first time point), the classes of typical and non-typical pneumonia were known, as there were instances in them.
- $E(\alpha_U \wedge F\alpha_{P*})$.
  Currently we have a patient who is known to have pneumonia the type of which is unknown, but which is classified into the third type at some future time.
- $F(\neg E\alpha_U)$.
  At some future time point, every pneumonia patient is an instance of either typical, non-typical or a third type of pneumonia.
- $E(\alpha_{Pn^*} \wedge F\alpha_{Pn^-})$.
  A patient who cannot be classified currently into the category pneumonia or non-pneumonia, may later be found not to have the disease.

Due to the presence of the global modal operator A and the descriptors in the language, one can define the following modal operators capturing Pawlak's lower and upper approximations [8] relative to different sets of attributes.

Let $S := \{b_1, b_2, \ldots, b_n\}$ be a subset of $\mathcal{A}$. Let $\mathcal{D}_S$ be the set of all S-*basic* wffs [8], i.e. wffs of the form $(b_1, v_1) \wedge (b_2, v_2) \wedge \ldots \wedge (b_n, v_n)$, $v_i \in V_{b_i}, i = 1, 2, \ldots n$. Then we define the operators

- $\Box_S \alpha := \bigwedge_{\beta \in \mathcal{D}_S} (\beta \rightarrow A(\beta \rightarrow \alpha))$.
- $\Diamond_S \alpha := \neg \Box_S \neg \alpha$.

The following wffs confirm that these operators indeed give the lower and upper approximations with respect to the equivalence relation Ind_S^t on W_t, that relates all objects assigned the same attribute-value for the attributes in S by f_t.

Proposition 4 *Let α be a wff not involving any temporal operator. Then*

- $[\![\Box_S \alpha]\!]_{\mathfrak{M},t} = \underline{[\![\alpha]\!]_{\mathfrak{M},t}}_{Ind_S^t}$.
- $[\![\Diamond \alpha]\!]_{\mathfrak{M},t} = \overline{[\![\alpha]\!]_{\mathfrak{M},t}}_{Ind_S^t}$.

Thus using these modal operators, we can make statements involving lower and upper approximations. For instance, the language can express the statement that if an object is in the lower approximation of a certain set X relative to a given set S of attributes, then the object has typical pneumonia. $\Box_S p \rightarrow \alpha_P$ captures this statement, where p is a propositional variable representing the set X.

4 Conclusions

We present certain temporal information systems that formalize an open world model with one category. A temporal logic $\mathcal{L}_{OW}$ with descriptors and the global modality is proposed for reasoning with these structures.

As observed in Sect. 1, the open world model may be generalized in a number of ways. We may include more than one category in the OWIS by considering a collection of special attribute sets B_i indexed over some set I. There may be more classifications inside a category B—these may be accommodated by considering a larger collection of distinguished values for the attributes in the set C_B. Accordingly, one can make the necessary changes in the logic $\mathcal{L}_{OW}$.

A tableaux-based proof procedure has been defined for the temporal logics for rough sets studied in [3]. One expects to obtain a sound and complete proof procedure for $\mathcal{L}_{OW}$ in the same line. The decidability problem of the logic with respect to the class of all $\mathcal{L}_{OW}$-models is open. However, we can see that decidability with respect to the class of models with domain of fixed cardinality, can be proved as in [1].

A Hilbert-style axiomatization for $\mathcal{L}_{OW}$ is also an open question. It is clear that any such axiom set would include, apart from the standard axioms and rules for the temporal and global modalities, axioms pertaining to the descriptors. For instance, one would have the following, the last three giving the open world conditions:

1. $(a, v) \rightarrow \neg(a, v')$, for $v \neq v'$,
2. $\bigvee_{v \in V_a}(a, v)$,
3. $(b, v_b) \rightarrow G(b, v_b)$, $b \in B$,
4. $(c, v'_c) \rightarrow G(c, v'_c)$ and
5. $(c, v''_c) \rightarrow G(c, v''_c)$, $c \in C_B$.

In fact, $\mathcal{L}_{OW}$ with the basic temporal modalities and without the global modality, can be shown to have a complete axiomatization, following standard techniques. However, the global modality helps us to refer to members of the world during the transition in the open world over time—consider the satisfiable wffs given in Observation 1 of Sect. 3.

More properties of open world models may be expressed if information *updates* could be included in the syntax of the logic. This may be explored, in the line of the work on dynamic logics of information systems in [5, 6].

References

1. Banerjee, M., Khan, M.A.: Rough set theory: a temporal logic view. In: Chakraborty, M.K., Löwe, B., Mitra, M.N., Sarukkai, S. (eds.) Studies in Logic. Proceedings of the Logic, Navya-Nyaya and Applications: Homage to Bimal Krishna Matilal, 3–6 January 2007, Kolkata, Vol. 15, pp. 1–20. College Publications, London (2008)
2. Ju, S., Liu, H.: The logical structure of open sets. In: Ju, S. (ed.) Logic and Cognition: The Progress in Logical Study in China, pp. 596–603. Social Sciences in China Press, Beijing (2003)
3. Khan, M.A., Banerjee, M.: Logics for some dynamic spaces—I and II. J. Log. Comput. (2014). doi:10.1093/logcom/exu055-56
4. Khan, M.A., Banerjee, M.: A logic for complete information systems. In: Sossai, C., Chemello, G. (eds.) The 10th European Conference on Symbolic and Quantitative Approaches to Reasoning with Uncertainty, Verona, Italy. LNAI, vol. 5590, pp. 829–840. Springer (2009)
5. Khan, M.A., Banerjee, M.: Logics for information systems and their dynamic extensions. ACM Trans. Comput. Log. **12**(4), 29 (2011)
6. Khan, M.A., Banerjee, M.: An update logic for information systems. Int. J. Approx. Reason. **55**(1), 436–456 (2014)
7. Orłowska, E.: Dynamic information system. Fundam. Inf. **5**, 101–118 (1982)
8. Pawlak, Z.: Rough Sets. Theoretical Aspects of Reasoning about Data. Kluwer Academic Publishers, Dordrecht (1991)
9. Synak, P.: Temporal feature extraction from temporal information systems. In: Zhong, Z., Ras, Z.W., Tsumoto, S., Suzuki, E. (eds.) Foundations of Intelligent Systems. Proceedings of ISMIS 2003, LNCS 2871, pp. 270–278. Springer, Berlin (2003)

Approximate Reasoning Under Type-2 Fuzzy Logics

Sudin Mandal, Namrata Bhattacharyya and Swapan Raha

Abstract In this paper, we have made a study of approximate reasoning based on a Type-2 fuzzy set theory. We have focused upon two typical rules of inference used mostly in ordinary approximate reasoning methodology based on Type-1 fuzzy set theory. Similarity is inherent in approximate reasoning. The concept of similarity between Type-2 fuzzy sets is discussed and a similarity-based approximate reasoning technique is proposed. The proposal is illustrated with a typical artificial example. Prediction is the causal basis for decision making. Different measures leading to prediction under uncertainty are proposed for a better understanding of the power of Type-2 fuzzy set theory.

Keywords Type-2 fuzzy set · Type-2 fuzzy logic · Approximate reasoning

1 Introduction

In 1965, the concept of a fuzzy set was introduced by Zadeh [13] and it has already established its usefulness through successful applications in different fields. Considering the importance of fuzzy logic as a basis for approximate reasoning, a systematic development of fuzzy set theory, the deductive aspects and structures of the underlying fuzzy logics were extensively studied [2]. In dealing with

This research has been financially supported by the UGC SAP (DRS) Phase-II Project under the Department of Mathematics, Visva-Bharati and the UGC Major Research Project No. F. 36-293/2008(SR).

S. Mandal · N. Bhattacharyya · S. Raha (✉)
Department of Mathematics, Visva-Bharati,
Santiniketan 731235, India
e-mail: swapan.raha@visva-bharati.ac.in

S. Mandal
e-mail: sudin.mandal@rediffmail.com

N. Bhattacharyya
e-mail: namratamathematics@gmail.com

M.K. Chakraborty et al. (eds.), *Facets of Uncertainties and Applications*,
Springer Proceedings in Mathematics & Statistics 125,
DOI 10.1007/978-81-322-2301-6_7

vagueness/impreciseness using fuzzy set theory, we come across situation, where it is difficult to find satisfactorily the degree of membership of an element of the universal set in a particular fuzzy set is found impossible. This motivated Zadeh to introduce a generalization of fuzzy set, a Type-2 fuzzy set, in 1975 [14]. The key point in this generalization is that membership degrees of elements in a Type-2 fuzzy set are the traditional fuzzy sets in [0, 1] while that for ordinary fuzzy sets are real numbers in [0, 1]. Accordingly, any Type-2 fuzzy logic is a generalization of some fuzzy logic as proposed by Zadeh [15].

Now, reasoning is a mental activity that allows us to derive new information on the object of study from existing knowledge with some degree of confidence. It is a fact that human beings are better at reasoning than machines as they have the ability to make effective decisions on the basis of imprecise information. A collection of imprecise information given by human experts often forms the basis of a fuzzy system which is represented by fuzzy sets, fuzzy relations, and subsequently manipulated using fuzzy set theory. The task of a fuzzy system is to exploit knowledge acquired by experts about the system. In a rule-based system, from a given rule (antecedent–consequent condition) and an observed state of the antecedent, we conclude something similar to the consequent by applying a method of inference which, we call approximate reasoning. Here, most of the reasoning is performed based on imperfect matching, e.g., "ripe" and "very ripe."

Approximate reasoning methodology was developed by Zadeh [16] to formulate complex problems of human reasoning. The main motivation of the theory of approximate reasoning is apparently, the desire to build up a qualitative framework that will allow one to derive an approximate conclusion from a collection of imprecise knowledge. Fuzzy logic is the basis of approximate reasoning. Fuzzy sets and fuzzy relations are used to represent simple and complex fuzzy propositions in fuzzy logic. Rules of inference are used to derive new propositions (fuzzy logical forms) from an observed data and given knowledge on the same.

Comparison of objects is a widespread operation necessary in many frameworks. It has been observed that relations play an important role in comparison. It is based on the idea that if two objects are sufficiently similar (expressed as a given relation) then, a transfer of knowledge is possible from one to the other (status of one is induced by the observed status of the other). This comparison is frequently achieved through a measure intended to determine the extent to which, the descriptions have similarities or differ from each other. For that many measures of comparison have been proposed and studied by researchers in different disciplines for different purposes. It has also been observed that a basic tool for human cognition is the ability to assess similarity. The process of thinking is dependent on a sense of sameness. Derivation of an imprecise statement from a set of imprecise statements is efficiently performed by humans, using analogy or similarity. For a better understanding of how human beings assess similarity in problem-solving, categorization, information retrieval, and reasoning, we have studied different aspects of similarity for a theory of similarity-based approximate reasoning with fuzzy sets of Type-2.

Prediction on the nature of transformational change of state of a dynamical system is determined by a knowledge of causation. Observation followed by rigorous analysis help scientists in decision making under uncertainty. The transformational change of one set of state to another is the dynamics of the system. This dynamics is modeled by experts using linguistic tools. Imprecision is inherent in natural language. We propose a few measures based on Type-2 fuzzy set theory leading to the prediction of the clinical behavior of a patient [3].

This research work is organized into six sections. In Sect. 2, we define a few terms in order to communicate using Type-2 fuzzy sets. Approximate reasoning is discussed in the context of Type-2 fuzzy set theory in Sect. 3. Two basic rules of inference have been developed. The computational procedure is presented in a subsection. Examples are considered to illustrate the problem. A few characteristic measures of the clinical dynamics are presented in Sect. 4. The work is briefly concluded in Sect. 5. This is followed by a list of references.

2 Definition and Basic Concepts

An attempt is made to study approximate reasoning based on Type-2 fuzzy logics. Accordingly, a brief study on the theory of Type-2 fuzzy sets is presented. We focus on the study of operations on Type-2 fuzzy sets. The concept of a Type-2 fuzzy relation and fuzzy connectives *not* ($\neg$), *and* ($\wedge$), and *or* ($\vee$) are also studied. Appropriate interpretation of connectives is one of the basic problems in any fuzzy logic and its application. Classes of negation functions (to model complement operators), continuous triangular norms (to model conjunction), and triangular co-norms (t-co-norms to model disjunction) are also examined. These classes of operations are found to be mathematically sound and contain a wide variety of particular members.

Definition 1 A fuzzy set of Type-2 A in a set U is characterized by a fuzzy membership function with the value at u being called a fuzzy grade of the point u, is a fuzzy subset of $J \subseteq [0, 1]$. More explicitly, A is characterized by a membership function $\mu_A(u, v)$, where $u \in U$ and $v \in J$, i.e.,

$$A = \{((u, v), \mu_A(u, v)) \mid \forall u \in U, \forall v \in J \subseteq [0, 1]\}.$$

Example 1 Let us consider U ={Lotus, Marigold, Tube rose, Dahlia, Rose} to be a set of flowers and that A is a fuzzy subset of Type-2 of *beautiful flowers* defined over U. We then express the same in set theoretic notation as,

A = beautiful flowers = $middle/Lotus$ + *not low*/$Marigold$ + *low*/*Tube rose* + *very high*/$Dahlia$ + $high/Rose$

where the fuzzy grades labeled *middle, low, high* are assumed to be fuzzy sets in $J = 0, 0.1, \ldots, 0.9, 1 \subseteq [0, 1]$ and, for example, are expressed as in the following:

$middle = 0.4/0.3 + 0.8/0.4 + 1/0.5 + 0.8/0.6 + 0.4/0.7$
$low = 1/0 + 0.9/0.1 + 0.6/0.2 + 0.3/0.3$
$high = 0.5/0.7 + 0.7/0.8 + 0.9/0.9 + 1/1.$

Moreover, fuzzy grades for elements that belong to fuzzy sets such as *not low* and *very high* are determined from fuzzy grades of the same as in *low* and *high*, respectively, by using the concept of linguistic hedges as in ordinary fuzzy sets

not low $= 0.1/0.1 + 0.4/0.2 + 0.7/0.3 + 1/(0.4 + 0.5 + \cdots + 1)$.

Let A and B be fuzzy sets of Type-2 over the universe of discourse X and $\mu_A(u)$ and $\mu_B(u)$ be the fuzzy grades of the point u in A and B, respectively, (that is, fuzzy subsets in $J \subseteq [0, 1]$), represented as

$$\begin{aligned}\mu_A(u) &= f(v_1)/v_1 + f(v_2)/v_2 + \cdots + f(v_n)/v_n.\\ &= \sum f(v_i)/v_i,\ v_i \in J,\\ \mu_B(u) &= g(w_1)/w_1 + g(w_2)/g_2 + \cdots + g(w_m)/w_m\\ &= \sum g(w_j)/w_j,\ w_j \in J,\end{aligned}$$

where f and g are membership functions of elements in $J \subseteq [0, 1]$ and the values $f(u_i)$ and $g(w_j)$ in $[0, 1]$ represent the grades of u_i and $w_j \in J$, respectively. At each value of u, the plane section determined by u and $\mu_A(u, v)$ for different values of v is the vertical slice of A at u. Vertical slices represent secondary membership of u in A; it is a Type-1 fuzzy set. The domain of secondary membership function is regarded as the primary membership of u.

Now, the operations on fuzzy sets of Type-2 are expressed as in the following:

Definition 2 The union of two Type-2 fuzzy sets A and B is given as

$$\begin{aligned}A \cup B \Leftrightarrow \mu_{A\cup B}(u) &= \mu_A(u) \sqcup \mu_B(u)\\ &= \sum_i f(u_i)/u_i \cup \sum_j g(w_j)/w_j\\ &= \sum_{i,j} (f(u_i) \wedge g(w_j))/(u_i \vee w_j)\end{aligned}$$

Definition 3 The intersection of two Type-2 fuzzy sets A and B is given as

$$\begin{aligned}A \cap B \Leftrightarrow \mu_{A\cap B}(u) &= \mu_A(u) \sqcap \mu_B(u)\\ &= \sum_i f(u_i)/u_i \sqcap \sum_j g(w_j)/w_j\\ &= \sum_{i,j} (f(u_i) \wedge g(w_j))/(u_i \wedge w_j)\end{aligned}$$

Definition 4 The complement of a Type-2 fuzzy set A

$$\begin{aligned}\bar{A} \Leftrightarrow \mu_A(u) &= \neg\mu_A(u)\\ &= \sum_i f(u_i)/1 - u_i,\end{aligned}$$

where $\bigvee$ and $\bigwedge$ represent max and min, respectively. We call the operations for fuzzy grades, that is, $\sqcup$ as *join*, $\sqcap$ as *meet*, and $\neg$ as *negation* hereafter.

Example 2 Here, let $J = \{0, 0.1, \ldots, 0.9, 1\}$ and let fuzzy grades $\mu_A(u)$ and $\mu_B(u)$ be given as

$$\mu_A(u) = 0.5/0 + 0.7/0.1 + 0.3/0.2,$$

$$\mu_B(u) = 0.9/0 + 0.6/0.1 + 0.2/0.2.$$

Then, we have

$$\begin{aligned}
\mu_A(u) \sqcup \mu_B(u) &= (0.5/0 + 0.7/0.1 + 0.3/0.2) \sqcup (0.9/0 + 0.6/0.1 + 0.2/0.2) \\
&= \tfrac{0.5\wedge 0.9}{0\vee 0} + \tfrac{0.5\wedge 0.6}{0\vee 0.1} + \tfrac{0.5\wedge 0.2}{0\vee 0.2} \\
&\quad + \tfrac{0.7\wedge 0.9}{0.1\vee 0} + \tfrac{0.7\wedge 0.6}{0.1\vee 0.1} + \tfrac{0.7\wedge 0.2}{0.1\vee 0.2} \\
&\quad + \tfrac{0.3\wedge 0.9}{0.2\vee 0} + \tfrac{0.3\wedge 0.6}{0.2\vee 0.1} + \tfrac{0.3\wedge 0.2}{0.2\vee 0.2} \\
&= 0.5/0 + 0.5/0.1 + 0.2/0.2 + 0.7/0.1 + 0.6/0.1 + 0.2/0.2 \\
&= 0.5/0 + \tfrac{(0.5\wedge 0.7\wedge 0.6)}{0.1} + \tfrac{(0.2\wedge 0.2\wedge 0.3\wedge 0.3\wedge 0.2)}{0.2} \\
&= 0.5/0 + 0.7/0.1 + 0.3/0.2
\end{aligned}$$

Similarly, we have

$$\begin{aligned}
\mu_A(x) \sqcap \mu_B(x) &= 0.7/0 + 0.6/0.1 + 0.2/0.2 \\
\neg\mu_A(x) &= 0.5/1 + 0.7/0.9 + 0.3/0.8.
\end{aligned}$$

Definition 5 Let U_1, U_2 be two universes of discourse, and A be a Type-2 fuzzy set defined over U_1. The *cylindrical extension* of A to $U_1 \times U_2$ is a Type-2 fuzzy relation *ce(A)* defined by

$$\Sigma_{U_1\times U_2}\mu_{ce(A)}(u_1, u_2)/(u_1, u_2), \text{ where } u_1 \in U_1, u_2 \in U_2 \text{ and;}$$

$$\mu_{ce(A)}(u_1, u_2) = \mu_A(u_1).$$

Thus, the cylindrical extension clearly produces the largest fuzzy relation that is compatible with the given projection. Such a relation is the least specific of all relations compatible with the projection.

Definition 6 Let U_1, U_2 be two universes of discourse, and R be a Type-2 fuzzy relation defined over $U_1 \times U_2$. The *Projection* of R on U_1 is a Type-2 fuzzy set $proj_{U_1}(R)$ defined by

$$\Sigma_{U_1}\mu_{proj_{U_1}(R)}(u_1)/u_1, \text{ where } u_1 \in U_1;$$

$$\mu_{proj_{U_1}(R)}(u_1) = \sup_{u_2\in U_2} \mu_R(u_1, u_2).$$

3 Approximate Reasoning

In this section, we demonstrate how conclusions can be obtained from given premises with the help of Type-2 fuzzy set theory. According to Zadeh, approximate reasoning using the scheme **from** 'X **is** A' **and** '(X, Y) **is** R' **infer** 'Y **is** B' is termed as the compositional rule of inference (Table 1).

Here, A is a type-2 fuzzy subset of U, B is a type-2 fuzzy subset of V, R is a type-2 fuzzy subset of $U \times V$. Explicitly, let

$$A = \sum_{u \in U} \mu_A(u)/u \; ; \quad \mu_A(u) = \sum_{x \in [0,1]} \mu_u(x)/x.$$

$$R = \sum_{(u,v) \in U \times V} \mu_R(u, v)/(u, v) \; ; \quad \mu_R(u, v) = \sum_{z \in [0,1]} \mu_R(z)/z.$$

$$B = AoR = \sum_{v \in V} \mu_B(v)/v \; ; \quad \mu_B(v) = \sup_{u \in U} \left\{ \mu_{ce(A)} \bigcap \mu_{R(u,v)}(u, v) \right\}.$$

i.e., $\mu_B(v) = \sup_{u \in U} \left\{ (\mu_A(u) \bigcap \mu_{R(u,v)}(u, v)) \right\}.$

Approximate reasoning using the scheme **from 'X is A^*' and 'if X is A then Y is B' infer 'Y is B^*'** is termed as the Generalized modus ponens. Here, A, A' are fuzzy subsets of type-2 defined over the universe of discourse U; B and B' are fuzzy subsets of type-2 defined over the universe of discourse V. Interpreting $(A \rightarrow B)$ as a type-2 binary fuzzy relation R, i.e., $(A \rightarrow B) = R(A, B)$ and using Compositional rule of inference deduce $A'oR = B'$. The scheme can be best described in Table 2.

$$A = \sum_{u \in U} \mu_A(u)/u \; ; \quad \mu_A(u) = \sum_{x \in [0,1]} \mu_u(x)/x.$$

$$B = \sum_{v \in V} \mu_B(v)/v \; ; \quad \mu_B(v) = \sum_{y \in [0,1]} \mu_v(y)/y.$$

$$R = \sum_{(u,v) \in U \times V} \mu_R(u, v)/(u, v) \; ; \quad \mu_R(u, v) = \sum_{z \in [0,1]} \mu_R(z)/z.$$

$$A' = \sum_{u \in U} \mu_{A'}(u)/u \; ; \quad \mu_{A'}(u) = \sum_{x \in [0,1]} \mu_u(x)/x.$$

Table 1 Compositional rule of inference

p :	X is A	
q :	(X, Y) is R	
r :		Y is B

Table 2 Generalized modus ponens

p :	if X is A	then	Y is B
q :	X is A′		
r :			Y is B′

$$B = A'oR = \sum_{v\in V} \mu_B(v)/v\ ;\quad \mu_B(v) = \sup_{u\in U}\left\{\mu_{ce(A')\bigcap \mu_{R(u,v)}}(u,v)\right\}.$$

i.e., $\mu_B(v) = \sup\limits_{u\in U}\left\{(\mu_{A'}(u)\bigcap \mu_{R(u,v)}(u,v))\right\}$.

3.1 Computational Procedure

We now present computations involved in the above inference mechanisms in the following algorithms one-by-one.

ALGORITHM OAR: Ordinary Approximate Reasoning
Step 1. Represent A as a Type-2 fuzzy subset of U and R as a Type-2 fuzzy relation over the universe of discourse $U \times V$.
Step 2. Compute $ce(A)$, the cylindrical extension of A over $U \times V$.
Step 3. Compose $ce(A)$ and R to form a fuzzy relation of Type-2 S using some conjunction operation meant for fuzzy sets of Type-2.
Step 4. Compute $B = Proj_V S$, i.e., $B = \sum\limits_{v\in V} \mu_B(v)/v$ where $\mu_B(v) = \bigcup\limits_{u\in U} \mu_S(u,v)$, some disjunction of a collection of fuzzy sets of type-1 over [0, 1].

Example 3 Let us consider U ={Riya, Rama, Rimi} to be a set of women and that A is a Fuzzy Set of Type-2 of *intelligent women* in U. Then, we may have A = intelligent = *more or less*/*Riya* + *highly*/*Rama* + *not at all*/*Rimi*.

$$= \frac{\{0.6/0.4 + 1/0.5 + 0.5/0.6 + 0.2/0.7\}}{Riya} + \frac{\{0.3/0.7 + 0.6/0.8 + 0.8/0.9 + 1/1\}}{Rama} + \frac{\{1/0.1 + 0.8/0.2 + 0.5/0.3 + 0.2/0.4\}}{Rimi}.$$

Let V ={Mita,Priti,Soumi} be another set of girls and that B is a fuzzy set of Type-2 of *smart girl* in V.
Then, B = smart = *highly*/*Mita* + *more or less*/*Prity* + *not at all*/*Soumi*

$$= \frac{\{0.2/0.7 + 0.5/0.8 + 0.8/0.9 + 1/1\}}{Mita} + \frac{\{0.3/0.4 + 0.8/0.5 + 1/0.6 + 0.4/0.7\}}{Priti} + \frac{\{1/0 + 0.8/0.1 + 0.5/0.2 + 0.2/0.3\}}{Soumi}.$$

R : **a fuzzy relation on** $U \times V$

	Mita	Priti	Soumi
Riya	0.2/0.7	0.3/0.4 + 0.8/0.5 + 0.5/0.6 + 0.2/0.7	0.0
Rama	0.2/0.7 + 0.5/0.8 + 0.8/0.9 + 1.0/1.0	0.3/0.7	0.0
Rimi	0.0	0.2/0.4	0.8/0.1 + 0.5/0.2 + 0.2/0.3

Let A' be an observation over the set U and represented by a fuzzy set of Type-2 as

A' = Very intelligent

$= \frac{\{0.36/0.4 + 1/0.5 + 0.25/0.6 + 0.04/0.7\}}{Riya} + \frac{\{0.09/0.7 + 0.36/0.8 + 0.64/0.9 + 1.0/1.0\}}{Rama} + \frac{\{1.0/0.1 + 0.64/0.2 + 0.25/0.3 + 0.04/0.4\}}{Rimi}$.

Now, we find the cylindrical extension of A' and let us set

S : **the cylindrical extension of** A'

	Mita	Priti	Soumi
Riya	0.36/0.4 + 1/0.5 + 0.25/0.6 + 0.04/0.4	0.36/0.4 + 1/0.5 + 0.25/0.6 + 0.04/0.4	0.36/0.4 + 1/0.5 + 0.25/0.6 + 0.04/0.4
Rama	0.09/0.7 + 0.36/0.8 + 0.64/0.9 + 1.0/1.0	0.09/0.7 + 0.36/0.8 + 0.64/0.9 + 1.0/1.0	0.09/0.7 + 0.36/0.8 0.64/0.9 + 1.0/1.0
Rimi	1.0/0.1 + 0.64/0.2 + 0.25/0.3 + 0.04/0.4	1.0/0.1 + 0.64/0.2 + 0.25/0.3 + 0.04/0.4	1.0/0.1 + 0.64/0.2 + + 0.25/0.3 + 0.04/0.4

Next, let $R_1 = RoS = \mathrm{R} \sqcap S$ and we obtain

$B' = \{0.09/0.7 + 0.36/0.8 + 0.64/0.9 + 1/1\}/Mita +$
$\{0.3/0.4 + 0.8/0.5 + 0.25/0.6 + 0.9/0.7\}/Priti$
$+ \{0.8/0.1 + 0.5/0.2 + 0.2/0.3\}/Soumi.$

Defuzzification of $B' = 1.0/Mita + 0.5/Priti + 0.1/Soumi.$

We conclude that Mita is *very highly* smart, Priti is *more or less* smart and Soumi is *not at all* a smart girl.

<u>**ALGORITHM RBAR**</u>: Rule-based Approximate Reasoning

Step 1. Represent A, A' as Type-2 fuzzy subsets of U and B as a Type-2 fuzzy subset V.

Step 2. Compute R as a Type-2 fuzzy relation over the universe of discourse $U \times V$.

Step 3. Compose *ce*(A') and R to form a fuzzy relation of Type-2 S using some conjunction operation meant for fuzzy sets of Type-2.

Step 4. Compute $B = Proj_V S$, i.e., $B = \sum_{v \in V} \mu_B(v)/v$ where $\mu_B(v) = \bigcup_{u \in U} \mu_S(u, v)$, some disjunction of a collection of fuzzy sets of type-1 over [0, 1].

3.2 Similarity Based Approximate Reasoning

Let $A = \sum_{u \in U} \mu_A(u)/u$ and $B = \sum_{u \in U} \mu_B(u)/u$.

be two fuzzy subsets of Type-2 defined over the universe of discourse U(,say). The similarity between A and B is defined as $S(A, B; U)$ or simply $S(A, B)$ and is defined as

$$S(A, B) = 1 - \left\{ \frac{\sum_{u \in U} (def\{\mu_A(u)\} - def\{\mu_B(u)\})^2}{n} \right\}^{\frac{1}{2}}.$$

ALGORITHM SBAR: Similarity-based Approximate Reasoning
Step 1. Translate premise if X is A then Y is B and compute $R(A, B)$ using any suitable translating rule possibly, a T-norm operator.
Step 2. Compute $S(A, A')$ according to some definition.
Step 3. Modify $R(A, B)$ with $S(A, A')$ to obtain the modified conditional relation $R(A \mid A', B)$ according to some scheme C.
Step 4. Use sup-projection operation on $R(A \mid A', B)$ to obtain B' as

$$\mu_{B'}(v) = \sup_u \mu_{R(A'|A,B)}(u, v). \tag{1}$$

Now, for a given fact, we need to deduce a schemes C for computation of the modified conditional relation $R(A \mid A', B)$ as given in Step 3. We need to choose translating rules and specify T-norm functions logically.

4 Measures of Prediction

Two fuzzy sets of Type-2, A and B, are defined over the universe of discourse U. Let $M(A)$ and A^c represent, respectively, the fuzzy cardinality and the complement of the fuzzy set A. The fuzzy entropy of a fuzzy set is denoted by $E(A)$.

Definition 7 The degree to which one set(,say) A belongs to the other set B is expressed by the formula

$$S(A, B) = \frac{M\left(A \bigcap B\right)}{M(A)}.$$

Definition 8 The degree to which the elements of one fuzzy set(,say) A are compatible with that of the other set B is expressed by the formula

$$C(A, B) = \left[\left(\frac{M(A \bigcap B)}{M(B)}\right)^2 \left(\frac{M(A \bigcap B)}{M(A)}\right)^2\right]^{\frac{1}{2}}.$$

Definition 9 The degree to which the elements of one fuzzy set(,say) A contradict with the other is expressed by the formula

$$D(A, B) = \left[\left(\frac{M(A^c \bigcap B)}{M(B)}\right)^2 \left(\frac{M(A \bigcap B^c)}{M(A)}\right)^2\right]^{\frac{1}{2}}.$$

The degree to which the symmetry of the elements of one fuzzy set(,say) A is broken by that of the other is expressed by the formula [3]

$$K(A,B) = \frac{\left(\frac{M(A\bigcap B)}{M(B)}\right)^2\left(\frac{M(A\bigcap B)}{M(A)}\right)^2}{\left[\left(\frac{M(A\bigcap B)}{M(B)}\right)^2\left(\frac{M(A\bigcap B)}{M(A)}\right)^2\right]^{\frac{1}{2}}}.$$

Definition 10 Analogy is the fuzzy complement of $K(A, B)$.

Compatibility and contrast/contradiction, taken together, define a degree of paradox between the pair of fuzzy sets A and B. Accordingly, considering information as fuzzy entropy, we define

Definition 11 The degree of paradox between two pieces of information on the same universe represented by two Type-2 fuzzy sets A and B is defined as

$$\text{P(A, B)} = \text{E(compatibility, contrast)}.$$

The higher the fuzzy entropy of the fuzzy set the greater is the paradox.

Definition 12 A measure of surprise is defined as the degree of paradox per symmetry assimilation between two fuzzy sets of Type-2 and is given explicitly as

$$\text{Surprise} = \text{E(paradox, analogy)}.$$

The higher the fuzzy entropy of the fuzzy set the greater is the surprise [3].

All these measures are, in principle, dynamic in nature—neither belong wholly to fuzzy set A nor to fuzzy set B. Accordingly, it is expected that the dynamic thought process could be modeled by using some or all of these measures that change themselves between the fuzzy sets. As a case in point, these measures carry causal implication for the clinical state of the medical patient. It is observed that cause is the transformative connective dynamic process between one dynamic state of measured elements and their context to another.

5 Conclusion

Any attempt to study Type-2 fuzzy logics and the corresponding theory of Type-2 fuzzy sets is still considered to be difficult. Moreover, any study on the possibility of using Type-2 fuzzy logic in handling uncertainties in rule-based systems is interesting and important. It has already been established that approximate reasoning is an important topic of research because of its scope of applications in different fields of research particularly, in fuzzy control. This research on modeling approximate reasoning using Type-2 fuzzy set theory will definitely help the research community. It is hoped that with modeling of Generalized Modus Ponens and Compositional Rule of Inference using Type-2 fuzzy set theory, approximate reasoning methodology can be made more versatile in so far as decision-making under uncertainty is concerned.

References

1. Chen, S.M.: A weighted fuzzy reasoning algorithm for medical diagnosis. Decis. Support. Syst. **11**, 37–43 (1994)
2. Hajěk, P.: Metamathematics of Fuzzy Logic. Kluwer Academic Press, Dordrecht (1998)
3. Helgason, C.M., Jobe, T.H.: Causal efficacy in the fuzzy hypercube: representation of concomitant mechanisms for stroke. Neural Netw. **11**, 548–555 (1998)
4. Mendel, J.M., John, R.I.B.: Type-2 fuzzy sets made simple. IEEE Trans. Fuzzy Syst. **10**(1), 117–127 (2002)
5. Mizumoto, M., Tanaka, K.: Some properties of fuzzy sets of type-2. Inf. Control. **31**, 312–340 (1976)
6. Nieminen, J.: On the Algebraic Structure of Fuzzy Sets of Type-2, vol. 13. Kybernetica, New York (1977)
7. Raha, S., Pal, N.R., Ray, K.S.: Similarity based approximate reasoning: methodology and application. IEEE Trans. Syst. Man Cybern.—Part A: Syst. Hum. **32**(4), 541–547 (2002)
8. Raha, S., Hossain, A., Ghosh, S.: Similarity based approximate reasoning: fuzzy control. J. Appl. Log. **6**(1), 47–71 (2008)
9. Xuechang, L.: Entropy, distance measure and similarity measures of fuzzy sets and their relations. Fuzzy Sets Syst. **52**, 171–174 (1992)
10. Yager R.R.: Measurement of properties on fuzzy sets and possibility distribution. In: Klement, E.P. (ed.) Proceedings of the 3rd International Seminar on Fuzzy Set Theory, pp. 211–222. Johannes Kepler University of Linz
11. Yager, R.R.: An Introduction to applications of possibility theory. Hum. Syst. Manag. **3**, 246–253 (1983)
12. Yang, M.S., Lin, D.C.: On similarity and inclusion measures between type-2 fuzzy sets with an application to clustering. Comput. Math. Appl. **57**, 896–907 (2009)
13. Zadeh, L.A.: Fuzzy sets. Inf. Control. **8**, 338–353 (1965)
14. Zadeh, L.A.: The concept of linguistic variable and its application to approximate reasoning—I. Inf. Sci. **8**, 199–249 (1975)
15. Zadeh, L.A.: A computational approach to fuzzy quantifiers in natural languages. Comput. Math. Appl. **9**, 149–184 (1983)
16. Zadeh, L.A.: A theory of approximate reasoning. In: Hayes, J.E., Mitchie, D., Mikulich, L.I. (eds.) Machine Intelligence. A theory of approximate reasoning, vol. 9, pp. 149–194 (1979)

Approximation Dialectics of Proto-Transitive Rough Sets

A. Mani

Abstract Rough Sets over generalized transitive relations like proto-transitive ones have been initiated by the present author in Mani (Inst. Math. Sci. ICLA'2013, 1–12, Chennai 2013, [1]) and detailed semantics have been developed in forthcoming papers. In this research paper, approximation of proto-transitive relations by other relations is investigated and the relation with rough approximations is developed toward constructing semantics that can handle *fragments of structure*. It is also proved that difference of approximations induced by some approximate relations need not induce rough structures.

Keywords Proto-transitive rough sets · Approximate relations · Nelson algebras · Axiomatic theory of granules · Contamination problem

1 Introduction

Proto-transitivity is one of the infinite number of generalizations of transitivity. The structure of definite objects and knowledge interpretation in *proto-transitive approximation spaces* (PRAX) have been investigated by the present author in [1]. Semantics of PRAX is hard because the representation of rough objects is involved [1]. Though as many as five different semantic approaches have already been developed by the present author, there is scope for further enhancement.

If R is a relation on a set S, then R can be approximated by a wide variety of partial/quasi-order relations in both classical and rough set perspective [2]. Though the methods are essentially equivalent for binary relations, the latter method is more

I would like to thank Prof Mihir Chakraborty for discussions on PRAX and referees for useful remarks.

A. Mani (✉)
Department of Pure Mathematics, University of Calcutta, 9/1B, Jatin Bagchi Road, Kolkata 700029, India
e-mail: a.mani.cms@gmail.com
URL: http://www.logicamani.in

M.K. Chakraborty et al. (eds.), *Facets of Uncertainties and Applications*, Springer Proceedings in Mathematics & Statistics 125,
DOI 10.1007/978-81-322-2301-6_8

general. When the relation R satisfies proto-transitivity, then many new properties emerge. This aspect is developed in some detail in the present paper.

When R is a quasi-order relation, then a semantics for the set of ordered pairs of lower and upper approximations $\{(A^l, A^u);\ A \subseteq S\}$ has recently been developed in [3, 4]. Though such a set of ordered pairs of lower and upper approximations are not rough objects in the PRAX context, we can use the approximations for an additional semantic approach to it. We prove that differences of consequent lower and upper approximations suggest partial structures for *measuring* structured deviation. The developed method should also be useful for studying correspondences between the different semantics [5, 6]. Because of this we devote some space to the nature of transformation of granules by the relational approximation process.

Rough objects as explained in [7, 8] are collections of objects in a classical domain (Meta-C) that appear to be indistinguishable among themselves in another rough semantic domain (Meta-R). But their representation in most RSTs in purely order theoretic terms is not known. For PRAX, this is solved in [1]. In a PRAX, these need not correspond to intervals (bounded by definite objects a, b) of the form $]a, b[$ with b covering (in the order on definite objects) a.

Definition 1 A binary relation R on a set S is said to be *weakly transitive, transitive, or proto-transitive*, respectively, on S iff S satisfies

- ⋆ If whenever $Rxy,\ Ryz$ and $x \neq y \neq z$ holds, then Rxz. (i.e., $(R \circ R) \setminus \Delta_S \subseteq R$ (where $\circ$ is relation composition), or
- ⋆ whenever Rxy & Ryz holds then Rxz (i.e. $(R \circ R) \subseteq R$), or
- ⋆ whenever $Rxy,\ Ryz,\ Ryx,\ Rzy$ and $x \neq y \neq z$ holds, then Rxz follows, respectively. Proto-transitivity of R is equivalent to $R \cap R^{-1} = \tau(R)$ being weakly transitive.

Note that *weak transitivity* of [9] is *proto-transitivity* here. $Ref(S)$, $w\tau(S)$, $p\tau(S)$, and $EQ(S)$ will, respectively, denote the set of reflexive, weakly transitive, proto-transitive, and equivalence relations on the set S, respectively. We can prove $w\tau(S) \subseteq p\tau(S)$ and $\forall R \in Ref(S)(R \in p\tau(S) \leftrightarrow \tau(R) \in EQ(S))$.

Example 1 It is easy to derive PRAX from population census, medical, gender studies, and other indeterminate information systems [1]. Let $\mathcal{I}$ be survey data in table form with column names being for sex, gender, sexual orientations, other personal data, and opinions on sexist contexts with each row corresponding to a person. We write Rab if and only if person a *agrees with* b's opinions (this predicate can be constructed empirically or from the data by a suitable heuristic). Often R is a proto-transitive, reflexive relation and this condition can be imposed to complete partial data as well (as a rationality condition). If a agrees with the opinions of b, then we will say that a is an *ally* of b—if b is also an ally of a, then they are *comrades*. Specifically, the relation P, defined by Pxy iff x thinks that y thinks that color of object O is *a maroon*, is proto-transitive and not transitive.

2 Approximations and Definite Elements in PRAX

The proofs of the results of this section can be found in [1].

Definition 2 By a *Proto-Approximation Space* S (PRAS for short), we will mean a pair of the form $\langle \underline{S}, R\rangle$ with $\underline{S}$ being a set and R being a proto-transitive relation on it. If R is also reflexive, then it will be called a *Reflexive Proto-Approximation Space* (PRAX) for short). $\underline{S}$ may be infinite.

If S is a PRAX, then we will, respectively, denote successor and symmetrized successor neighborhoods generated by an element $x \in S$ by $[x] = \{y;\ Ryx\}$ and $[x]_o = \{y;\ Ryx \,\&\, Rxy\}$, respectively. Taking these as granules, the associated granulations will be denoted by $\mathcal{G} = \{[x] : x \in S\}$ and $\mathcal{G}_o$, respectively. In all that follows, S will be a PRAX unless indicated otherwise. Our motivation for considering the following approximations are in application contexts, generative value (for other generalized transitivity), simplicity, and granularity in foundations.

Definition 3 Definable approximations on S include ($A \subseteq S$):

Upper Proto: $A^u = \bigcup_{[x]\cap A\neq\emptyset} [x]$.
Lower Proto: $A^l = \bigcup_{[x]\subseteq A} [x]$.
Symmetrized Upper Proto: $A^{uo} = \bigcup_{[x]_o\cap A\neq\emptyset} [x]_o$.
Symmetrized Lower Proto: $A^{lo} = \bigcup_{[x]_o\subseteq A} [x]_o$.
Point-wise Upper: $A^{u+} = \{x : [x]\cap A \neq \emptyset\}$.
Point-wise Lower: $A^{l+} = \{x : [x] \subseteq A\}$.

Example 2 In the context of our example 1, $[x]$ is the *set of allies* x, while $[x]_o$ is the set of comrades of x. A^l is the *union of the set of all allies of at least one of the members of* A if they are all in A. A^u is the union of the set of all allies of persons having at least one ally in A. A^{l+} is the set of all those persons in A all of whose allies are within A. A^{u+} is the set of all those persons having allies in A.

Proposition 1 *For any subset* $A \subseteq S$, *all of the following hold:*

- ⋆ *It is possible that* $A^l \neq A^{l+}$ *and in general,* A^l *is not comparable with* A^{lo}.
- ⋆ $A \subseteq S$, $A^{uo} \subseteq A^u$.

Definition 4 If X is an approximation operator, then by a X-*definite element*, we will mean a subset A satisfying $A^X = A$. The set of all X-definite elements will be denoted by $\delta_X(S)$, while the set of X and Y-definite elements (Y being another approximation operator) will be denoted by $\delta_{XY}(S)$. In particular, we will speak of *lower proto-definite*, *upper proto-definite*, and *proto-definite* elements (those that are both lower and upper proto-definite).

Theorem 1 *In a* PRAX S, *the following hold:*

- ⋆ $\delta_u(S) \subseteq \delta_{uo}(S)$, *but* $\delta_{lo}(S) = \delta_{uo}(S)$ *and* $\delta_u(S)$ *is a complete sublattice of* $\wp(S)$ *with respect to inclusion.*

- ⋆ $\delta_l(S)$ *is not comparable with* $\delta_{lo}(S)$ *in general.*
- ⋆ *It is possible that* $\delta_u \nsubseteq \delta_{uo}$.

A^{u+}, A^{l+} *like operators have been more commonly used in the literature and the only kind of approximation studied in* [10] for example.

Theorem 2 *All of the following hold in PRAX:*

- ⋆ $(\forall x)\, x^{cl+} = x^{u+c},\ x^{cu+} = x^{l+c}$*—that is* $l+$ *and* $u+$ *are mutually dual.*
- ⋆ $u+$ *(*$l+$ *resp.) is a monotone* $\vee$*- (complete* $\wedge$*- resp.) morphism.*
- ⋆ $\partial(x) = \partial(x^c)$*, where partial stands for the boundary operator.*
- ⋆ $\Im(u+)$ *is an interior system while* $\Im(l+)$ *is a closure system.*
- ⋆ $\Im(u+)$ *and* $\Im(l+)$ *are dually isomorphic lattices.*

Theorem 3 *In a PRAX,* $(\forall A \in \wp(S))\, A^{l+} \subseteq A^l,\ \ A^{u+} \subseteq A^u$ *and*

Bi $(\forall A \in \wp(S))\, A^{ll} = A^l \,\&\, A^u \subseteq A^{uu}$,
l-Cup $(\forall A, B \in \wp(S))\, A^l \cup B^l \subseteq (A \cup B)^l$,
l-Cap $(\forall A, B \in \wp(S))\, (A \cap B)^l \subseteq A^l \cap B^l$,
u-Cup $(\forall A, B \in \wp(S))\, (A \cup B)^u = A^u \cup B^u$,
u-Cap $(\forall A, B \in \wp(S))\, (A \cap B)^u \subseteq A^u \cap B^u$ *and*
Dual $(\forall A \in \wp(S))\, A^{lc} \subseteq A^{cu}$ *hold.*

Proof

l-Cup For any $A, B \in \wp S$, $x \in (A \cup B)^l$

$\Leftrightarrow$ $(\exists y \in (A \cup B))\, x \in [y] \subseteq A \cup B$.
$\Leftrightarrow$ $(\exists y \in A)\, x \in [y] \subseteq A \cup B$ or $(\exists y \in B)\, x \in [y] \subseteq A \cup B$.
$\Leftrightarrow$ $(\exists y \in A)\, x \in [y] \subseteq A$ or $(\exists y \in A)\, x \in [y] \subseteq B$ or $(\exists y \in B)\, x \in [y] \subseteq A$ or $(\exists y \in B)\, x \in [y] \subseteq B$—this is implied by $x \in A^l \cup B^l$.

l-Cap For any $A, B \in \wp S$, $x \in (A \cap B)^l$

$\Leftrightarrow$ $x \in A \cap B$
$\Leftrightarrow$ $(\exists y \in A \cap B)\, x \in [y] \subseteq A \cap B$ and $x \in A,\ x \in B$
$\Leftrightarrow$ $(\exists y \in A)\, x \in [y] \subseteq A$ and $(\exists y \in B)\, x \in [y] \subseteq B$—Clearly this statement implies $x \in A^l \& x \in B^l$, but the converse is not true in general.

u-Cup $x \in (A \cup B)^u$

$\Leftrightarrow$ $x \in \bigcup_{[y]\cap(A\cup B)\neq\emptyset}[y]$
$\Leftrightarrow$ $x \in \bigcup_{([y]\cap A)\cup([y]\cap B)\neq\emptyset}$
$\Leftrightarrow$ $x \in \bigcup_{[y]\cap A\neq\emptyset}[y]$ or $x \in \bigcup_{[y]\cap B\neq\emptyset}[y]$
$\Leftrightarrow$ $x \in A^u \cup B^u$.

u-Cap By monotonicity, $(A \cap B) \subseteq A^u$ and $(A \cap B) \subseteq B^u$, so $(A \cap B)^u \subseteq A^u \cap B^u$.

Dual If $z \in A^{lc}$, then $z \in [x]^c$ for all $[x] \subseteq A$ and either, $z \in A \backslash A^l$ or $z \in A^c$. If $z \in A^c$ then $z \in A^{cu}$. If $z \in A \backslash A^l$ and $z \neq A^{cu} \setminus A^c$ then $[z] \cap A^c = \emptyset$. But this contradicts $z \notin A^{cu} \backslash A^c$. So $(\forall A \in \wp(S))\, A^{lc} \subseteq A^{cu}$.

Theorem 4 *In a* PRAX *S, all of the following hold:*

1. $(\forall A, B \in \wp(S))\,(A \cap B)^{l+} = A^{l+} \cap B^{l+}$
2. $(\forall A, B \in \wp(S))\, A^{l+} \cup B^{l+} \subseteq (A \cup B)^{l+}$
3. $(\forall A \in \wp(S))\,(A^{l+})^c = (A^c)^{u+}\, \&\, A^{l+} \subseteq A^{lo}\, \&\, A^{uo} \subseteq A^{u+}\, \& A^{l+} \subseteq A^{lo}$.

From the above, we have the following relation between approximations in general ($u+ \longrightarrow u$ should be read as *the* $u+$*- approximation of a set is included in the* u*-approximation of the same set*):

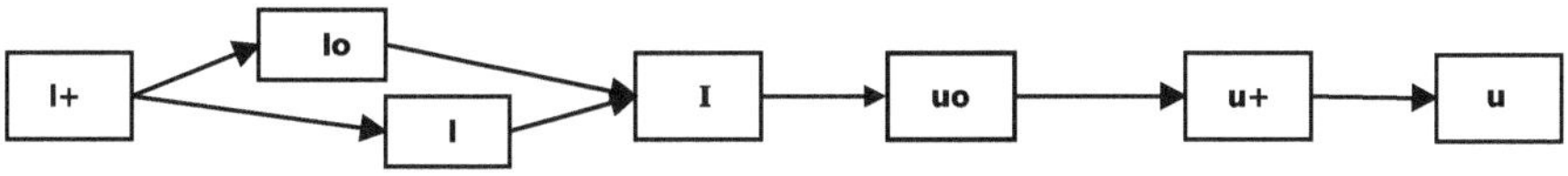

If a relation R is reflexive and not proto-transitive on a set S, then the relation $\tau(R) = R \cap R^{-1}$ will not be an equivalence and for a $A \subset S$, it is possible that $A^{uol} \subseteq A$ or $A^{uol} \parallel A$ or $A \subseteq A^{uol}$.

Theorem 5 *On the set of proto-definite elements* $\delta_{lu}(S)$ *of a* PRAX *S, we can define the following:*

1. $x \wedge y \stackrel{\Delta}{=} x \cap y;\ x \vee y \stackrel{\Delta}{=} x \cup y$
2. $0 \stackrel{\Delta}{=} \emptyset\ ;\ 1 \stackrel{\Delta}{=} S\ ;\ x^c \stackrel{\Delta}{=} S \backslash x$.

The resulting algebra $S_{proto}(S) = \langle \delta_{lu}(S), \vee, \wedge, c, 0, 1\rangle$ *is a Boolean lattice.*

3 Approximate Relations

If R is a binary relation on a set X, then we let $R^o \stackrel{\partial}{=} R \cup \Delta_X$. The weak transitive closure of R will be denoted by $R^{\#}$. If $R^{(i)}$ is the i-times composition $R \circ R \ldots \circ R$ (i-times), then $R^{\#} = \bigcup R^{(i)}$. R is *acyclic* iff $(\forall x)\, \neg R^{\#}xx$. The relation $R^{\cdot}$ is defined by $R^{\cdot}ab$ iff $Rab\, \&\, \neg(R^{\#}ab\, \&\, R^{\#}ba)$.

Definition 5 If R is a relation on a set S, then the relations $R^{\leftthreetimes}$, R^{cyc}, and R^h will be defined via

- $R^{\leftthreetimes}ab$ iff $[b]_{R^o} \subset [a]_{R^o}\, \&\, [a]_{iR^o} \subset [b]_{iR^o}$.
- $R^{cyc}ab$ iff $R^{\#}ab$ and $R^{\#}ba$.
- $R^h ab$ iff $R^{\leftthreetimes}ab$ and $R^{\cdot}ab$.

In case of PRAX, $R^o = R$, so the definition of $R^{\leftthreetimes}$ would involve neighborhoods of the form $[a]$ and $[a]_i$ alone. $R^{\leftthreetimes} \subset R$ and $R^{\leftthreetimes}$ is a partial order.

Example 3 In our example 1, $R^{\#}ab$ happens when a is an ally of an ally of b. $R^{\leftthreetimes}ab$ happens iff every ally of b is an ally of a and if a is ally of c, then b is an ally of c—this can happen, for example, when b is a Marxist feminist and a is a socialist feminist. $R^{cyc}ab$ happens when *a is an ally of an ally of b and b is an ally of an ally of a*. $R^{\cdot}ab$ happens whenever a is an ally of b, but b is not an ally of anybody who is an ally of a.

Theorem 6 $R^h = \emptyset$.

Proof $R^h ab \Leftrightarrow R^{\leftthreetimes}ab \,\&\, R^{\cdot}ab$
$\Leftrightarrow \tau(R)ab \,\&\, (R\backslash\tau(R))ab.$
But $\neg(\exists a)(R\backslash\tau(R))aa$.
So $R^h = \emptyset$.

Proposition 2 *All of the following hold in a PRAX S:*

- ⋆ $R^{\cdot}ab \leftrightarrow (R\backslash\tau(R))ab.$
- ⋆ $(\forall a, b)\neg(R^{\cdot}ab \,\&\, R^{\cdot}ba).$
- ⋆ $(\forall a, b, c)(R^{\cdot}ab \,\&\, R^{\cdot}bc \longrightarrow \neg R^{\cdot}ac).$

Proof $R^{\cdot}ab \leftrightarrow Rab \,\&\, \neg(R^{\#}ab\, R^{\#}ba)$. But $\neg(R^{\#}ab\, R^{\#}ba)$ is possible only when both Rab and Rba hold. So $R^{\cdot}ab \leftrightarrow Rab \,\&\, \neg(\tau(R)ab) \leftrightarrow (R\backslash\tau(R))ab.$

Theorem 7

1. $R^{\#\cdot} = R^{\#}\backslash\tau(R).$
2. $R^{\cdot\#} = (R\backslash\tau(R))^{\#}.$
3. $(R\backslash\tau(R))^{\#} \subseteq R^{\#}\backslash\tau(R).$

Proof

1. $R^{\#\cdot}ab \leftrightarrow R^{\#}ab \,\&\, \neg(R^{\#\#}ab \,\&\, R^{\#\#}ba) \leftrightarrow$
$R^{\#}ab \,\&\, \neg(R^{\#}ab \,\&\, R^{\#}ba). \leftrightarrow R^{\#}ab \,\&\, \neg\tau(R)ab \leftrightarrow (R^{\#}\backslash\tau(R))ab.$
2. $R^{\cdot\#}ab \leftrightarrow (R^{\cdot})^{\#}ab \leftrightarrow (R\backslash\tau(R))^{\#}ab.$
3. Can be checked by a contradiction or a direct argument.

We now look at possible properties that approximations of proto-transitive relations may/should possess. If $<$ is a strict partial order on S and R is a relation, then consider the conditions:

PO1 $(\forall a, b)(a < b \longrightarrow R^{\#}ab).$
PO2 $(\forall a, b)(a < b \longrightarrow \neg R^{\#}ba).$
PO3 $(\forall a, b)(R^{\leftthreetimes}ab \,\&\, R^{\cdot}ab \longrightarrow a < b.$
PO4 If $a \equiv_R b$, then $a \equiv_< b$.
PO5 $(\forall a, b)(a < b \longrightarrow Rab).$

As per [2], $<$ is said to be a *partial order approximation* POA (resp. *weak partial order approximation* WPOA) of R iff **PO1, PO2, PO3, PO4** (resp. **PO1, PO3, PO4**) hold. A POA $<$ is *inner approximation* IPOA of R iff **PO5** holds. **PO4** has a role beyond that of approximation and depends on both successor and predecessor neighborhoods. R^h, $R^{\cdot\leftthreetimes}$ are IPOA, while $R^{\cdot\#}$, $R^{\#\cdot}$ are POAs. By a *lean quasi-order approximation* $<$ of R, we will mean a quasi-order satisfying **PO1** and **PO2**— the set of such approximations of R will be denoted by $LQO(R)$.

Theorem 8 *For any $A, B \in LQO(R)$, we can define the operations &, $\vee$, $\top$:*

- $\star$ $(\forall x, y)(A \& B)xy$ *iff* $(\forall x, y) Axy \,\&\, Bxy$.
- $\star$ $(A \vee B) = (A \cup B)^{\#}$, $\top = R^{\#}$.

Proof If Aab then $R^{+}ab$ and if Bab then $R^{+}ab$. But if $(A\&B)ab$, then both Aab and Bab. So $R^{+}ab$.

Similarly it can be shown that $A \vee B \in LQO(R)$. It is always defined and contained within $R^{\#}$ as it is the transitive completion of $A \cup B$. $\top = R^{\#}$ as transitive closure is a closure operator.

Theorem 9 *In a PRAX,* $R^{\cdot\#} \& R^{\#\cdot} xy \leftrightarrow (R \backslash \tau(R))^{\#} xy$.

3.1 Granules of Derived Relations

The behavior of approximations and rough objects corresponding to derived relations is investigated in this subsection.

Definition 6 The relation $R^{\#\cdot}$ will be termed the *trans ortho-completion* of R. The following granules will be associated with each $x \in S$:

$$[x]_{ot} = \{y\,;\ R^{\#\cdot} yx\} \tag{1}$$

$$[x]_{ot}^{i} = \{y\,;\ R^{\#\cdot} xy\} \tag{2}$$

$$[x]_{ot}^{o} = \{y\,;\ R^{\#\cdot} yx \,\&\, R^{\#\cdot} xy\}. \tag{3}$$

Let the corresponding approximations be l_{ot}, u_{ot} and so on.

Theorem 10 *In a PRAX S,* $(\forall x \in S)\, [x]_{ot}^{o} = \{x\}$.

Proof $R^{\#\cdot} xy \,\&\, R^{\#\cdot} yx$ means that the pair (x, y) is in the transitive completion of R and not in $\tau(R)$. So $y \in [x]_{ot}^{o}$ iff

$(\exists a, b) Rxa \,\&\, Ray \,\&\, (\neg Rax \vee \neg Rya) \,\&\, (Ryb \,\&\, Rbx) \,\&\, (\neg Rby \vee \neg Rxb)$.

If we assume that $x \neq y$, then each of the possibilities leads to a contradiction as is shown below. In the context of the above statement:

Case-1: $Rxa \,\&\, Ray \,\&\, \neg Rax \,\&\, Rya \,\&\, Ryb \,\&\, Rbx \,\&\, \neg Rby \,\&\, Rxb$. This yields $R^{\#}xa \,\&\, R^{\#}bb \,\&\, R^{\#}ba \,\&\, R^{\#}ab$. So, $R^{\#}xb \,\&\, R^{\#}ya \,\&\, R^{\#}ax$ and we have contradicted our original assumption.

Case-2: $Rxa \,\&\, Ray \,\&\, Rax \,\&\, \neg Rya \,\&\, Ryb \,\&\, Rbx \,\&\, Rby \,\&\, \neg Rxb$. So $R^{\#}ab$.

Case-3: $Rxa \,\&\, Ray \,\&\, \neg Rax \,\&\, Rya \,\&\, Ryb \,\&\, Rbx \,\&\, Rby \,\&\, \neg Rxb$.

So $R^{\#}ba \,\&\, R^{\#}ab \,\&\, R^{\#}aa \,\&\, R^{\#}bb$ and $R^{\#}yy \& R^{\#}xy \& R^{\#}yx \& Rya \& R^{\#}xa$. But such a $R^{\#}$ is not possible.

Somewhat similarly the other cases can be seen to lead to contradictions.

By the *symmetric center* of a relation R, we will mean the set $K_R = \bigcup e_i(\tau(R)\setminus \Delta_S)$—basically the union of elements in either component of $\tau(R)$ minus the diagonal relation on S.

Proposition 3 *$(\forall x)\, [x]\triangle[x]_{ot} \neq \emptyset$ as $(x \notin K_R \longrightarrow [x] \subset [x]_{ot})$ and $(x \in K_R \longrightarrow [x] \nsubseteq [x]_{ot} \,\&\, \{x\} \subset [x] \cap [x]_{ot})$.*

Proof

$$z \in [x]_{ot} \leftrightarrow R^{\#\cdot}zx \leftrightarrow R^{\#}zx \,\&\, \neg\tau(R)zx$$
$$\leftrightarrow (Rzx \,\&\, \neg Rxz) \text{ or } (\neg Rzx \,\&\, \neg Rxz \,\&\, (R^{\#}\setminus R)zx).$$

K_R can be used to partially categorize subsets of S based on intersection.

Proposition 4 *$(R\setminus\tau(R))^{\#} \cup \tau(R)$ is not necessarily a quasi-order.*

Proof $(x, y) \in (R\setminus\tau(R))^{\#} \cup \tau(R)$ and $(x, y) \notin \tau(R)$ and $x \in K_R \,\&\, y \notin K_R$ and $\exists z \in K_R \,\&\, z \neq x \,\&\, Rzx$ do not disallow Rzy. So $(R\setminus\tau(R))^{\#} \cup \tau(R)$ is not necessarily a quasi-order. We leave the missing part to the reader.

Proposition 5 *$((R\setminus\tau(R))^{\#} \cup \tau(R))^{\#} = R^{\#}$.*

Proof Clearly $R \subseteq ((R\setminus\tau(R))^{\#} \cup \tau(R))^{\#}$ and it can be directly checked that if $a \in ((R\setminus\tau(R))^{\#} \cup \tau(R))^{\#}\setminus R$ then $a \in R^{\#}\setminus R$ and conversely.

The main conclusion of this section is that we should select our approximate relations and approximations based on our context (of course avoiding the redundant ones). In case of our main example, all of $R^{\#}$, $R^{\#\cdot}$, $R^{\cdot\#}$, $R^{\curlywedge}$ are sensible in different perspectives.

4 Transitive Completion and Approximate Semantics

The interaction of the rough approximations in a PRAX and the rough approximations in the transitive completion can be expected to follow some order. *The definite or rough objects most closely related to the difference of lower approximations and those related to the difference of upper approximations can be expected to be related in a nice way*. We show that this *nice way* is not really a *rough way*. But the results proved remain relevant for the formulation of semantics that involves that of the transitive completion as in [3, 4]. A rough theoretical alternative is possible by simply starting from sets of the form $A^* = (A^l\setminus A^{l\#}) \cup (A^{u\#}\setminus A^u)$ and taking their lower ($l_\#$) and upper ($u_\#$) approximations—the resulting structure would be a partial algebra derived from a Nelson algebra over an algebraic lattice.

Proposition 6 *For an arbitrary proto-transitive reflexive relation R on a set S, (we use # subscripts for neighborhoods, approximation operators and rough equalities of the weak transitive completion) all of the following hold:*

$$(\forall x \in S)\, [x]_R \subseteq [x]_{R^{\#}} \tag{Nbd}$$

$$(\forall A \subseteq S)\, A^l \subseteq A^{l\#} \,\&\, A^u \subseteq A^{u\#} \tag{App}$$

$$(\forall A \subseteq S)(\forall B \in [A]_{\approx})(\forall C \in [A]_{\approx_{\#}})\, B^l \subseteq C^{l\#} \,\&\, B^u \subseteq C^{u\#} \tag{REq}$$

The reverse inclusions are false in general in the second assertion in a specific way. Note that the last condition induces a more general partial order $\preceq$ over $\wp(\wp(S))$ via $A \preceq B$ iff $(\forall C \in A)(\forall E \in B)\, C^l \subseteq E^{l\#} \,\&\, C^u \subseteq E^{u\#}$.

Proof The first of these is direct. For simplicity, we will denote the successor neighborhoods of x by $[x]$ and $[x]_{\#}$, respectively. We look at the possibility tracking in the first part of the second assertion. If $z \in A^{l\#}$ then $z \in A^l$ as $[x]_{\#} \subseteq A$ implies $[x] \subseteq A$. If $z \in A^l$ then $(\exists x)\, z \in [x] \subseteq A^l$. For this $x, z \in [x]_{\#}$, but it is possible that $[x]_{\#} \subseteq A$ or $[x]_{\#} \nsubseteq A$. If $[x]_{\#} \nsubseteq A$, and $(\exists b \notin A)\, R_{\#}ax \,\&\, Rab \,\&\, Rbx$ then we have a contradiction as Rbx means $b \in [x]$. If $[x]_{\#} \nsubseteq A$, and $(\exists b \in A)\, R_{\#}ax \,\&\, Rab \,\&\, Rbx$ all we need is a $c \notin A \,\& Rcb$ that is compatible with $R_{\#}cx$ and $A^l \nsubseteq A^{l\#}$.

Definition 7 By the *l-scedastic approximation* $\hat{l}$ and the *u-scedastic approximation* $\hat{u}$ of a subset $A \subseteq S$ we will mean the following approximations:

$$A^{\hat{l}} = (A^l \setminus A^{l\#})^l, \quad A^{\hat{u}} = (A^{u\#} \setminus A^u)^{u\#}.$$

The above cross-difference approximation is the best possible from closeness to properties of rough approximations.

Theorem 11 *For an arbitrary subset $A \subseteq S$ of a PRAX S, the following statements and diagram of inclusion ($\rightarrow$) hold:*

- ★ $A^{l\#l} = A^{l\#} = A^{ll\#} = A^{l\#l\#}$
- ★ *If $A^u \subset A^{u\#}$ then $A^{uu\#} \subseteq A^{u\#u\#}$.*

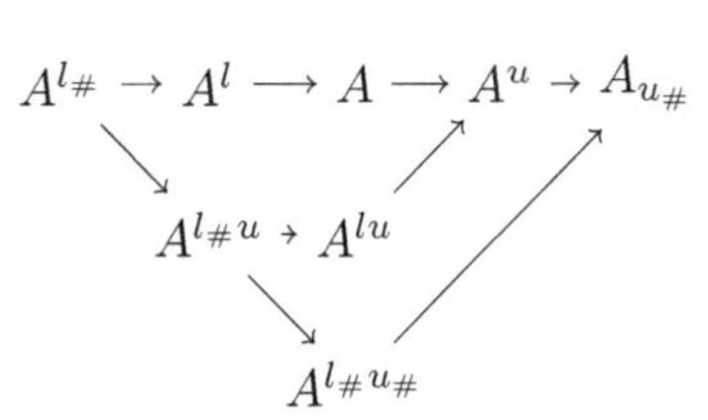

Proof It is clear that $A^l \subseteq A^u \subseteq A^{u\#}$. So $A^l \nsubseteq A^{u\#} \setminus A^u$.

- ⋆ $x \in (A^l \setminus A^{l\#})^l \Rightarrow (\exists y)\,[y]_\# \nsubseteq A \,\&\, x \in [y] \subset A \,\&\, x \in [y]_\#$
- ⋆ $\Rightarrow x \in A^{u\#} \,\&\, x \in A^u \Rightarrow x \notin A^{u\#} \setminus A^u$.
- ⋆ But $[y]_\# \subset A^{u\#}$, $(\exists z)\, z \in A^{u\#} \,\&\, z \notin A^u \,\&\, z \in [y]_\#$.
- ⋆ So $[y]_\# \subset (A^{u\#} \setminus A^u)^{u\#}$ and it is possible that $[y]_\# \nsubseteq (A^{u\#} \setminus A^u)^u$.

Theorem 12 *For an arbitrary subset $A \subseteq S$ of a PRAX S,*

$$(A^l \setminus A^{l\#})^l \nsubseteq (A^{u\#} \setminus A^u)^{u\#} \longrightarrow A^{u\#} = A^u.$$
$$A^{u\#} \neq A^u \longrightarrow A^l \setminus A^{l\#})^l \subseteq (A^{u\#} \setminus A^u)^{u\#}.$$

Proof Let $S = \{a, b, c, e, f\}$ and R be the transitive completion satisfying Rab, Rbc, Ref. If $B = \{a, b\}$, $B^{\hat{l}} = B$, but $B^{u\#} = \{a, b, c\} = B^u$. So $B^{\hat{u}} = \emptyset$. The second part follows from the proof of the above proposition under the restriction in the premise.

Theorem 13 *Key properties of the scedastic approximations follow:*

1. $(\forall B \in \wp(S))(B^{\hat{l}} = B \nrightarrow B^{\hat{u}} = B)$.
2. $(\forall B \in \wp(S))(B^{\hat{u}} = B \rightarrow B^{\hat{l}} = B)$.
3. $(\forall B \in \wp(S))\, B^{\hat{l}\hat{l}} = B^{\hat{l}}$.
4. $(\forall B \in \wp(S))\, B^{\hat{u}\hat{u}} \neq B^{\hat{u}}$.
5. *It is possible that* $(\exists B \in \wp(S)\, B^{\hat{u}\hat{u}} \subset B^{\hat{u}})$.

Proof 1. The counterexample in the proof of the above theorem works for this statement.

2. $x \in B \leftrightarrow x \in (B^{u\#} \setminus B^u)^{u\#} \leftrightarrow (\exists y \in B^{u\#})(\exists z \in B^{u\#} \setminus B^u)\, x, z \in [y]_\# \,\&\, z \in B^{u\#} \,\&\, z \notin B^u$. But this situation requires that elements of the form z be related to x and so we should have $B^{u\#} = B^u$.
3. $B^{\hat{l}\hat{l}} = (B^{\hat{l}l} \setminus B^{\hat{l}l\#})^l = ((B^l \setminus B^{l\#})^l \setminus \emptyset)^l = B^{\hat{l}}$. The missing step is of proving $(B^l \setminus B^{l\#})^{ll\#} = \emptyset$.

4–5. We prove the last two assertions together. We provide a counterexample and also show the essential pattern of deviation.

Let $S = \{a, b, c, e, f\}$ and R be a reflexive relation s.t. Rab, Rbc, Ref.
If $A = \{a, e\}$, then $A^{u\#} = \{a, b, c, e\}$ and $A^u = \{a, b, e\}$.
Therefore $A^{\hat{u}} = \{c\} \,\&\, A^{\hat{u}\hat{u}} = \emptyset \,\&\, A^{\hat{u}\hat{u}} \subset A^{\hat{u}}$.
In general if B is some subset, then $x \in B^{\hat{u}} = (A^{u\#} \setminus A^u)^{u\#} \Rightarrow (\exists y \in A^{u\#})(\exists z)\, y \in [z]_\# \,\&\, y \notin A^u \,\&\, y \notin A \,\&\, z \in A \,\&\, y \notin [z] \,\&\, y \in [x]_\#$.

An interesting problem can be given A for which $A^{u\#} \neq A^u$, when does there exist a B such that

$$B^l = (A^l \setminus A^{l\#})^l = A^{\hat{l}} \,\&\, B^u = (A^{u\#} \setminus A^u)^{u\#} = A^{\hat{u}}?$$

5 Remarks

In this research, we have developed the relation between approximation of proto-transitive relations and approximations in PRAX. The relation of transitive completions to such relations is examined in detail for further dialectical semantics relying on Nelson algebras over algebraic lattices—this internalized semantics will appear separately. It is shown that transitive completions of proto-transitive relations do not happen in any uniform rough way unless we start from sets of the form A^*. The methods will be of relevance for other RSTs as well.

References

1. Mani, A.: Dialectics of knowledge representation in a granular rough set theory. Inst. Math. Sci. ICLA'2013. Chennai, 1–12 (2013) http://arxiv.org/abs/1212.6519. Refereed Conference Paper
2. Janicki, R.: Approximation of arbitrary binary relations by partial orders: classical and rough set models. Trans. Rough Sets XIII (LNCS 6499) 17–38 (2011)
3. Jarvinen, J., Radeleczki, S.: Representation of Nelson algebras by rough sets determined by quasi-orders. Algebra Univers. **66**, 163–179 (2011)
4. Jarvinen, J., Pagliani, P., Radeleczki, S.: Information completeness in Nelson algebras of rough sets induced by quasiorders. Studia Logica 1–20 (2012)
5. Mani, A.: Axiomatic approach to granular correspondences. In: Li, T., et al. (eds.) Proceedings of RSKT'2012, LNAI 7414. vol. LNAI 7414, pp. 482–487. Springer (2012)
6. Mani, A.: Contamination-free measures and algebraic operations. In: Pal, N., et al. (eds.) 2013 IEEE International Conference on Fuzzy Systems, Fuzzy Systems (FUZZ), vol. F-1438, pp. 1–8 Hyderabad (2013)
7. Mani, A.: Choice inclusive general rough semantics. Inf. Sci. **181**(6), 1097–1115 (2011)
8. Mani, A.: Dialectics of counting and the mathematics of vagueness. Trans. Rough Sets XV (LNCS 7255) 122–180 (2012)
9. Chajda, I., Haviar, M.: Induced pseudo orders. Acta Univ. Palack. Olomou **30**(1), 9–16 (1991)
10. Jarvinen, J.: Lattice theory for rough sets. In: Peters, J.F. et al. (eds.) Transactions on Rough Sets VI. LNCS vol. 4374, pp. 400–498. Springer, New York (2007)

Part III
Hybridization of Uncertainties

A Probabilistic Approach to Information System and Rough Set Theory

Md. Aquil Khan

Abstract We propose a generalization of information systems which provides the probability of an object to take an attribute-value for an attribute. Notions of distinguishability relations and corresponding notions of approximations are proposed and studied in comparison with the existing one.

Keywords Rough sets · Approximation operators · Information system · Indiscernibility relation · Similarity relation

1 Introduction

Rough set theory, introduced by Pawlak, is based on the concept of *approximation space* [10] which is defined as a tuple (U, R), where R is an equivalence relation on the set U. Any concept represented as a subset (say) X of the partitioned domain U, is then approximated from "within" and "outside," by its *lower* and *upper approximations* given as $\underline{X}_R := \{x : [x]_R \subseteq X\}$ and $\overline{X}_R := \{x : [x]_R \cap X \neq \emptyset\}$, respectively. Here, $[x]_R$ denotes the equivalence class of $x \in U$. With time, Pawlak's simple rough set model has seen many generalizations due to demands from different practical situations (e.g. [2, 6, 11–13, 17]). A useful natural generalization is where the relation R is not necessarily an equivalence. For instance, in [3, 12], a *tolerance approximation space* is considered, where R is a tolerance relation. The notion of lower and upper approximations of a set in these generalized approximation spaces is then defined in a natural way.

There is another way to look at generalizations of Pawlak's rough set theory, viz. from the point of view of *information systems* (e.g. [1, 7, 8, 15]). Most applications of rough set theory are based on these attribute-value representation models.

The author would like to thank the referees and editors for their valuable comments.

Md. A. Khan (✉)
Discipline of Mathematics, Indian Institute of Technology, Indore 452 017, India
e-mail: aquilk@iiti.ac.in

M.K. Chakraborty et al. (eds.), *Facets of Uncertainties and Applications*,
Springer Proceedings in Mathematics & Statistics 125,
DOI 10.1007/978-81-322-2301-6_9

Definition 1 A *deterministic information system* (DIS) $\mathcal{K} := (U, \mathcal{A}, \{\mathcal{V}_a\}_{a\in\mathcal{A}}, f)$, comprises a nonempty set U of objects, finite set $\mathcal{A}$ of attributes, finite set $\mathcal{V}_a$ of attribute-values for each $a \in \mathcal{A}$, and information function $f : U \times \mathcal{A} \to \bigcup_{a\in\mathcal{A}} \mathcal{V}_a$ such that $f(x, a) \in \mathcal{V}_a$.

Given a deterministic information system $\mathcal{K} := (U, \mathcal{A}, \{\mathcal{V}_a\}_{a\in\mathcal{A}}, f)$ and a set $B \subseteq \mathcal{A}$, the *indiscernibility relation* $Ind_{\mathcal{K},B}$ is an equivalence relation on U defined by:

$$(x, y) \in Ind_{\mathcal{K},B}, \text{ if and only if } f(x, a) = f(y, a) \text{ for all } a \in B.$$

Thus, given a DIS $\mathcal{K}$ and a set B of attributes, we obtain an approximation space $(U, Ind_{\mathcal{K},B})$.

From Definition 1 of DISs, it is clear that for each object of the domain, we have information about each attribute of the system. However, we could have situation where some attribute-values for an object may be missing. A distinguished attribute-value $*$ is used to depict this absence of information.

Definition 2 An *incomplete information system* (IIS) is a tuple $\mathcal{K} := (U, \mathcal{A}, \{\mathcal{V}_a\}_{a\in\mathcal{A}} \cup \{*\}, f)$, where $f : U \times \mathcal{A} \to \bigcup_{a\in\mathcal{A}} \mathcal{V}_a \cup \{*\}$ such that $f(x, a) \in \mathcal{V}_a \cup \{*\}$.

In [4, 5], instead of an indiscernibility relation, a *similarity* relation (defined below) is considered as the distinguishability relation in the context of an IIS. The assumption here is that the real value of missing attributes is one from the attribute domain.

$(x, y) \in Sim_{\mathcal{S},B}$ if and only if, $f(x, a) = f(y, a)$ or $f(x, a) = *$, or $f(y, a) = *$, for all $a \in B$.

DISs are deterministic in the sense that objects take a single value for each attribute. Thus, a natural generalization of DISs is obtained by allowing an object to take a *set of values* for an attribute.

Definition 3 A tuple $\mathcal{K} := (U, \mathcal{A}, \{\mathcal{V}\}_{a\in\mathcal{A}}, f)$ is called a *non-deterministic information system* (NIS), where $f : U \times \mathcal{A} \to 2^{\bigcup_{a\in\mathcal{A}} \mathcal{V}_a}$ such that $f(x, a) \subseteq \mathcal{V}_a$.

One may attach different interpretations with '$f(x, a) = V$', for $V \subseteq \mathcal{V}_a$. For instance, one could interpret $f(x, a) = V$ as object x takes precisely one attribute-value from V, and under this interpretation the following similarity relations are found to be useful.

Similarity: $(x, y) \in Sim_{\mathcal{K},B}$ if and only if $f(x, a) \cap f(y, a) \neq \emptyset$ for all $a \in B$.

Weak similarity: $(x, y) \in Sim^{w}_{\mathcal{K},B}$ if and only if $f(x, a) \cap f(y, a) \neq \emptyset$ for some $a \in B$.

Let us consider an object x, an attribute a, and attribute-value $v \in \mathcal{V}_a$. Consider the event

$$E\text{: object } x \text{ takes the attribute-value } v \text{ for } a.$$

Under a DIS, we have precise information whether this event occurs or not. But situation is not so simple in the case of IIS when $f(x, a) = *$. In this case, we can at most say that the event E has a probability $\frac{1}{|\mathcal{V}_a|}$ to occur, where $|X|$ denotes the cardinality of the set X. Situation is similar in the case of NIS. Under the interpretation of $f(x, a) = V$ given above, if $v \notin f(x, a)$, then we are certain that event E will not occur, but if $v \in f(x, a)$, then we do not have precise information about the event, and again we can only assign a probability to the occurrence of this event.

The above observation shows that we can have a situation where we only know the probability of an object to take an attribute-value for an attribute. Therefore, in this article, we propose and study a generalization of information systems called *probabilistic information system (PIS)*, which provides only the probability of an object to take an attribute-value for an attribute. A few similarity relations are defined on PIS, and it is shown that the indiscernibility relation defined on DISs, and similarity relations defined on IISs and DISs are all originated from a single similarity relation defined on PISs. We would like to add here that several work has been done on the applications of probabilistic approaches to rough set theory (cf. e.g. [9, 16]), but most of these works are based on the proposals of approximations of sets in approximation spaces keeping in view the overlap of the equivalence classes with the set. In this article, instead we take into account the source of approximation spaces, that is, information systems.

The remainder of this article is organized as follows. In Sect. 2, we present the notion of the PISs, and study the notion of approximations on PISs. In Sect. 3, we present a comparative study of PISs with the DISs, IISs, and NISs. Section 4 concludes the article.

2 Probabilistic Information Systems

Let U be a set of objects, and $\mathcal{A}$ be a set of attributes of the objects of U. For each $a \in \mathcal{A}$, let $\mathcal{V}_a$ be the set of possible attribute-values that the objects from U can take for the attribute a. For $x \in U, a \in \mathcal{A}$ and $v \in \mathcal{V}_a$, let us use the tuple (x, a, v) to denote the event that the object x takes the value v for the attribute a. In many practical situations, we may not have the precise information for the event (x, a, v). For instance, in an election, we may not know precisely to whom a voter x is going to vote, but we may know the probabilities of x voting to different candidates. A *probabilistic information system* with domain U, attribute set $\mathcal{A}$, and attribute-value set $\bigcup_{a\in\mathcal{A}} \mathcal{V}_a$ is a structure which assigns probabilities to these events. Formally, we have the following definition.

Definition 4 A *probabilistic information system (PIS)* is defined as a tuple $\mathcal{K} := (U, \mathcal{A}, \{\mathcal{V}_a\}_{a\in\mathcal{A}}, F)$, where $U, \mathcal{A}, \mathcal{V}_a$ are as in Definition 1, and $F : \mathcal{D}_\mathcal{K} \to [0, 1]$, $\mathcal{D}_\mathcal{K}$ being the set $\{(x, a, v) : x \in U, a \in \mathcal{A}, \text{ and } v \in \mathcal{V}_a\}$ such that $\sum_{v\in\mathcal{V}_a} F(x, a, v) = 1$.

Let $\mathcal{K} := (U, \mathcal{A}, \{\mathcal{V}_a\}_{a\in\mathcal{A}}, F)$ be a PIS. Corresponding to $x, y \in U, x \neq y$, and $a \in \mathcal{A}$, we obtain a sample space $E_{(x,y,a)}$ defined as

$$E_{(x,y,a)} := \{\langle(x, a, v), (y, a, u)\rangle : v, u \in \mathcal{V}_a\},$$

and a probability mass function $P_{(x,y,a)} : E_{(x,y,a)} \to [0, 1]$ such that

$$P_{(x,y,a)}\langle(x, a, v), (y, a, u)\rangle := F(x, a, v)F(y, a, u).$$

One can easily verify the following property of the probability mass function.

Proposition 1 $\sum_{\beta\in E_{(x,y,a)}} P_{(x,y,a)}(\beta) = 1.$

The element $\langle(x, a, v), (y, a, u)\rangle$ of the sample space $E_{(x,y,a)}$ represents the event that the objects x and y take the attribute-value v and u, respectively, for the attribute a. Moreover, $P_{(x,y,a)}\langle(x, a, v), (y, a, u)\rangle$ gives the probability of this event to occur based on the information provided by the PIS $\mathcal{K}$.

Recall that an event Q of the sample space $E_{(x,y,a)}$ is a subset of $E_{(x,y,a)}$, and its probability is given by

$$P_{(x,y,a)}(Q) = \sum_{\beta\in Q} P_{(x,y,a)}(\beta).$$

We use this fact to define the following fuzzy relations on U.

Definition 5 Let $\mathcal{K} := (U, \mathcal{A}, \{\mathcal{V}_a\}_{a\in\mathcal{A}}, F)$ be a PIS. For each $a \in \mathcal{A}$, we define the mappings $R_a : U \times U \to [0, 1]$ as follows:

$$R_a(x, y) := \begin{cases} P_{(x,y,a)}\{\langle(x, a, v), (y, a, v)\rangle : v \in \mathcal{V}_a\}, & \text{if } x \neq y \\ 1, \text{otherwise.} \end{cases}$$

The mappings defined above are not indexed with the underlying PIS to make the notation simple, and should not create any confusion. We note that $R_a(x, y)$ gives the probability of the event that the objects x and y take the same attribute-value for the attribute a. On unfolding the definition of R_a, we obtain the following result.

Proposition 2 $R_a(x, y) = \sum_{v\in\mathcal{V}_a} F(x, a, v)F(y, a, v), \ x \neq y.$

For a given PIS $\mathcal{K} := (U, \mathcal{A}, \{\mathcal{V}_a\}_{a\in\mathcal{A}}, F)$, we now use the relation R_a to define the following fuzzy and crisp similarity relations on U. Let $B \subseteq \mathcal{A}$, and $x, y \in U$.

Definition 6

Similarity: $S_{\mathcal{K},B}(x, y) := \prod_{a\in B} R_a(x, y).$

Weak Similarity: $S^w_{\mathcal{K},B}(x, y) := 1 - \prod_{a\in B} (1 - R_a(x, y)).$

Crisp Similarity: For $\lambda \in [0, 1)$, $(x, y) \in S^c_{\mathcal{K},B}(\lambda)$ if and only if for all $a \in B$, $R_a(x, y) > \lambda$.

A generalization of the above-defined crisp similarity relation $S^c_{\mathcal{K},B}$ would be the case where a different threshold λ_a is provided for different $a \in B$. But in this article, we shall consider only the above-defined crisp similarity relation to make the presentation simple.

Let us observe the following facts about these relations.

- $S_{\mathcal{K},B}(x, y)$ gives the probability of the event that the objects x and y take the same attribute-value for each attribute in B.
- $S^w_{\mathcal{K},B}(x, y)$ gives the probability of the event that the objects x and y take the same attribute-value for some attributes in B.
- $(x, y) \in S^c_{\mathcal{K},B}(\lambda)$ if and only if for all $a \in B$, the probability of the event that the objects x and y take the same attribute-value for a is more than λ.

In Sect. 3, we shall see the close connections of the above-defined relations with some of the indistinguishability relations defined on information systems. But before that, we propose the following notion of lower and upper approximations. Let $\mathcal{K} := (U, \mathcal{A}, \{\mathcal{V}_a\}_{a\in\mathcal{A}}, F)$ be a PIS, and $x \in U$, $B \subseteq \mathcal{A}$, $\lambda \in [0, 1)$. We will use the following notation:

$$[x]^\lambda_{S_{\mathcal{K},B}} := \{y \in U : S_{\mathcal{K},B}(x, y) > \lambda\}; \quad [x]^\lambda_{S^w_{\mathcal{K},B}} := \{y \in U : S^w_{\mathcal{K},B}(x, y) > \lambda\};$$
$$[x]^\lambda_{S^c_{\mathcal{K},B}} := \{y \in U : (x, y) \in S^c_{\mathcal{K},B}(\lambda)\}.$$

Corresponding to each $R \in \{S^c_{\mathcal{K},B}, S_{\mathcal{K},B}, S^w_{\mathcal{K},B}\}$, and $\lambda \in [0, 1)$, we obtain the lower and upper approximation operators L_R and U_R defined as follows:

$$L_R(X, \lambda) := \{x \in U : [x]^\lambda_R \subseteq X\}; \; U_R(X, \lambda) := \{x \in U : [x]^\lambda_R \cap X \neq \emptyset\}.$$

Note that the relation $S^c_{\mathcal{K},B}$ is a crisp tolerance relation, and hence all the results that hold for tolerance relation based approximation operators follow automatically for the approximation operators based on $S^c_{\mathcal{K},B}$. On the other hand, the relations $S_{\mathcal{K},B}$, $S^w_{\mathcal{K},B}$ are fuzzy and hence the theory develop on these relations will take the course of fuzzy-rough sets. Therefore, it seems to be interesting to see how the theory develops for these relations. In the rest of this section, we explore a few properties of the approximation operators defined above. In this direction, we first note that the fuzzy relations $S_{\mathcal{K},B}$ and $S^w_{\mathcal{K},B}$ satisfy the reflexivity and symmetry conditions, but fail to satisfy transitivity condition: $\sigma(x, y) > \lambda \,\&\, \sigma(y, z) > \lambda \Rightarrow \sigma(x, z) > \lambda$. As a consequence of it, the lower and upper approximation operators defined on $S_{\mathcal{K},B}$ and $S^w_{\mathcal{K},B}$ satisfy all the standard properties of Pawlak's lower and upper approximation operators except the *idempotence*.

For each of the relation $R \in \{S^c_{\mathcal{K},B}, S_{\mathcal{K},B}, S^w_{\mathcal{K},B}\}$, the following holds:

Proposition 3 *1.* $L_R(X, \lambda) = \left(U_R(X^c, \lambda)\right)^c$, *where* X^c *denotes the complement of the set* X *relative to* U.

Table 1 PIS $\mathcal{K}$

	a				b		
	v_1	v_2	v_3	v_4	u_1	u_2	u_3
x_1	1	0	0	0	$\frac{1}{3}$	$\frac{1}{3}$	$\frac{1}{3}$
x_2	0	$\frac{1}{2}$	$\frac{1}{2}$	0	$\frac{1}{2}$	$\frac{1}{2}$	0
x_3	$\frac{1}{3}$	$\frac{1}{3}$	0	$\frac{1}{3}$	1	0	0

2. *For* $\lambda_2 \geq \lambda_1$, $L_R(X, \lambda_1) \subseteq L_R(X, \lambda_2)$ *and* $U_R(X, \lambda_2) \subseteq U_R(X, \lambda_1)$.

The following proposition gives the connection between different lower approximation operators defined above.

Proposition 4

1. $L_{S^c_{\mathcal{K},B}}(X, 0) = L_{S_{\mathcal{K},B}}(X, 0)$.
2. $L_{S^c_{\mathcal{K},B}}(X, \lambda) \subseteq L_{S_{\mathcal{K},B}}(X, \lambda)$, $\lambda \in [0, 1)$.
3. $L_{S^w_{\mathcal{K},B}}(X, \lambda) \subseteq L_{S_{\mathcal{K},B}}(X, \lambda)$, $\lambda \in [0, 1)$.

Example 1 Let us consider a PIS $\mathcal{K} := (U, \{a, b\}, \{\mathcal{V}_a, \mathcal{V}_b\}, F)$ with $U := \{x_1, x_2, x_3\}$, $\mathcal{V}_a := \{v_1, v_2, v_3, v_4\}$, $\mathcal{V}_b := \{u_1, u_2, u_3\}$ given by the Table 1. Thus $F(x_1, a, v_1) = 1$, $F(x_1, a, v_2) = 0$ and so on. The relations R_a and R_b giving the probability of two objects to take same attribute-value for a and b, respectively, are obtained as follows:

$$R_a : (x_1, x_2) \mapsto 0 \qquad R_b : (x_1, x_2) \mapsto \tfrac{1}{3}$$
$$(x_1, x_3) \mapsto \tfrac{1}{3} \qquad (x_1, x_3) \mapsto \tfrac{1}{3}$$
$$(x_2, x_3) \mapsto \tfrac{1}{6} \qquad (x_2, x_3) \mapsto \tfrac{1}{2}$$

This, in turn, determines the mapping $S_{\mathcal{K},B}$ and $S^w_{\mathcal{K},B}$, $B = \{a, b\}$, and are given as follows:

$$S_{\mathcal{K},B} : (x_1, x_2) \mapsto 0 \qquad S^w_{\mathcal{K},B} : (x_1, x_2) \mapsto \tfrac{2}{3}$$
$$(x_1, x_3) \mapsto \tfrac{1}{9} \qquad (x_1, x_3) \mapsto \tfrac{4}{9}$$
$$(x_2, x_3) \mapsto \tfrac{1}{12} \qquad (x_2, x_3) \mapsto \tfrac{4}{12}$$

The Table 2 gives the lower approximations of some subsets of U, relative to different relations corresponding to threshold $\lambda = 0, \frac{1}{10}, \frac{1}{3}$.

Note that when we fix $\lambda = 0$ so that two objects x and y are considered to be indistinguishable relative to the attribute set B if indistinguishability probability $S_{\mathcal{K},B}(x, y) > 0$. Therefore, x_3 does not lie in the lower approximation of the set $\{x_1, x_3\}$ relative to $S_{\mathcal{K},B}$. On the other hand, if we raise the indistinguishability threshold, and take $\lambda = \frac{1}{10}$, then we obtain x_3 in this lower approximation of $\{x_1, x_3\}$. This is due to the fact that the probability of x_2, x_3 to be indistinguishable relative to B is $\frac{1}{12} \lneq \frac{1}{10}$, so that $[x_3]^{\frac{1}{10}}_{S_{\mathcal{K},B}} = \{x_1, x_3\}$.

Table 2 Lower approximations for different thresholds

		X		
		$\{x_1, x_2\}$	$\{x_1, x_3\}$	$\{x_2, x_3\}$
$\lambda = 0$	$L_{S^c_{\mathcal{K},B}}(X, \lambda)$	$\emptyset$	$\{x_1\}$	$\{x_2\}$
	$L_{S_{\mathcal{K},B}}(X, \lambda)$	$\emptyset$	$\{x_1\}$	$\{x_2\}$
	$L_{S^w_{\mathcal{K},B}}(X, \lambda)$	$\emptyset$	$\emptyset$	$\emptyset$
$\lambda = \frac{1}{10}$	$L_{S^c_{\mathcal{K},B}}(X, \lambda)$	$\emptyset$	$\{x_1\}$	$\{x_2\}$
	$L_{S_{\mathcal{K},B}}(X, \lambda)$	$\{x_2\}$	$\{x_1, x_3\}$	$\{x_2\}$
	$L_{S^w_{\mathcal{K},B}}(X, \lambda)$	$\emptyset$	$\emptyset$	$\emptyset$
$\lambda = \frac{1}{3}$	$L_{S^c_{\mathcal{K},B}}(X, \lambda)$	$\{x_1, x_2\}$	$\{x_1, x_3\}$	$\{x_2, x_3\}$
	$L_{S_{\mathcal{K},B}}(X, \lambda)$	$\{x_1, x_2\}$	$\{x_1, x_3\}$	$\{x_2, x_3\}$
	$L_{S^w_{\mathcal{K},B}}(X, \lambda)$	$\{x_2\}$	$\{x_3\}$	$\emptyset$

3 PISs and Information Systems

In this section, we shall give a comparative study of PISs with different types of information systems.

3.1 Deterministic Information Systems

Let $\mathcal{K} := (U, \mathcal{A}, \{\mathcal{V}_a\}_{a\in\mathcal{A}}, f)$ be a deterministic information system (DIS). Then it can also be viewed as a PIS $T(\mathcal{K}) := (U, \mathcal{A}, \{\mathcal{V}_a\}_{a\in\mathcal{A}}, F)$, where

$$F(x, a, v) = \begin{cases} 1, & \text{if } f(x, a) = v \\ 0, & \text{otherwise.} \end{cases}$$

Observe that the above defined F satisfies the required condition of probability distribution viz. $\sum_{v\in\mathcal{V}_a} F(x, a, v) = 1$. Moreover, as $F(x, a, v) \in \{0, 1\}$, it follows that under a PIS $T(S)$, the probability of an object x to take an attribute-value v for an attribute a is either 0 or 1. This reflects the fact that in a DIS, we have the precise information regarding the attribute-values of the objects.

We note the following facts about the PIS $T(\mathcal{K})$.

Proposition 5 *The range of the mappings R_a, $S_{T(\mathcal{K}),B}$ and $S^w_{T(\mathcal{K}),B}$ is* $\{0, 1\}$.

The Proposition 5 captures the fact that in a PIS $T(\mathcal{K})$, relative to any set of attributes, two objects will be considered as distinguishable or indistinguishable. There is no intermediate grading of distinguishability relation.

The following proposition gives the precise connection between the approximation operators defined on DISs and PISs.

Proposition 6 *Consider a DIS $\mathcal{K}$ and corresponding PIS $T(\mathcal{K})$. Then the following holds:*

1. $(x, y) \in Ind_{\mathcal{K},B}$ *if and only if* $S_{T(\mathcal{K}),B}(x, y) > 0$.
2. $Ind_{\mathcal{K},B} = S^c_{\mathcal{K},B}(\lambda)$, *for all* $\lambda \in [0, 1)$.
3. $\underline{X}_{Ind_{\mathcal{K},B}} = L_{S_{T(\mathcal{K}),B}}(X, 0) = L_{S^c_{T(\mathcal{K}),B}}(X, 0)$.
   $\overline{X}_{Ind_{\mathcal{K},B}} = U_{S_{T(\mathcal{K}),B}}(X, 0) = U_{S^c_{T(\mathcal{K}),B}}(X, 0)$.

3.2 Incomplete Information Systems

Recall that in an incomplete information system (IIS) $\mathcal{K} := (U, \mathcal{A}, \{\mathcal{V}_a\}_{a\in\mathcal{A}} \cup \{*\}, f)$, $f(x, a) = *$ denotes the absence of information about x regarding the attribute a. Moreover, in that case, each of the attribute-value $v \in \mathcal{V}_a$ has the equal probability to be assigned to the object x for the attribute a. Due to this fact, it is natural to assign the probability $\frac{1}{|\mathcal{V}_a|}$ to the event of taking attribute-value v for a by the object x. Under this observation, we can view an IIS $\mathcal{K} := (U, \mathcal{A}, \{\mathcal{V}_a\}_{a\in\mathcal{A}} \cup \{*\}, f)$ as a PIS $T(\mathcal{K}) := (U, \mathcal{A}, \{\mathcal{V}_a\}_{a\in\mathcal{A}}, F)$, where

$$F(x, a, v) = \begin{cases} 1, \text{ if } f(x, a) = v \\ \frac{1}{|\mathcal{V}_a|}, \text{ if } f(x, a) = * \\ 0, \text{ otherwise.} \end{cases}$$

One can again easily verify that $\sum_{v\in\mathcal{V}_a} F(x, a, v) = 1$. From the definition of F, it follows that under a IIS $T(\mathcal{K})$, the probability of an object x to take an attribute-value v for an attribute a is 0 or 1, or each of the attribute-value from $\mathcal{V}_a$ has equal probability to be assign to x for the attribute a.

The following proposition captures the relationship between the approximation operators defined on IISs and PISs.

Proposition 7 *Consider an IIS $\mathcal{K}$ and corresponding PIS $T(\mathcal{K})$. Then the following holds:*

1. $(x, y) \in Sim_{\mathcal{K},B}$ *if and only if* $S_{T(\mathcal{K}),B}(x, y) > 0$.
2. $\underline{X}_{Sim_{\mathcal{K},B}} = L_{S_{T(\mathcal{K}),B}}(X, 0) = L_{S^c_{T(\mathcal{K}),B}}(X, 0)$,
   $\overline{X}_{Sim_{\mathcal{K},B}} = U_{S_{T(\mathcal{K}),B}}(X, 0) = U_{S^c_{T(\mathcal{K}),B}}(X, 0)$.

Observe that in an IIS $\mathcal{K}$ if $f(x, a) = f(x, b) = *$, it does not mean that $T(\mathcal{K})$ will assign equal probability to the events (x, a, v) and (x, b, u). This is due to the fact that probability distribution in $T(\mathcal{K})$ also depends on the size of the attribute-value set. Moreover, PISs can also express more general situation where one does not know the exact attribute-value, but can exclude some values. For instance, let $V_a := \{v_1, v_2, v_3\}$, and suppose that we do not have information about the attribute-value of x for the attribute a, but we have the information that it cannot be v_1. This fact cannot be captured in a IIS, but can be represented in a PIS by assigning the probability $F(x, a, v_1) = 0$, and $F(x, a, v_2) = F(x, a, v_2) = \frac{1}{2}$.

We would like to add here that the lower approximation operator $L_{S^c_{T(\mathcal{K}),B}}$ defined on $T(\mathcal{S})$ for IIS $\mathcal{S}$ coincides with the one defined on IIS using valued-tolerance relation in [14].

3.3 Nondeterministic Information Systems

Let us consider a nondeterministic information system (NIS) $\mathcal{S} := (U, \mathcal{A}, \{\mathcal{V}_a\}_{a\in\mathcal{A}}, f)$ under the assumption that $f(x, a) = V$, for $V \subseteq \mathcal{V}_a$, represents a situation where we do not know what attribute-value the object x takes for the attribute a, but we know that it is one of the member of V. Under this assumption, the probability of the event (x, a, v) is zero for $v \notin V$, and for $v \in V$, the probability of the event (x, a, v) is $\frac{1}{|V|}$. This observation suggests that a NIS $\mathcal{S} := (U, \mathcal{A}, \{\mathcal{V}_a\}_{a\in\mathcal{A}}, f)$ can be viewed as a PIS $T(\mathcal{K}) := (U, \mathcal{A}, \{\mathcal{V}_a\}_{a\in\mathcal{A}}, F)$, where

$$F(x, a, v) := \begin{cases} \frac{1}{|f(x,a)|}, & \text{if } v \in f(x, a) \\ 0, & \text{otherwise.} \end{cases}$$

We again note that F satisfies the required condition $\sum_{v\in\mathcal{V}_a} F(x, a, v) = 1$ of a probability distribution.

The following proposition provides the precise connection between different indistinguishability relations and corresponding lower and upper approximation operators defined on NISs and PISs.

Proposition 8 *Consider a NIS $\mathcal{K}$ and corresponding PIS $T(\mathcal{K})$. Then the following holds:*

1. (a) $(x, y) \in Sim_{\mathcal{K},B}$ *if and only if* $S_{T(\mathcal{K}),B}(x, y) > 0$;
   (b) $(x, y) \in Sim^w_{\mathcal{K},B}$ *if and only if* $S^w_{T(\mathcal{K}),B}(x, y) > 0$;
2. (a) $\underline{X}_{Sim_{\mathcal{K},B}} = L_{S_{T(\mathcal{K}),B}}(X, 0) = L_{S^c_{T(\mathcal{K}),B}}(X, 0)$
   $\overline{X}_{Sim_{\mathcal{K},B}} = U_{S_{T(\mathcal{K}),B}}(X, 0) = U_{S^c_{T(\mathcal{K}),B}}(X, 0)$;
   (b) $\underline{X}_{Sim^w_{\mathcal{K},B}} = L_{S^w_{T(\mathcal{K}),B}}(X, 0)$, *and* $\overline{X}_{Sim^w_{\mathcal{K},B}} = U_{S^w_{T(\mathcal{K}),B}}(X, 0)$.

Example 2 Let us consider the nondeterministic information system $\mathcal{K}_1$ given by the Table 3. The corresponding PIS $T(\mathcal{K}_1)$ is given by Table 1. From Example 1, we obtain $\underline{\{x_1, x_3\}}_{Sim_{\mathcal{K}_1,B}} = L_{\mathcal{K}_1,B}(X, 0) = \{x_1\}$. The object x_3 does not belong to $\underline{\{x_1, x_3\}}_{Sim_{\mathcal{K}_1,B}}$ due to the fact that the object x_3 and x_2 has some possibility, although it could be very small, to take the common value v_2. But, if we also consider the measure of this possibility, then situation could be different. For instance, as illustrated in Example 1, if we fix $\lambda = \frac{1}{10}$, then we obtain x_3 in the lower approximation of $\{x_1, x_3\}$.

Table 3 PIS $\mathcal{K}_1$

	a	b
x_1	$\{v_1\}$	$\{u_1, u_2, u_3\}$
x_2	$\{v_2, v_3\}$	$\{u_1, u_2\}$
x_3	$\{v_1, v_2, v_4\}$	$\{u_1\}$

From Propositions 6–8, it follows that lower (and hence upper) approximations defined on deterministic information systems (relative to indiscernibility relation), nondeterministic, and incomplete information systems (relative to similarity relation) are all actually instances of only one notion of lower (upper) approximation defined on PISs namely $L_{S_{\mathcal{K},B}}(X, 0)$ corresponding to threshold $\lambda = 0$. Moreover, as illustrated in Example 1, by assigning different values for threshold λ, we obtain approximation operators which are different from the standard one defined on nondeterministic and incomplete information systems.

4 Conclusions

In order to capture the situation where information regarding the attribute-values of the objects are not precise, but given in terms of probability, we propose the notion of probabilistic information system (PIS). Notions of distinguishability relations and corresponding approximation operators are proposed and studied. It is shown that the DISs, IISs, and NISs are all special instances of PISs. Moreover, the approximation operators defined on DIS (relative to indiscernibility), IISs, and NISs (relative to similarity relations) are all originated from a single approximation operator defined on PISs. It may be noted here that this may not be the case for the other types of relation defined on NISs (cf., e.g., [1, 7, 15]), and we may need to come up with a different set of relations defined on PISs to capture these relations. We would also like to add here that we have the proposal of a sound and complete logic for PISs where one can express the notions of approximations defined here. But this issue is outside the scope of the current article.

References

1. Düntsch, I., Gediga, G., Orlowska, E.: Relational attribute systems II: reasoning with relations in information structures. In: J.F. Peters et al. (eds.) Transactions on Rough Sets VII, LNCS 4400, pp. 16–35. Springer-Verlag Berlin Heidelberg (2007)
2. Dubois, D., Prade, H.: Rough fuzzy sets and fuzzy rough sets. Int. J. Gen. Syst. **17**, 191–200 (1990)
3. Komorowski, J., Pawlak, Z., Polkowski, L., Skowron, A.: Rough sets: a tutorial. In: Pal, S.K., Skowron, A. (eds.) Rough Fuzzy Hybridization: A New Trend in Decision-Making, pp. 3–98. Springer, Singapore (1999)

4. Kryszkiewicz, M.: Rough set approach to incomplete information systems. Inf. Sci. **112**, 39–49 (1998)
5. Kryszkiewicz, M.: Rules in incomplete information systems. Inf. Sci. **113**, 271–292 (1999)
6. Lin, T.Y., Yao, Y.Y.: Neighborhoods system: measure, probability and belief functions. In: Proceedings of the 4th International Workshop on Rough Sets and Fuzzy Sets and Machine Discovery, pp. 202–207. Nov 1996
7. Orlowska, E., Pawlak, Z.: Representation of nondeterministic information. Theor. Comput. Sci. **29**, 27–39 (1984)
8. Pawlak, Z.: Systemy Informacyjne. Podstawy Teoretyczne. Wydawnictwa Naukowo-Techniczne, Warszawa (1983)
9. Pawlak, Z.: Rough probability. Bull. Pol. Acad. Sci. (Math) **32**(9–10), 607–612 (1984)
10. Pawlak, Z.: Rough Sets. Theoretical Aspects of Reasoning about Data. Kluwer Academic Publishers, Dordrecht (1991)
11. Pomykala, J.A.: Approximation, Similarity and Rough Constructions. ILLC Pre-publication Series for Computation and Complexity Theory CT-93-07. University of Amsderdam, Amsderdam (1993)
12. Skowron, A., Stepaniuk, J.: Tolerance approximation spaces. Fundamenta Informaticae **27**, 245–253 (1996)
13. Ślęzak, D., Ziarko, W.: The investigation of the Bayesian rough set model. Int. J. Approx. Reason. **40**, 81–91 (2005)
14. Stefanowski, J., Tsoukiàs, A.: On the extension of rough sets under incomplete information. In: Zhong, N., Skowron, A., Ohsuga, S. (eds.) RSFDGrC '99: Proceedings of the 7th International Workshop on New Directions in Rough Sets. Data Mining, and Granular-Soft Computing, pp. 73–81. Springer, London, UK (1999)
15. Vakarelov, D.: Modal logics for knowledge representation systems. Theor. Comput. Sci. **90**, 433–456 (1991)
16. Yao, Y.Y.: Probabilistic rough set approximations. Int. J. Approx. Reason. **49**(2), 255–271 (2008)
17. Ziarko, W.: Variable precision rough set model. J. Comput. Syst. Sci. **46**, 39–59 (1993)

Uncertainty Analysis of Contaminant Transportation Through Ground Water Using Fuzzy-Stochastic Response Surface

Subrata Bera, D. Datta and A.J. Gaikwad

Abstract The process of contaminant transportation through ground water can be varied with different parameters such as soil characteristics, ground water flow velocity, longitudinal and transverse dispersion coefficients, self-degradation of contaminant etc. The precise definition of these parameters is very difficult due to various factors such as measurement error, sampling error, dependence of complex physical phenomena, etc. The analytical solution of transient advection–diffusion equation is being used to assess the ground water contamination due to the industrial discharge. The paper describes a methodology to estimate the hybrid uncertainty, i.e., combination of aleatory and epistemic using the fuzzy-stochastic system. Aleatory uncertainty due to random variation of input parameter is estimated using polynomial chaos expansion method. To take into account the effect of imprecise variation (i.e., epistemic uncertainty) of input parameter, a fuzzy α-cut technique has been used. The large sample space of concentration reduction factor (CRF) have been generated using fuzzy-stochastic response surface to arrive the upper uncertainty bound corresponds to the 95th percentile value at a specified distance from the source and period of time. The methodology will be very useful to assess the safety margin or discharge limit from the industry.

Keywords Uncertainty · Stochastic response surface · Polynomial chaos expansion · Fuzzy set theory · Ground water contamination

1 Introduction

Ground water is being contaminated due to the industrial effluent discharge to the environment. The process of contamination is governed by two physical phenomena, viz. advection and diffusion. In the advection process, contamination is trans-

S. Bera (✉) · A.J. Gaikwad
Nuclear Safety Analysis Division, AERB, Mumbai 400094, India
e-mail: sbera@aerb.gov.in

D. Datta
Computational Radiation Physics Section, HPD, BARC, Mumbai 400085, India

M.K. Chakraborty et al. (eds.), *Facets of Uncertainties and Applications*,
Springer Proceedings in Mathematics & Statistics 125,
DOI 10.1007/978-81-322-2301-6_10

ported due to the fluid bulk motion, whereas the transport of contaminant in diffusion process is due to the concentration difference. Various methodologies are being used to estimate the concentration of a contaminant, e.g., analytical solution of different dimension, numerical solution of advection, and diffusion equation through computational fluid dynamics, etc. The complexity of problem is decided by many factors such as steady-state, transient, consideration of uncertainty of input parameters, e.g., soil properties, self-decay of contaminant, dispersion coefficients along the flow, and its perpendicular directions, etc. The nature of variation of input parameters can be modeled using a probabilistic frame work [1–3] or a fuzzy computing framework [4–8]. The source of uncertainty may be due to the random variation of input parameter (aleatory) and lack of knowledge (epistemic) [1]. An innovative blending methodology of probabilistic framework with fuzzy computing framework has been developed using formulation of stochastic response surface with fuzzy argument [9, 10]. This fuzzy-stochastic response surface is used to create a large random sample space of concentration reduction factors to be used for statistical analysis. The polynomial chaos expansion methodology has been adopted to formulate the stochastic response surface with Harmite polynomials.

In the present analysis, the uncertainty associated with the contaminant transported through groundwater has been carried out using polynomial chaos expansion method and interval Fuzzy α-cut techniques. Contaminant concentration estimation at a distance from source has been calculated using analytical solution of transient advection and diffusion equation. Analytical solution as derived by Domenico is used for estimation of contaminant concentration [11]. Two parameters, i.e., hydraulic conductivity and weighting factor for longitudinal dispersivity are considered as uncertain due to their randomness and fuzziness. Uniform and imprecise variation is considered in analysis for hydraulic conductivity and weighting factor for longitudinal dispersivity, respectively. For uniform variation of hydraulic conductivity, a stochastic response surface has been developed using polynomial chaos expansion. Uncertainty of the fuzzy variable is selected on the basis of the lack of measurement. Measurement process is generally addressed as mean $\pm$ standard deviation. Therefore, the least-bounded uncertain variable can be expressed by considering the extreme bounds as mean $-$ standard deviation and mean $+$ standard deviation. So, considering these extreme bounds with the most likely value as mean value of the uncertain variable composes a triangular imprecise variable, which we have addressed as triangular fuzzy variable. For this reason, triangular membership functions are selected. The results shown in this paper using the method as described are already applied for liquid effluent discharged due to routine operation of nuclear plant. The similar kind of results can be achieved due to the routine operation of other industry. Therefore, obviously results are useful for real-life applications. Moreover, the method can be applied very easily for many other process in our real-life applications, for example, uncertainty occurred in any pharmacokintetic process can be explained by using the same method. Moreover, these results as well as the methodology will be very useful to assess the safety margin or discharge limit from the industry.

2 Theoretical Model of Ground Water Transport

The Advection-dispersion process is the main physical phenomena that governs contaminant transport process in ground water. The two-dimensional partial differential equation for advection-diffusion equation for groundwater transport is given below [5]

$$\frac{\partial c}{\partial t} = D_x \frac{\partial^2 c}{\partial x^2} + D_y \frac{\partial^2 c}{\partial y^2} - v \frac{\partial c}{\partial x} \tag{1}$$

where C is the contaminant concentration in groundwater; t is the time (day); v is the groundwater seepage velocity (m/day); x, y are the coordinates to two dimensions (m); D_x, D_y are the dispersion coefficients for the x and y dimension (m), respectively. Analytical solution of advection-diffusion equation introduced by Domenico (1987) has been used to evaluate transient plume behavior under conditions of a continuous source and finite source dimensions with one-dimensional groundwater velocity, longitudinal, transverse, and vertical dispersion and first-order degradation rate constant.

The analytical solution [5] of the two-dimensional advection-diffusion equation is given below:

$$\mathrm{C(x, y, t)} = \frac{c_0}{4} exp\left\{\frac{x}{2\alpha_x}\left[1 - \sqrt{(1 + \frac{4\lambda\alpha_x}{v})}\right]\right\} erfc\left\{\frac{x - vt\sqrt{(1 + \frac{4\lambda\alpha_x}{v})}}{2\sqrt{\alpha_x vt}}\right\}$$
$$\left\{erf\left[\frac{y + \frac{Y}{2}}{2\sqrt{\alpha_y x}}\right] - erf\left[\frac{y - \frac{Y}{2}}{2\sqrt{\alpha_y x}}\right]\right\} \tag{2}$$

where C(x, y, t) is the contaminant concentration (mg/l) in at time t and position (x, y); C_0 is the source concentration,; α_x, α_y are the longitudinal and traverse dispersion coefficient (m), respectively; λ is the degradation rate constant and equal to $0.693/T_{1/2}$; v is the groundwater velocity (m/day); erf(x) and erfc(x) are the error and complementary error functions, respectively.

The calculation of groundwater velocity is given below:

$$v = \frac{k_{hydro} H_{grad}}{n} \tag{3}$$

where K_{hydro} is the hydraulic conductivity (m/day); H_{grad} is the hydraulic gradient (m/m); and n is soil porosity (%).

The literature survey yields various empirical formula for longitudinal and transverse dispersivity. Following empirical formulae which is commonly used is given below:

$$\alpha_x = w_l x \tag{4}$$

$$\alpha_y = w_t \alpha_x \tag{5}$$

where W_l is longitudinal weighting factor and as per the literature survey its value is around 0.1; W_t is the transverse weighting factor and taken as a 33 % of α_x, according to the literature survey. But, in this analysis W_l is imprecisely defined, and hence it is considered as a fuzzy parameter with triangular membership function.

3 Polynomial Chaos Expansion (PCE) Methodology

The PCE [1–3] is the representation of a random variable, more generally a stochastic process, with an infinite series of orthogonal polynomials that take a vector of independent and identically distributed (IID) random variables as arguments. Mathematically, the stochastic response surface is represented with one uncertain parameter with Hermite polynomial of degree n can be represented by

$$y = \sum_{i=1}^{n+1} a_i \Psi_i(\xi) \tag{6}$$

where y is the model estimation; a_i is the coefficient of stochastic response surface; ψ_i (ξ) is the Hermite polynomial of order I as a function of standard normal random number (ξ). Hermite polynomials represent the Gaussian process, which is further represented as a normal distribution. Uncertain parameters of the environmental models basically follow normal and normal-like distributions such as log normal distribution, and gamma distribution. All these distributions including uniform distribution can be transformed in the domain of Hermite polynomials. This signifies the usage of Hermite polynomial in this paper, and as far as the degree of the Hermite polynomial is concerned, higher the degree more is the possibility of overshooting the fitting process due to the inherent nature of inconsistency least square fitting method. That is why more correct degree from the suitability of a response surface for fitting is the order of three which has been chosen for the present work. The terms of Hermite polynomial of order up to three have been given below:

$$\Psi_0(\xi) = 1;\quad \Psi_1(\xi) = \xi;\ \Psi_2(\xi) = \xi^2 - 1;\ \Psi_3(\xi) = \xi^3 - 3\xi \tag{7}$$

An explicit representation of third-order polynomial chaos expansion for one variable can be written as:

$$y = a_0 + a_1\xi + a_2(\xi^2 - 1) + a_4(\xi^3 - 3\xi) \tag{8}$$

The coefficients of the stochastic response surface have been determined by solving a set of linear equation generated with standard normal random number. The large sample space of concentration reduction function has been generated for statistical analysis. Therefore, for four unknown coefficients, four model outputs are to be generated for the specified model. Sampling points for the generation of these outputs will be obtained from the model uncertain inputs for which inputs are to be transformed into standard normal variables (SRV). Standard transformation of normal probability distribution functions with mean (μ) and standard deviation (σ) of model inputs in terms of SRVs can be written as:

$$\text{Normal}(\mu, \sigma) : \mu + \sigma\xi \tag{9}$$

4 Fuzzy Set Theory

Fuzzy set theory is the combination of classical set theory and proportional logic. In contrast to the binary membership function used in classical set theory, FST allows to use partial membership function. The partial membership function can be represented by triangular membership function with interval [l, m, r] as shown in Fig. 1. where, l, m, and r are the left, middle, and right points of triangular membership function. Mathematical formulation of the membership function is given in (10).

$$\mu_A(x) = \begin{cases} \frac{x-l}{m-l}, l \le x \le m \\ \frac{r-x}{r-m}, m \le x \le r \end{cases} \tag{10}$$

Alpha-cut (α-cut) technique of a fuzzy set [5–8] provides the interval range corresponding to a specific value of membership function. Mathematically, α-cut techniques is represented by

$$Cut_\alpha(A) = \{x | \mu_A(x) \ge \alpha\} \tag{11}$$

$$^{\alpha}X = [l + \alpha(m - l), r - \alpha(r - m)] \tag{12}$$

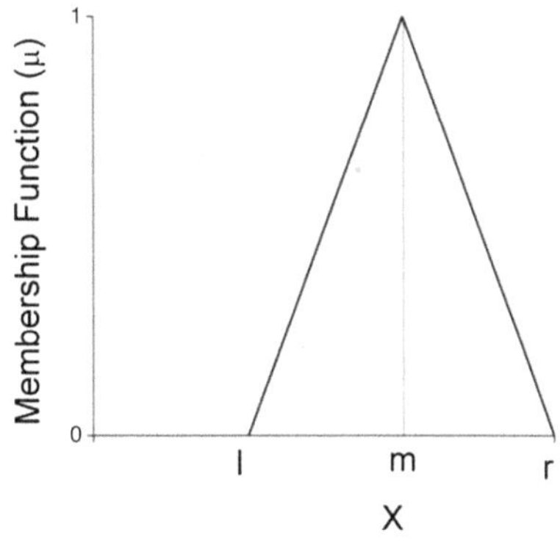

Fig. 1 Triangular fuzzy membership function

In the uncertainty analysis, various nomenclatures are used to represent variation of a fuzzy parameter depending on the value of alpha. These are "core" ($\alpha = 1.0$), "upper quartile" ($\alpha \geq 0.75$), "cross over point" ($\alpha = 0.50$), "middle quartile" ($\alpha \geq 0.50$), and "lower quartile" ($\alpha \geq 0.25$).

5 Cumulative Density Function in Fuzzy-Stochastic System

In the probabilistic space, cumulative density function is derived from the probability density function of a random variable to estimate the portion of population with a value less than a specified value. In the other way, it may be represented as the probability having a value less than a specified value. CDF does not depend on the probability density function of the input variable. If a model contains both random and imprecise variables, CDF can be formulated with the random variable for a specific value of imprecise parameter. Since, degree of impreciseness of the imprecise parameter is represented by the α-cut value and uncertainty with interval, lower bound and upper bound of the interval for a specific α-cut value generates two CDFs, which represent the uncertainty due to the imprecise variable in the fuzzy-stochastic system.

6 Result and Discussion

Stochastic response surface with third-order Hermite polynomial has been formed for one uncertain parameter (hydraulic conductivity). This response surface is used to get CDF of concentration. The CDF of concentration has been estimated for both spatial profile with fixed time and temporal profile with fixed location from source. Variation of CDF at a distance of 500 m from source and time period of 180 days has been shown in the Fig. 2. During the estimation of CDF, the fuzzy variable, weighting factor of longitudinal dispersion coefficient (W_1) value kept at its core value at which membership function equal to unity ($\alpha = 1.0$). This variation is due to the random variation of hydraulics conductivity alone. Various percentile values of concentration reduction factor have been estimated and given in the Table 1.

The CDF of concentration of contaminant has been calculated for various alpha cut value of W_1 at a distance 500 m from source and time period of 180 days. Calculations are carried out for α-cut value equal to 0.05, 0.25, 0.50, 0.75, and 0.95. The nature of variation of CDF with concentration reduction factor (CRF) for different α-cut values has been shown in the Fig. 3.

From the Fig. 3, it is found that uncertainty in cumulative distribution of concentration is increasing with higher α-cut value of W_1. For a single α-cut value of W_1, there are two CDFs around mean CDF. These two CDFs around the mean CDF are the representation of uncertainty due to fuzzy variable for a specific membership value. The uncertainty due fuzzy-stochastic system is confined in between two CDFs corresponding to a particular α-cut value of W_1. The uncertainty boundary

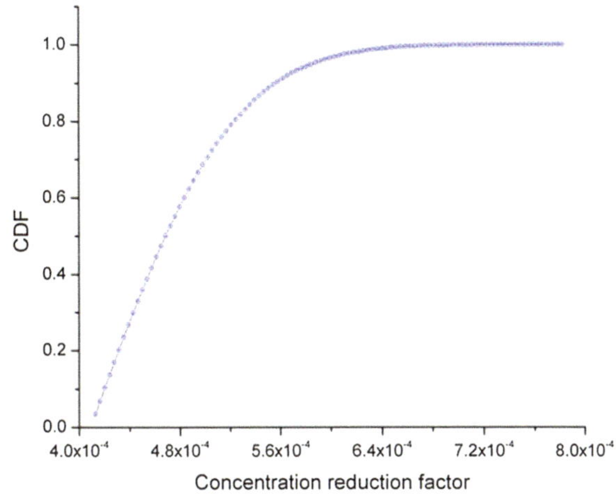

Fig. 2 CDF of concentration reduction factor at distance 500 m and time 180 days CDF

Table 1 Uncertainty estimation in terms of percentile value

Percentile	Concentration reduction factor
5th	4.1648E-04
25th	4.3886E-04
50th (Mean value)	4.6869E-04
75th	5.1345E-04
95th	5.5821E-04

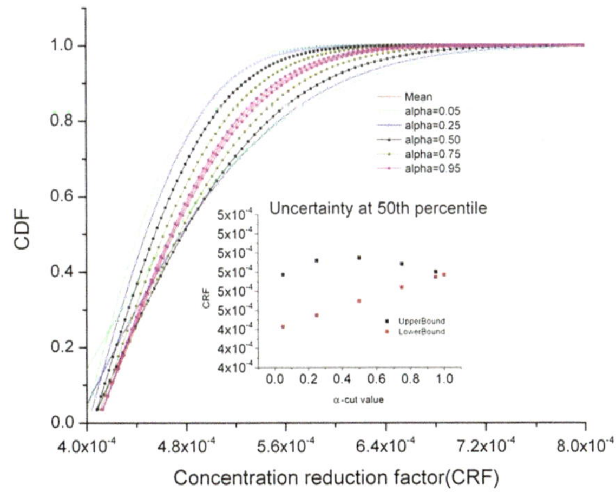

Fig. 3 Variation of CDFs with different α-cut values of fuzzy parameter

of CRF for various α-cut value at the 50th percentile has been shown in the inset figure of Fig. 3. From the inset figure it is found that the uncertainty reduces with increasing α-cut value. Uncertainty boundary becomes zero at the α-cut value equal to 1.0 corresponding to the crisp value.

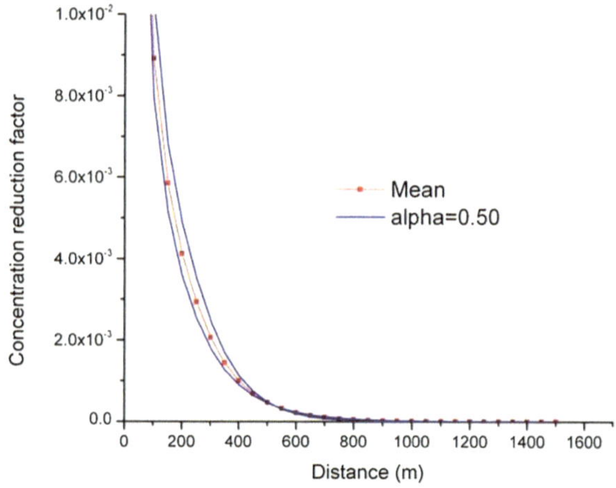

Fig. 4 Spatial variation of concentration reduction factor with fuzzy uncertainty

Concentration of contaminant has been estimated at 30 locations starting from 50 to 1500 m from the source with equal interval of 50 m for time period of 6 months or 180 days. Mean value of concentration corresponding to the 50th percentile of cumulative distribution and fuzzy membership function equal to unity has been compared with interval parameters corresponding to fuzzy α-cut value equal to 0.5 (shown in Fig. 4).

It is found from Fig. 4, that the concentration of contaminant in groundwater is drastically reduced to a very low value at a distance 600 m from the source within the time period of 18 days. The variation of concentration reduction factor is found to be about 10 % of mean value due to the degree of impreciseness of 0.5 of W_1. It is found that uncertainty is significant up to the distance of 500 m. If the degree of

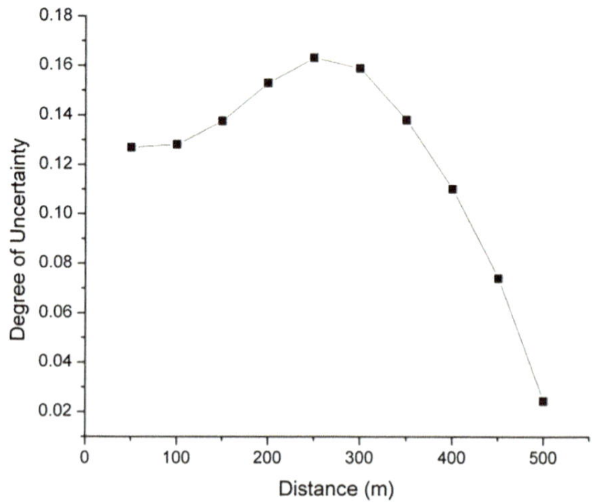

Fig. 5 Variation of the degree of uncertainty with distance

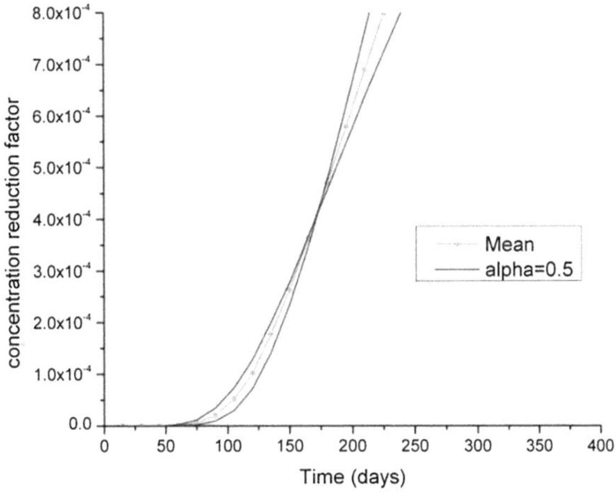

Fig. 6 Temporal variation of concentration reduction factor with fuzzy uncertainty

uncertainty at a particular distance is defined as the ratio of difference between upper value and lower value to the addition of upper and lower value, then the variation of the degree of uncertainty with distance up to 500 m has been shown in the Fig. 5.

From the Fig. 5, it is found that the degree of uncertainty is initially increasing with distance and then decreases from the distance around 300 m. This result infers that the maximum uncertainty arises at a distance of 300 m.

The variation of concentration reduction factor with time has been carried out for time period of one year with interval of fifteen days. Temporal variation at a distance of 500 m from the source has been shown in the Fig. 6. It is found that concentration is increasing with time. The important observation is that at a time period of 180 days, fuzzy-based uncertainty band becomes zero. It is happened because of the nature of variation of complementary error function. At that particular period of time, complementary error function becomes unity. However, after that particular period of time, the uncertainty boundary lines follow the continuity and become diverged from the mean value. Analysis shows that if we increase the time of observation beyond 400 days, the concentration of the contaminant at 500 m gets saturated.

7 Conclusion

Fuzzy-stochastic response surface has been developed using one normally distributed parameter and one fuzzy parameter. Stochastic response surface has been developed using Hermite polynomial expansion of random variable (hydraulic conductivity) up to order of three; however, fuzzy variable is kept at a desire α-cut value. Fuzzy-stochastic response surface has been used to get cumulative density function (CDF) of contamination reduction factor. The CDF for contamination reduction factor for a specific distance from source and period of time has been estimated with 100,000 sample calculation of stochastic response surface. Various percentile value of concentration reduction function has been estimated. The uncertainty due to one imprecise parameter (weighting factor for longitudinal dispersivity) is included through implementation of Fuzzy α-cut technique. Uncertainty involved with various degree of impreciseness of parameter has been demonstrated. Finally, the variation of mean concentration reduction factor for various distances and time periods have been estimated with fuzzy uncertainty bounds. It is found that the uncertainty associated with the spatial variation of concentration reduction factor is significant up to the distance of 500 m for period of time of 180 days. The degree of uncertainty is found to be maximum at a distance of 300 m from the source. In the time profile of concentration reduction factor, it is found that the complementary error function plays an important role for the quantification of uncertainty.

References

1. Kushwaha, H.S.: Uncertainty Modeling and Analysis. Bhabha Atomic Research Centre, Mumbai (2009). ISBN 978-81-90726-0-8
2. Datta, D., Kushwaha, H.S.: Uncertainty quantification using stochastic response surface method case study-transport of chemical contamination through ground water. Int. J. Energy, Inf. Commun. **2**(3), 49–58 (2011)
3. Datta, D.: Uncertainty analysis using polynomial chaos theory case study- transport of contamination through ground water and temperature profile of a rectangular bar. Int. J. Math. Arch. **2**(6), 935–943 (2011)
4. Zadeh, L.A.: Fuzzy sets. Inf. Control **8**, 338–353 (1965)
5. Ross, T.J.: Fuzzy Logic with Engineering Applications, 3rd edn. Wiley, New York (2010)
6. Kaufmann, A., Gupta, M.: Introduction to Fuzzy Arithmetic: Theory and Application. van Nonstrand Reinhold Company Inc., New York (1985)
7. Dutta, P., Boruah, H., Ali, T.: Fuzzy arithmetic with and without alpha-cut method: a comparative study. Int. J. Latest Trends Comput. **2**(1), 99–108 (2011)
8. Bera, S., Datta, D., Gaikwad, A.J.: Role of Fuzzy set theory in air pollution estimation. In: International Conference on Advance Engineering Optimization Through Intelligent Techniques (AEOTIT), SVNIT, Surat, India, 01–03 July 2013
9. Li, J., Huang, G.H., Zeng, G., Maqsood, I., Huang, Y.: An integrated fuzzy-stochastic modeling approach for risk assessment of ground water contamination. J. Environ. Manag. **82**, 173–188 (2007)
10. Isukapalli, S.S., Roy, A., Georgopoulos, P.G.: Efficient sensitivity/uncertainity analysis using the combined stochastic response surface method and automated differentiation: Application to environmental and biological system. Risk anal. **20**(5), 591 (2000)
11. Lesser-Carrillo, L.E.: Modeling of a contaminant plume in a tidally influenced river using Domenico's equation. Int. J. Geosci. **4**, 529–536 (2013)

Development of a Fuzzy Random Health Risk Model

D. Datta, S. Kar and H.S. Kushwaha

Abstract This work focuses on the development of a fuzzy random health risk model. The concept of fuzzy random variable is used to develop this health risk model. Health risk is addressed as the risk due to exposure to uranium through ingestion of food grown in and around a high background area which is rich in groundwater and has a substantial amount of phosphate deposits that constitute uranium. Lack of data about the activity concentration of uranium and its variability at many locations of that area justifies its fuzziness and randomness. A similar reason is valid for the consumption of food from the area. Therefore, these input parameters of the exposure model are the fuzzy random variable. Exposure model computes the daily average ingestion and risk is computed by multiplying this with the corresponding cancer slope factor. Fuzziness of daily average ingestion is computed at a specified percentile of the lower and upper fuzzy cumulative distribution of daily average ingestion. Randomness is computed at every alpha cut of the fuzzy random daily average ingestion. Fuzzy random daily average ingestion is used to compute the fuzzy random risk. It has been shown that risk due to the consumption of naturally occurred uranium through ingestion of food is insignificant. Support, uncertainty index, possibility, necessity, and credibility of the fuzzy random risk are also computed to explore the role of fuzzy random variable in uncertainty of risk estimation.

Keywords Fuzziness · Randomness · Health risk model · Ingestion · Activity concentration

---

D. Datta (✉)
Health Physics Division, Bhabha Atomic Research Centre, Mumbai 400085, India
e-mail: ddatta@barc.gov.in

S. Kar
Department of Mathematics, Nit Durgapur, Durgapur, India

H.S. Kushwaha
Former Director, Health, Safety & Environment Group, Raja Rammanna Fellow, Bhabha Atomic Research Centre, Mumbai 400085, India

M.K. Chakraborty et al. (eds.), *Facets of Uncertainties and Applications*, Springer Proceedings in Mathematics & Statistics 125,
DOI 10.1007/978-81-322-2301-6_11

1 Introduction

Risk models involve two important sources of uncertainty: randomness and fuzziness. Randomness relates to the stochastic variability of all possible outcomes of a situation. Fuzziness, on the other hand, relates to the unstructured boundaries of the parameters of the risk model and can be traced to incomplete knowledge of the situation. Thus, randomness is more an instrument of normative analysis that focuses on the future, while fuzziness is an instrument of descriptive analysis reflecting the past and its implications. Randomness and fuzziness can be merged to formulate a fuzzy random variable (FRV), that is, a function that assigns a fuzzy subset to each possible random outcome [1]. Therefore, it is obvious that randomness and fuzziness are complementary. One important facet of this relationship is the FRV, which is a measurable function from a probability space to the set of fuzzy variables. FRVs have also been referred to as random fuzzy sets and random upper semicontinuous functions [2].

Anecdotal evidence suggests that risk analysts are receptive to the notion of FRVs. They generally recognize that there are sources of uncertainty that random variables cannot capture and that fuzziness is a key component of that uncertainty. Consequently, since random variables are at the core of exposure concepts and fuzziness permeates every aspect of health risk modeling and analysis, one would expect to see FRVs implemented often, in the literatures of risk analysis.

This article explores these FRVs. We address: distinction between fuzziness and randomness, fundamental concepts, FRVs, a comparison of the FRVs defined by Kawakernakk [3, 4] and Puri and Ralescu [5]. Finally, fuzzy random variable concept is applied to analyze the health risk due to ingestion of contaminated or toxic elements.

2 Fuzziness and Randomness—How They Are Different?

Before introducing the fuzzy random variable, let us recall the difference between randomness and fuzziness. Randomness addresses the variability of the uncertain variable, whereas fuzziness describes the ignorance of the variable. Fuzziness can be reduced, whereas randomness cannot be reduced. Randomness is described by the probability distribution, whereas fuzziness is represented by possibility distribution. Therefore, it can be envisaged that there exists a distributional difference between fuzziness and randomness. A simple example to demonstrate the distributional differences is as follows. For a representative probability distribution, based on dosimetric survey of occupational workers in any nuclear plant, a probability of 0.94 can be assigned as zero overexposure, a 0.06 probability as one overexposure, and a 0.004 probability as two overexposures.

In contrast, one can see that the representative possibility distribution as, overexposure of zero and one occupational worker each have a high possibility of occurrence, and there is some possibility that the two occupational workers will be overexposed. Therefore, it can be concluded that, a probable event is always possible, while a possible event need not be probable. Zadeh [6] called this heuristic connection between possibilities and probabilities the probability/possibility consistency principle. This informal principle may be translated as: the degree of possibility of an event is greater than or equal to its degree of probability, which must be itself greater than or equal to its degree of necessity [7].

3 Probability, Possibility, and Credibility

In view of the difference between randomness and fuzziness, let us define formally the probability, possibility and credibility spaces. The basic features of these spaces are presented in Table 1 and following this we define them in the following way:

3.1 Probability

As indicated in Table 1, a probability space is defined as the 3-tuple (Ω, A, Pr), where $\Omega = \{\omega_1, \omega_2, \omega_3, \ldots, \omega_N\}$ is a sample space. A is the σ-algebra of subsets of Ω and Pr, a probability measure on Ω, such that it satisfies:

$\Pr(\Omega) = 1, \Pr\{\phi\} = 0, 0 \le \Pr\{A\} \le 1$ for any $A \in A$

For every countable sequence of mutually disjoint events $\{A_i\}$, i = 1, 2,....

$$\Pr\left\{\bigcup_{i=1}^{\infty} A_i\right\} = \sum_{i=1}^{\infty} \Pr\{A_i\} \tag{1}$$

Probability measure satisfies the law of excluded middle (which requires that a proposition be either true or false), the law of contradiction (which requires that a proposition cannot be both true and false), and the law of truth conservation (which requires that the truth values of a proposition and its negation should sum to unity) [8].

Table 1 Probability, possibility, and credibility spaces

Probability space	Possibility space	Credibility space
(Ω, A, Pr) is a probability space	(Θ, P(Θ), Pos) is a possibility space	(Θ, P(Θ), Cr) is a credibility space
Ω: sample space	Θ: sample space	Θ: sample space
A: σ-algebra of subsets of Ω	P(Θ): power set of Θ	P(Θ): power set of Θ
Pr: probability measure on Ω	Pos: possibility measure on Θ	Cr: credibility measure on Θ

3.2 Possibility

A possibility space from Table 1 is defined as the 3 tuple ((Θ, P(Θ), Pos), where $\Theta = \{\theta_1, \theta_2, \ldots, \theta_N\}$ is a sample space, P(Θ), also denoted as 2^{Θ}, is the power set of Θ, that is, the set of all subsets of Θ, and Pos is a possibility measure defined on Θ. Pos{A}, the possibility that A will occur, satisfies:

$$\text{Pos}\{\Theta\} = 1, \text{Pos}\{\phi\} = 0, 0 \le \text{Pos}\{A\} \le 1, \text{ for any A in P}(\Theta)$$

$\text{Pos}\left\{\bigcup_i A_i\right\} = \sup_i \text{Pos}\{A_i\}$ for any collection $\{A_i\}$ in P(Θ). The heavy red line shown in Fig. 1 represents the possibility of a fuzzy event characterized by $\zeta \ge$ x, where ζ = (2, 5, 9) is a triangular fuzzy variable given by the mathematical form

$$Pos\{\zeta \ge x\} = \begin{cases} 1, & x \le 5, \\ \dfrac{9-x}{9-5}, & 5 \le x \le 9, \\ 0, & x \ge 9 \end{cases} \tag{2}$$

It can be stated that the possibility of an event is determined by its most favorable case only, in contrast to the probability of an event where all favorable cases are accumulated. By its very nature, the possibility measure is inconsistent with the law of excluded middle and the law of contradiction and does not satisfy the law of truth conservation [8].

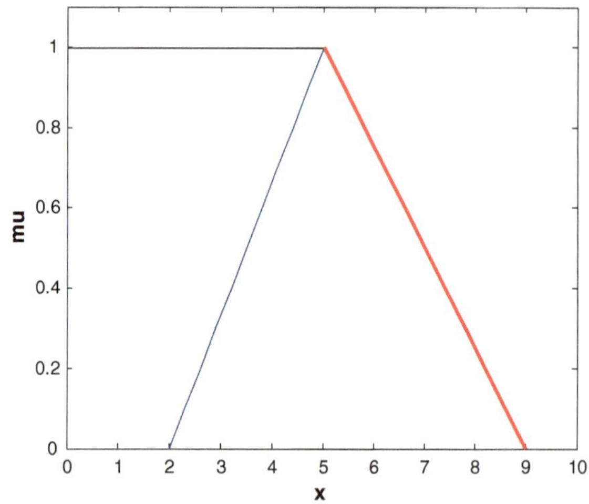

Fig. 1 Possibility that ζ is greater than x

3.3 Necessity

The necessity measure of a set A often is defined as the impossibility of the complement set A^c [9]. Formally, let (Θ, P(Θ), Pos) be a possibility space, and A be a set in P(Θ). Then the necessity measure of A is defined as

$$\mathrm{Nec}\{A\} = 1 - \mathrm{Pos}\{A^c\}$$

Considering the triangular fuzzy variable $\zeta = (2, 5, 9)$, we can represent Nec $\{\zeta \geq x\}$ by the mathematical equation, (The red line in Fig. 2 shows the necessity measure).

$$Nec\{\zeta \geq x\} = \begin{cases} 1, & x \leq 2, \\ \dfrac{5-x}{5-2}, & 2 \leq x \leq 5, \\ 0, & x \geq 5 \end{cases} \tag{3}$$

It can be noted from Fig. 2 that $\mathrm{Nec}\{\zeta \geq x\} = 1 - \mathrm{Pos}\{\zeta < x\}$.

3.4 Credibility

Given the limitations of the possibility measure, Liu and Liu [10] suggested replacing it with what they termed as credibility measure. The credibility measure takes the form

$$\mathrm{Cr}\{X \leq r\} = 0.5\,(\mathrm{Pos}\{X \leq r\} + \mathrm{Nec}\{X \leq r\})$$

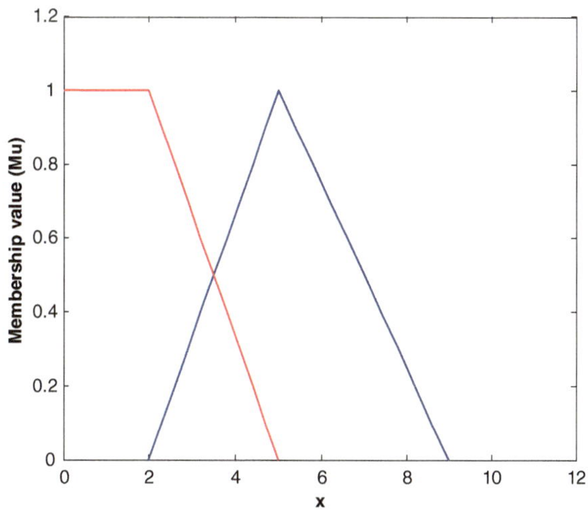

Fig. 2 Necessity that ζ is greater than or equal to x

or, equivalently,

$$Cr\{X \leq r\} = \frac{1}{2}\left(\sup_{t \leq r} \mu_x(t) + 1 - \sup_{t > r} \mu_x(t)\right) \tag{4}$$

$$Cr\left\{\bigcup_i A_i\right\} = \underset{i}{Sup}\, Cr\{A_i\}$$

The set Cr on the power set P is called a credibility measure if it satisfies the following four axioms [11]:

(i) $\mathrm{Cr}\{\Theta\} = 1$
(ii) $\mathrm{Cr}\{A\} \leq \mathrm{Cr}\{B\}$ whenever $A \subset B$
(iii) $\mathrm{Cr}\{A\} + \mathrm{Cr}\{Ac\} = 1$ for any event A
(iv) $Cr\left\{\bigcup_i A_i\right\} = \underset{i}{Sup}\, Cr\{A_i\}$ for any events $\{A_i\}$ with $\underset{i}{Sup}\, Cr\{A_i\} < 0.5$

It can be stated that the credibility measure is a special type of nonadditive measure with self-duality. In this context, a fuzzy event may fail even though its possibility achieves 1, and may hold even though its necessity is 0. However, the fuzzy event must hold if its credibility is 1 and fail if its credibility is 0. The mathematical representation of the credibility of $\zeta \geq x$ can be written with the help of the given triangular fuzzy number $(a, b, c)^{\cdot} \equiv (2, 5, 9)$ as

$$Cr(\zeta \geq x) = \begin{cases} 1 & x \leq a, \\ \dfrac{2b - a - x}{2(b - a)} & a \leq x \leq b, \\ \dfrac{c - x}{2(c - b)} & b \leq x \leq c, \\ 0 & x \geq c \end{cases} \tag{5}$$

The solid red line shown in Fig. 3 represents the credibility value of the fuzzy event characterized by $\zeta \geq x$.

4 Fundamental Concepts

Before we define the fuzzy random variable let us give a very short review of the fundamental concepts: fuzzy numbers, α-cuts, and Borel sets.

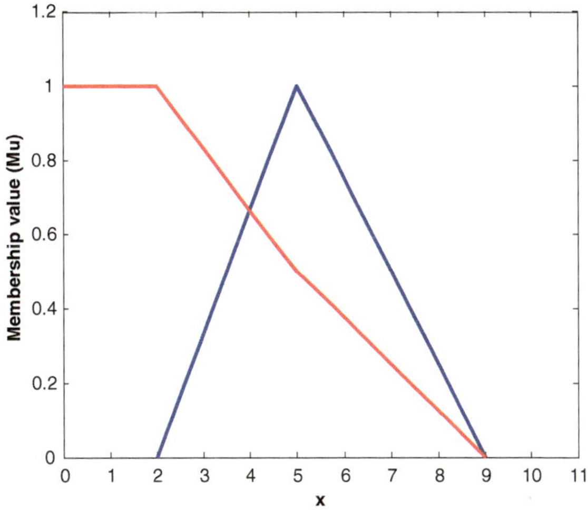

Fig. 3 Credibility that ζ is greater than or equal to x

4.1 Fuzzy Numbers

Fuzzy numbers are convex normalized paired numbers, (x, μ(x)), where x represents the crisp numbers and μ(x) represents its membership value, in which one of the membership values has to be unity (If the height of a fuzzy number (maximum membership value) is one, then it is said to be a normalized fuzzy set). The general characteristic of a fuzzy number ($a_1 = 1, a_2 = 5, a_3 = 7, a_4 = 8$) is as shown in Fig. 4.

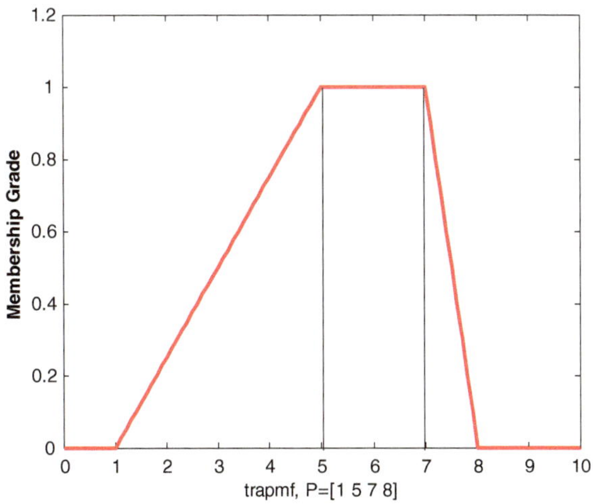

Fig. 4 Trapezoidal fuzzy number

The shape of this fuzzy number is referred to as trapezoidal fuzzy number. If $a_2 = a_3$, then the fuzzy number is referred to as triangular fuzzy number (TFN).

4.2 α-cuts

The set of crisp elements of a fuzzy set having membership grade equal to or greater than α is called the α-cut and is defined as

$$A_\alpha = \{(x, \mu(x)|x \in R, \mu(x) \geq \alpha\} \tag{6}$$

The importance of the α-cut is that it limits the domain under consideration to the set of elements with degree of membership of at least alpha, that is, α-level set. Therefore, α-cut is an interval and interval arithmetic operation can be performed while dealing with α-level set representation of two fuzzy numbers.

4.3 Borel Sets

If F is a collection of subsets of the sample space, Ω, then F is said to be a σ-algebra if the following conditions hold: Ω ∈ F; if A ∈ F then Ac ∈ F; and if $A = \bigcup_{i=1}^{\infty} A_i$ and $A_i \in F$ for $i \in I^+$, then A ∈ F. The Borel σ-algebra, B is the smallest σ-algebra that contains the set of all open intervals in R, the set of real numbers. Elements of B are called Borel sets and (R, B) is called Borel measurable space. An example can be given to clear the concept of Borel set. Let us consider the conflict case of consumption of rice in a typical family in the southern region of India. One group gave this figure as [100, 200] kg/year and another group said [150, 300] kg/year. If at least one of these groups is correct, the consumption of rice will fall within the union of two, i.e., [100, 300]. But, if both the groups are correct, the consumption of rice will fall in the intersection of their estimates, that is, the interval [150, 200]. Borel sets are used to describe such kind of data.

5 Fuzzy Random Variable Model

Two kinds of fuzzy random variables are cited in the literature [3–5]. One is due to Kwakernaak [3, 4], who coined the term "fuzzy random variable" and interpreted as FRVs as "random variables whose values are not real but fuzzy numbers." The other is due to Puri and Ralescu [5], who regarded FRVs as random fuzzy sets. Accordingly, we have two kinds of fuzzy random variable models and we present both the FRV models as follows:

5.1 Kwakernaak FRV Model

In this model, an FRV is a mapping $\zeta\colon \Omega\rightarrow$ F(R) such that for any $\alpha\in$ [0,1] and all $\omega \in \Omega$, the real-valued mapping is as follows:

$$\inf \zeta_\alpha : \Omega \rightarrow R, \text{ satisfying: } \inf \zeta_\alpha(\omega) = \inf(\zeta(\omega))_\alpha, \text{ and}$$
$$\sup \zeta_\alpha : \Omega \rightarrow R, \text{ satisfying: } \sup \zeta_\alpha(\omega) = \sup(\zeta(\omega))_\alpha.$$

These real valued mappings are real valued random variables, that is, Borel-measurable real-valued functions. These α-level constraints on ζ may be summarized as $\zeta_\alpha(\omega) = [\inf(\zeta(\omega))_\alpha, \sup(\zeta(\omega))_\alpha]$. In short, the Kwakernaak FRV takes the form of mappings from Ω to the left- and right-hand sides of the fuzzy target F(R), where the latter are real-valued random variables. If X is an FRV and $\Pi_A$is the collection of all A-measurable random variables of Ω, then the kth moment of Kwakernaak FRV x, E(x^k) is a fuzzy set on R with

$$\mu_{E(x^k)}(x) = \sup\left\{\mu_x(U)|U \in \Pi_A, \mathrm{E}U^k = x\right\}, x \in \mathbf{R} \tag{7}$$

The fuzzy variance of X is a fuzzy set Var[k(x)] on [0, ∞) with

$$\mu_{\mathrm{var}_k(x)}(\sigma^2) = \sup\left\{\mu_x(U)|U \in \Pi_A, D^2U = \sigma^2\right\}, \sigma^2 \in [0,\infty) \tag{8}$$

5.2 Puri and Ralescu FRV Model

Prior to presenting Puri and Ralescu's [5] FRV model, it is required to briefly describe Banach space. Banach space is a normed linear space which is complete as a metric space. Banach spaces are used to extend the domain of FRVs from the real line to Euclidean n-space. Puri and Ralescu [5] conceptualized an FRV as a fuzzification of a random set. If (B, |.|) is a separable Banach space, K(B) a nonempty compact subset of B, this model is addressed an FRV as a mapping $\zeta\colon \Omega\rightarrow$F(B) such that for any $\alpha \in$ [0,1] the set-valued mapping $\zeta\alpha : \Omega \rightarrow$K(B) (with $\zeta\alpha(\omega) = (\zeta(\omega))\alpha$ for all $\omega \in \Omega$) is a compact random set, that is, it is Borel-measurable with the Borel σ-field generated by the topology associated with the Hausdorff metric on K(B) [12]

$$d_H(P,Q) = \max\left\{\sup_{p\in P}\inf_{q\in Q}|p-q|, \sup_{q\in Q}\inf_{p\in P}|p-q|\right\} \tag{9}$$

If P and Q are bounded, then the Hausdorff metric becomes

$$d_H(P,Q) = \max\left\{|\inf p - \inf q|,\ |\sup p - \sup q|\right\} \tag{10}$$

6 Fuzzy Random Exposure and Health Risk Model

The dose evaluation for humans exposed to ionizing radiation is an important factor for risk assessment, consisting of a base criterion for decision-making in situations that may require the intervention of the regulatory agencies in many countries. Around the world there are many areas identified as high background area due to deposits of high concentrations of natural radionuclide such as uranium and thorium. People living in such areas are exposed to this radionuclide. Uranium is a significant radioactive element from a radiological point of view, particularly due to its chemical toxicity. The entry of uranium in the human body takes place through the exposure route of "ingestion of contaminated food grown in those areas and drinking of water." In order to estimate the health risk associated with the ingestion of food contaminated with uranium or any other naturally occurred radionuclide, it is required to first have an exposure model. Assuming that exposure and risk are directly proportional to each other, the exposure estimated by the model is multiplied by the cancer slope factor or risk factor of the radionuclide for the specified exposure route to obtain the proportional risk. Our present model dictating the exposure is due to the ingestion of food contaminated with naturally occurred uranium. The model estimates the daily average ingestion (DAI) of uranium in mBq/kg day.

The scenario for model formulation is as follows. The region around which the model is conceptualized is rich in groundwater and many other aquatic environments (river, estuaries and lakes). The inhabitants consume food products grown in several farms located in and around the zone. Body weight (BW), consumption rate (intake) of food (I), lifetime (LF), and exposure duration (ED) are the parameters considered for the given situation. Data for uranium activity concentration (C) are collected by sampling foods from the affected region and processing them in a radiochemical laboratory. The parameter, diet fraction (DF) which indicates the amount of contaminant transferred from the food consumed by humans after consumption of food is considered. Our formal model can be mathematically represented as

$$DAI = \frac{C \times I \times ED \times DF}{BW \times LF \times 365} \tag{11}$$

In this model, uranium activity concentration (C) and food consumption rate (I) are considered as fuzzy random variables. Due to lack of food samples collected for measuring activity concentration of uranium, mean value of the estimation is fuzzy which is represented as a triangular fuzzy number. However, standard deviation of the measurement being the same it is obviously a crisp number. But the variability of the activity concentration of uranium from place to place over the region of interest informs that the random nature of the same follows a lognormal distribution. So, the conclusion is that even through the variability of the activity concentration of uranium characterizes a lognormal distribution, but fuzziness exists in the mean value of the activity concentration and hence the activity concentration of uranium is justified as a fuzzy random variable. The similar reason is valid for the parameter, food consumption rate, which is also a fuzzy random variable, in which mean value of the lognormal distribution is taken as a triangular fuzzy number and standard deviation as crisp number. Variability of the exposure duration (ED) is best fitted as normal distribution. For every other parameter, the data collected through relevant questionnaires are best fitted by a probability distribution. Accordingly, the randomness of the parameters, DF and LF, is presented by a lognormal distribution. Randomness of the parameter, BW is specified by uniform distribution. The probability distributions were determined by fitting distribution functions to measured/surveyed data with the help of goodness-of-fit tests such as Chi-square, Kolmogorov-Smirnov, and Anderson-Darling (AD). Definition of distribution functions can be found elsewhere (Oracle 2007) [13]. Overall, the model output DAI becomes a fuzzy random variable and hence we call this model a fuzzy random exposure model. The next step is to formulate the fuzzy random health risk (Cancer risk) model associated with ingestion exposure [14], which is given as

$$\text{FRRisk} = \text{DAI} \times \text{CSF}, \tag{12}$$

where FRRisk, is the excess probability of developing cancer over a lifetime as a result of exposure to a contaminant and CSF is the cancer slope factor.

7 Computational Methodology and Results

The input values of the random and fuzzy random parameters of the present model are tabulated in Table 2 and these are taken from Rajkumar and Guesgen [15]. In the computation, random parameters, BW (uniform distribution), DF (log normal distribution), LF (log normal distribution), and ED (normal distribution) are

Table 2 Randomness and fuzziness of the input variables in the formulation of Fuzzy Random Exposure Model

Parameters	Probability distribution	L*	U*
BW, Body weight (kg)	Uniform	52	92
ED, Exposure Duration (day)	Normal	14741	10 % of (L)
DF, Diet fraction	Log normal	0.2	1.2
LF, Lifetime (years)	Log normal	62	1.2
C, Activity Concentration (mBq/kg)	Fuzzy log normal	(12,42,82)	50.4
I, Consumption of food rate (kg/day)	Fuzzy log normal	(10,32,54)	38.4

L* and U*: Lower and upper limit : Uniform distribution,
Mean and standard deviation : Normal distribution,
Mean and standard deviation : Lognormal distribution,
Triangular fuzzy number for mean and standard deviation : Fuzzy lognormal

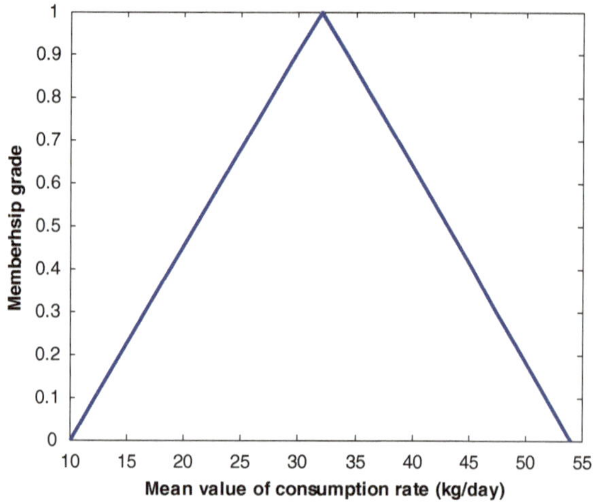

Fig. 5 Membership grade for mean value of consumption rate of food

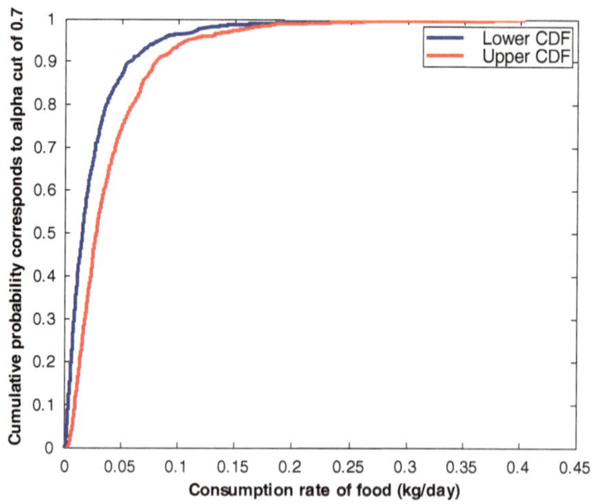

Fig. 6 *Lower* and *upper* cumulative probability for 0.7 α-cut of fuzzy random consumption rate of food containing uranium

generated by simple Monte Carlo simulation using Latin hypercube sampling of size 1,000. Fuzziness part of the fuzzy random parameter C (activity concentration) is as shown in Fig. 5 and I (consumption rate of food) is represented by α-cut and for each α-cut representation of the mean with standard deviation of the appropriate random distribution is simulated by Monte Carlo simulation. Fuzzy cumulative distribution of the parameters I and C are constructed for each α-cut levels. Lower and upper cumulative probabilities of the parameters I and C at an α-cut of 0.7 are presented in Figs. 6 and 7. Since α-cut is an interval, we have lower and upper bounds of the specified probability distribution for each α-cut of the fuzzy random parameters. Each such lower and upper bounds of the probability distribution of the fuzzy random parameters of the model are applied in the model along with the respective probability distribution of the remaining parameters to generate the lower and upper

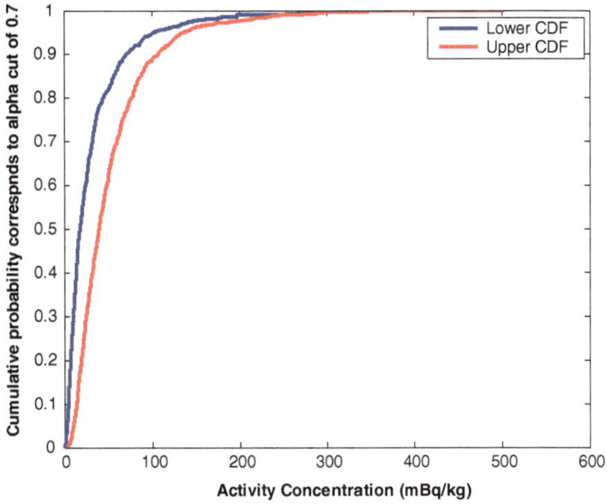

Fig. 7 *Lower* and *upper* cumulative probability for 0.7 α-cut of fuzzy random activity concentration of uranium

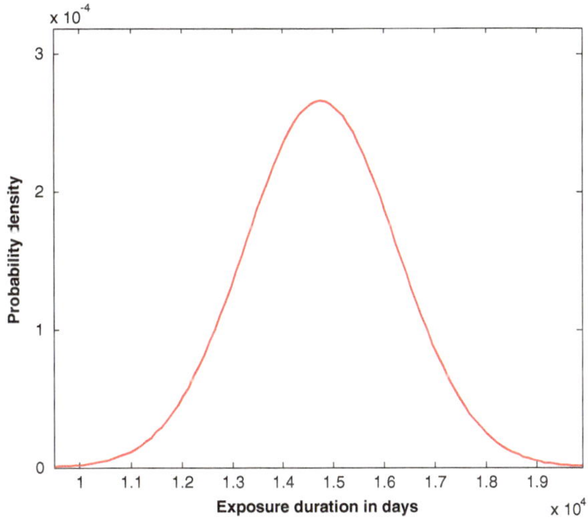

Fig. 8 Probability density of exposure Duration (ED)

bounds of the probability distribution of the model output, which is DAI. Lower and upper cumulative probability distribution of the model output, DAI are constructed for each α-cut levels. The same of the daily average ingestion (DAI) for an α-cut of 0.7 is as shown in Fig. 8. For a given α-cut level, say 0.7, corresponding lower and upper cumulative probability distributions have been used to generate various percentiles of the lower and upper bounds of the probability distribution of the model output. This scheme specifies the randomness of the model output at a specified α-cut level. On the other hand, at some specific percentiles, say 75 and 85th %le, of the lower and upper cumulative probability bounds of the DAI, various α-cut levels (0 to 1 with an increment of 0.1) are marginalized to obtain the membership function of the model output, DAI (Fig. 9). This scheme provides the knowledge of fuzziness

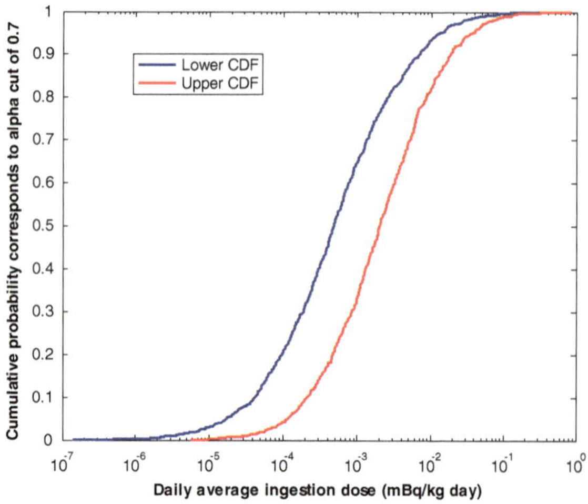

Fig. 9 Lower and upper cumulative probability for 0.7 α-cut of fuzzy random daily average ingestion

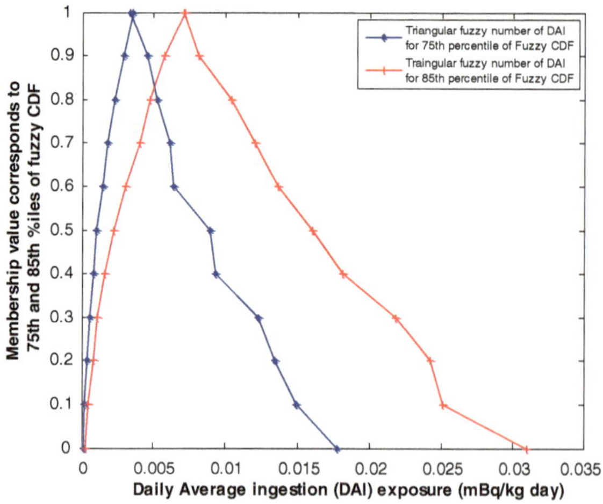

Fig. 10 Triangular fuzzy number for 75 and 85th percentiles of Fuzzy CDF of daily average ingestion (mBq/kg day)

of the model output at a specific percentile and by changing the order of the percentile we can obtain another fuzziness of the model output. Knowing the fuzziness of the model output at a specific percentile of the randomness, we have computed the support, uncertainty, possibility, necessity, and credibility of the model output. This scheme then updates the knowledge toward the randomness-fuzziness consistency. Fuzzy random risk due to exposure of uranium through ingestion of contaminated food has been computed by multiplying DAI with cancer slope factor [16] and the corresponding lower and upper cumulative probabilities at alpha cuts of 0.5, 0.7, 0.8, and 1 are presented in Fig. 10. Membership function of fuzzy random risk at 75th and 85th percentiles of corresponding fuzzy CDF is as shown in Fig. 11. It can be seen from Fig. 11 the risk from exposure to naturally occurred uranium through ingestion

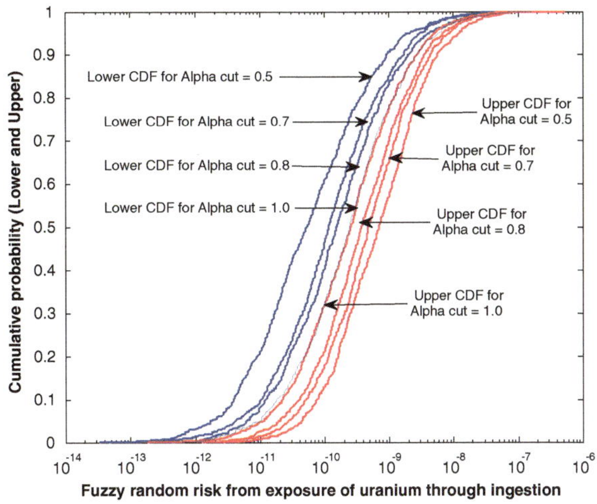

Fig. 11 *Lower* and *upper* cumulative probability of fuzzy random risk from exposure to uranium through ingestion of food

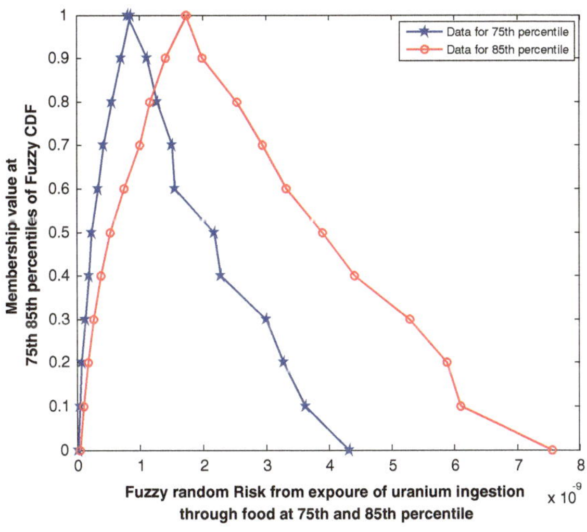

Fig. 12 Triangular fuzzy number for 75 and 85th percentiles of fuzzy CDF of risk from exposure to uranium through ingestion of food

of food grown in the region whose uranium deposits are very low. So the inhabitants of that region are always safe (Fig. 12).

7.1 Estimation of Support, Uncertainty, Possibility, and Necessity Measures

Two indices of uncertainty are computed in this study. The first is support that is widely used in the fuzzy literature and the other is the uncertainty index. The support for membership functions for fuzzy random risk, FRRisk, is defined as the range (spread between maximum and minimum) containing risk values with a nonzero membership in the set FRRisk. The support contains those elements of the universe for which μFRRisk(fr) $>$ 0. The uncertainty index is computed by dividing the support value with the most likely value of fuzzy random risk. For the present situation of fuzzy random risk estimation, the support value and uncertainty index of fuzzy random risk at 75 and 85th percentiles are tabulated in Table 3. The result obtained for fuzzy random risk can be compared with compliance guideline by the theory of possibility. For this purpose, possibility theory uses two measures, namely possibility measure and necessity measure [17]. The two measures are used to validate the proposition "the fuzzy random risk FRRisk, is lower or equal to the compliance guideline C."

The possibility measure *Pos* is defined as [18]

$$Pos\{FRRisk \leq C\} = \sup_u \min\{\mu_{FRRisk}(u), \mu_C(u)\}, \tag{13}$$

where μ_{FRRisk}(u) is the membership of FRRisk at any value u; μ_C(u) is the membership of C at any value u; Sup = largest value; min = minimization operator. Therefore initially, minimum values between the membership function of the risk and compliance criteria are calculated and then the largest value among them is calculated. This value then corresponds with the possibility value that fuzzy random risk is less than or equal to the compliance criteria C. For the present model, we have found that the highest value representing the minimum value between the membership function of risk and compliance criteria is at membership value of 0.7 which is the required possibility measure for the given compliance criteria. The possibility measure of 0.7 is estimated for both the 75 and 85th percentiles.

The necessity measure is defined as [18]

$$Nec\{FRRisk \leq C\} = \operatorname*{Inf}_u \max\{1 - \mu_{FRRisk}(u), \mu_C(u)\} \tag{14}$$

Here, the maximum values between the membership value of the complement of the risk value and the compliance criteria are first calculated. Then we find the minimum values among these values. This value then corresponds to the necessity measure satisfying the condition that, *FRRisk*, is less than or equal to the compliance criteria. We obtain the necessity measure of fuzzy random risk at 75 and 85th percentiles of fuzzy CDF of risk near to 0.5. Thus the greater the compliance criteria cover of both the arms of the risk profile, the higher the chance for the risk profile to satisfy both possibility and necessity measures.

Table 3 Support and uncertainty index of fuzzy random risk

Percentiles of fuzzy CDF of risk	Support	Uncertainty index
75th percentile	4.3E-9	3.31
85th percentile	7.5E-9	3.75

8 Conclusions

This paper presents the application of fuzzy random variable in formulating the fuzzy random health risk model. The use of fuzzy random concept does not replace the existing deterministic or probabilistic methods; however, it provides in-depth knowledge of human health risk assessment in the framework of nonprobabilistic uncertainties. There exist many examples of such uncertainties. One of them is the subjectivity present in the answers in the questionnaires providing the data that represent the fuzziness of the specific parameters. Subjectivity is due to imprecise information about the profile of the inhabitants from the uneducated population. Support, uncertainty index, possibility, and necessity measures have been computed. The possibility measure was performed using the left arm of the triangular membership function, while the necessity measure was calculated using the right arm of the triangular membership function. Possibility can be thought of as the criteria of an optimistic decision-maker, while necessity measure criteria can be thought of as the criteria of the pessimistic decision maker. A quantity satisfying possibility measure criteria may or may not satisfy necessity measure criteria but the reverse is always true.

The results indicate that fuzzy random radiation exposure and fuzzy random health risk model give the possibility to work out the nonprobabilistic uncertainties based on random sets. It also opens an area of research on the design of knowledge base or rule base on the basis of fuzzy random variable. Our future work will be on the implementation of this similar concept to assess the health risk due to inhalation of ^{222}Rn in the uranium mining environment.

References

1. Shapiro, A.F.: Fuzzy logic in insurance. Insur. Math. Econom. **35**, 399–424 (2004)
2. Gonazalez-Rodriguez, G., Coulbi, A., Gil, M.A.: A fuzzy representation of random variables: an operational tool in exploratory analysis and hypothesis testing. Int. J. Comput. Stat. Data Anal. **51**, 163–176 (2006)
3. Kwakernaak, H.: Fuzzy random variables—I. Definitions and theorems. Inf. Sci. **15**(1), 1–29 (1978)
4. Kwakernaak, H.: Fuzzy random variables—II. Algorithms and examples for the discreet case. Inf. Sci. **17**(3), 253–278 (1979)
5. Puri, M.L., Ralescu, D.A.: Fuzzy random variables. J. Math. Anal. Appl. **114**, 409–422 (1986)
6. Zadeh, L.A.: Probability measures of fuzzy events. J. Math. Anal. Appl. **23**, 421–427 (1968)
7. Dubois, D., Prade, J.: Mathematics in Science and Engineering. Fuzzy Sets and Systems: Theory and Applications, vol. 144. Academic Press, New York (1980)

8. Liu, B.: Uncertainty Theory, 4th edn., Uncertainty Theory Laboratory, Tsinghua University, Beijing
9. Liu, B.: Foundations of Uncertainty Theory. Tsinghua University, Department of Mathematical Sciences, Beijing (2005)
10. Liu, B., Liu, Y.-K.: Expected value of fuzzy variable and fuzzy expected value models. IEEE Trans. Fuzzy Syst. **10**, 445–450 (2002)
11. Liu, B.: Uncertainty Theory, 2nd edn. Springer, Berlin (2007)
12. Gil, M.A.: Fuzzy random variables: development and state of the art. In: Mathematics of Fuzzy Systems, Linz Seminar on Fuzzy Set Theory, Linz (2004)
13. Oracle, : Crystal Ball Reference Manual. Oracle Inc. (2007)
14. Patrik, D.R.: Risk assessment and risk management. In: Patrick, D.R. (ed.) Toxic Air Pollution Handbook, p. 347. Van Nostrand Reinhold, New York (1994)
15. Rajkumar, T., Guesgen, H.W.: Fuzzy exposure model, IEEE Xplore 1051–1056 (1996)
16. United Nation Scientific Committee on the effect of atomic radiation (UNSCEAR): Sources and Effects of Ionizing Radiation. United Nations, New York (2000)
17. Guoynnet, D., Bourgine, B., Dubois, D., Fargier, H., Come, B., Chiles, J.P.: Hybrid approach for addressing uncertainty in risk assessments. J. Environ. Eng. (ASCE) **129**, 68–78 (2003)
18. Kentel, E., Aral, M.M.: Probabilistic-fuzzy health risk modeling. Stoch. Environ. Res. Risk Assess. **18**, 324–338 (2004)

Uncertainty Analysis of Retardation Factor Using Monte Carlo, Fuzzy Set and Hybrid Approach

T.K. Pal, V. Arumugam and D. Datta

Abstract Uncertainty analysis of physical parameters present in the groundwater model is important from the point of safety measures in the field of nuclear science and technology. Researchers have carried out this uncertainty analysis using traditional Monte Carlo simulations. However, in practice, Monte Carlo simulation may not be possible because of lack of data obtained from field experiments. Therefore, the demand is to investigate uncertainty using imprecise-based method. In order to fulfill this demand, we have carried out uncertainty analysis of groundwater model parameter using fuzzy set and hybrid methods. Monte Carlo-based uncertainty is also presented in this paper. Overall, this paper highlights the various methodologies of uncertainty analysis. In the hybrid approach, the concept of fuzzy random variable and its computational details have been explored. Retardation factor is our representative groundwater model parameter on which illustration of the said methodologies of uncertainty modeling is presented.

Keywords Uncertainty · Monte Carlo · Fuzzy set · Fuzzy random variable · Retardation factor

1 Introduction

The rate at which a chemical constituent moves through soil is determined by several transport mechanisms. These mechanisms often act simultaneously on the chemical and may include processes such as convection, diffusion and dispersion, linear

T.K. Pal (✉) · V. Arumugam
Technology Development Division, Bhabha Atomic Research Centre,
Mumbai 400085, India
e-mail: tkpal@barc.gov.in

D. Datta
Health Physics Division, Bhabha Atomic Research Centre,
Mumbai 400085, India

M.K. Chakraborty et al. (eds.), *Facets of Uncertainties and Applications*,
Springer Proceedings in Mathematics & Statistics 125,
DOI 10.1007/978-81-322-2301-6_12

equilibrium adsorption, zero-order or first-order production, and decay. Because of the many mechanisms affecting solute transport, a complete set of analytical solutions should be available, not only for predicting actual solute transport in the field but also for analyzing the transport mechanisms themselves, for example, in conjunction with column displacement experiments. Numerous analytical solutions of the convective dispersive solute transport equation have been published in recent years, both in well-known and widely distributed scientific journals and in lesser known reports and conference proceedings. A large number of physical processes are involved in solute transport, for example, solute transport takes place by molecular diffusion wherein the phenomenon involved is due to the random thermal motion of solute ions and/or molecules. Each process involves parameters that are uncertain due to their variability and imprecision. Variability type of uncertainty is classified as aleatory uncertainty, which is irreducible, whereas imprecision-based uncertainty is classified as epistemic uncertainty, which is reducible. Aleatory uncertainty is quantified by traditional Monte Carlo method [1, 2] and epistemic uncertainty is estimated by Fuzzy set theory [3–5]. However, if the model presents both types of uncertain variables, then uncertainty can be analyzed using the hybrid method. Monte Carlo method-based uncertainty analysis is dependent on the probability density function of the uncertain parameters, while fuzzy set theoretic approach of uncertainty quantification is based on the alpha cut value of the corresponding uncertain fuzzy parameter and the associated interval arithmetic. Hybrid uncertainty is quantified using the concept of fuzzy random variable [6–8].

The objective of this paper is to demonstrate the various types of uncertainty analysis. Typical examples are given for illustration of various methods of uncertainty analysis. Finally, uncertainty analysis of convective–dispersive solute transport equation is presented in detail. The remainder of the paper is organized as follows: Sect. 2 presents the Monte Carlo method-based uncertainty analysis of physical models. Section 3 presents the Fuzzy set theoretic approach of uncertainty analysis of physical models. Section 4 presents the method of hybrid uncertainty analysis. Conclusion of the paper is presented in Sect. 5.

2 Uncertainty Analysis Using Monte Carlo Method

The goal of the Monte Carlo method in uncertainty analysis is to propagate parameter uncertainty through the specified model. Each individual parameter of a specific model is characterized by its probability distribution. Generation of random values of these uncertain parameters by sampling from this probability distribution is the basic aim of Monte Carlo method. Latin hypercube sampling scheme [9] is used to generate samples of random parameters of a model. Parametric uncertainty is propagated through the specific model to generate its probability density function (PDF) and the corresponding cumulative distribution function (CDF). Uncertainty of the model in this case is expressed in terms of an interval, the lower bound of which is the 5th percentile of the CDF and the upper bound of which is the 95th percentile of the CDF.

Consider the following example as illustration.

2.1 Illustration: Uncertainty Analysis of Retardation Factor (R)

Retardation factor [10] represents the partition of the solute between solid and liquid phases. When soil water moves under steady-state flow conditions and when diffusion and dispersion are zero, all of the solute moves at the same velocity and the plume front arrives in one discontinuous jump to the final concentration. This ideal condition is known as piston flow and the retardation factor using this piston flow can be mathematically written as

$$R = 1 + (\rho_b/\theta_v)\frac{dS}{dC_l} \tag{1}$$

where ρ_b = bulk density of the soil (mg/m^3), θ_v = volumetric soil water content (m^3/m^3), S is the amount of solute in the solid or adsorbed phase (kg solute/kg of soil), and C_l is the solute concentration in solution (kg of solute per m^3 of soil water). Now, the bulk density of the soil and volumetric soil water content are random parameters and their randomness is described by the probability density function given below:

$\rho_b \sim N(\mu = 1.5, \sigma = 0.15)$: Normal distribution

$\theta_v \sim U(Lower = 0.15, Upper = 0.35)$: Uniform distribution

The deterministic value of $\frac{dS}{dC_l} = 2\,\text{cm}^3/\text{g}$

Using the Latin hypercube sampling scheme we generate 1,000 samples of each of these uncertain parameters. Histogram of the uncertainty of the parameters is as shown in Figs. 1 and 2. PDF of Eq. (1) is constructed by the uncertain values of the parameters (ρ_b and θ_v) and $\frac{dS}{dC_l}$. Frequency distribution and cumulative distribution function of retardation factor as given in Eq. (1) are shown in Figs. 3 and 4.

From Fig. 4, the 5th and 95th percentiles of retardation factor are obtained as 9.46 and 19.85. Therefore, uncertainty of retardation factor, R is expressed as an interval [9.46, 19.85]. Degree of uncertainty of R can be estimated as (19.85 − 9.46)/(19.85 + 9.46) = 0.35 (35 %) and obviously this is nothing but the degree of uncertainty at 90 % confidence level.

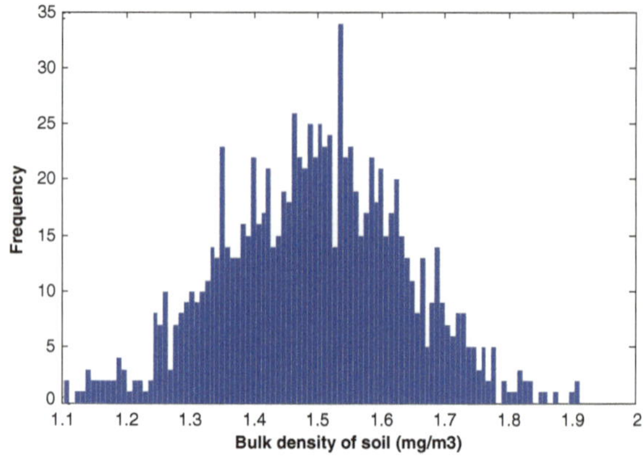

Fig. 1 Frequency distribution of the bulk density of soil

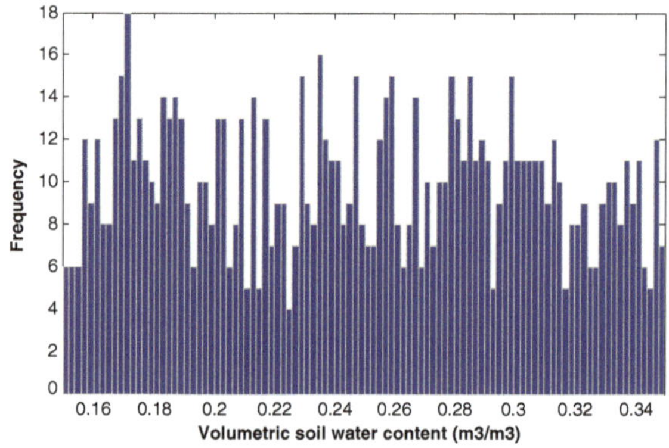

Fig. 2 Frequency distribution of soil water content

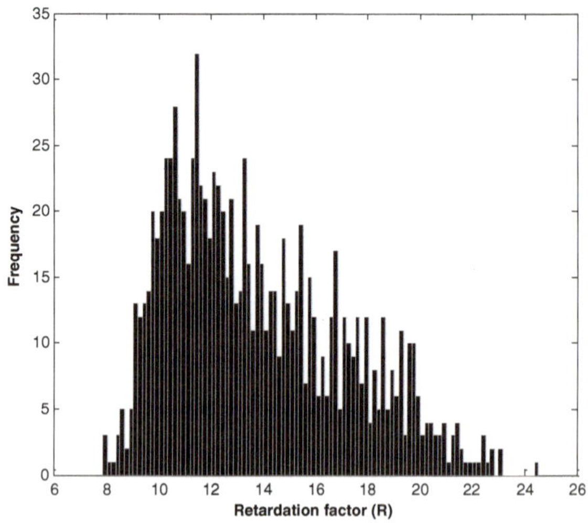

Fig. 3 Frequency distribution of retardation factor

Fig. 4 Cumulative distribution function of R

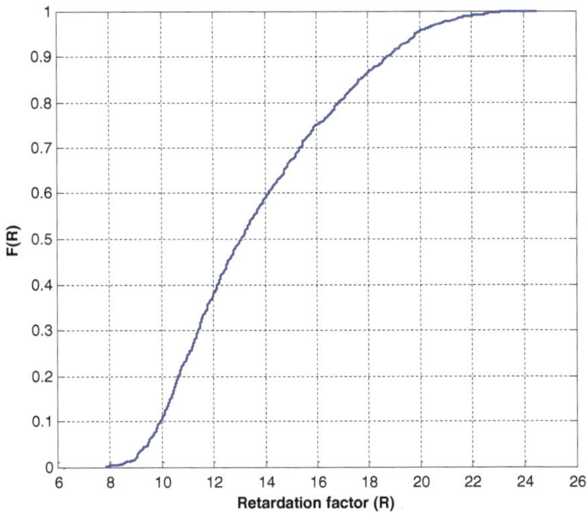

3 Uncertainty Analysis Using Fuzzy Set Theory

In this case, imprecision or incompleteness or vagueness are the cause for uncertainty and such uncertainty is quantified by fuzzy set theory [11]. Definition of a fuzzy set can be found elsewhere in the literature [12, 13]. However, before its implementation into uncertainty analysis we define a fuzzy set and its alpha cut representation as the alpha cut representation of a fuzzy set plays an important role in quantification of uncertainty of an imprecise system. A fuzzy set is defined as the set of a pair of numbers, such as A = {(x, μ(x)|, x $\in$ R, μ(x)$\in$[0,1]}, where μ(x) represents the membership grades of the crisp value, x. Alpha cut representation of this fuzzy set, A is defined as the set of values of x, whose membership values are greater than or equal to alpha. Basically, alpha cut of a fuzzy set is an interval and all the primary interval arithmetic operations {'+', '−', '*', '/'} are carried out depending on the model under quest for expressing its uncertainty in terms of a fuzzy set. If a fuzzy set is normal and bounded [13] then that fuzzy set is convex and is called the convex normal fuzzy number. Further, if the membership values of the fuzzy set are triangular in shape, then that fuzzy number is represented as a triangular fuzzy number. Mathematical structure of a triangular fuzzy number is governed by Eq. (2).

$$\mu(x) = \begin{cases} \dfrac{x_L - a}{b - a}, & a \le x_L \le b \\ \dfrac{c - x_R}{c - b}, & b \le x_R \le c \\ 0 & otherwise \end{cases} \tag{2}$$

Alpha cut representation of the fuzzy set, A(x) is given as

$$A_\alpha = [x_L^\alpha, x_R^\alpha] = [a + (b - a)\alpha,\ c - (c - b)\alpha] \tag{3}$$

In order to have the uncertainty analysis of the model as given in Eq. (1), all the uncertain parameters are assumed as triangular fuzzy numbers for convenience of computation. However, one can also have different fuzzy numbers (trapezoidal, Gaussian, etc.) for different parameters. Fuzzy set theory-based uncertainty analysis is demonstrated with the same example of retardation factor.

3.1 Illustration: Uncertainty Analysis of Retardation Factor (R) Using Fuzzy Set

Here, the uncertain parameters, the bulk density of the soil, and volumetric soil water content are expressed as triangular fuzzy numbers and their attributes are given below:

$\rho_b \sim TFN(1.35,\ 1.5,\ 1.65)$: TFN stands for triangular fuzzy number

$\theta_v \sim TFN(0.15,\ 0.25,\ 0.35)$

Membership plots of these fuzzy numbers are as shown in Figs. 5 and 6 respectively. Alpha cut representation of these fuzzy numbers provide an interval; substituting these interval representations of the uncertain fuzzy numbers, bulk density of the soil, and volumetric soil water content in Eq. (1) we obtain the corresponding interval representation of the retardation factor. Membership function of the retardation factor is as shown in Fig. 7.

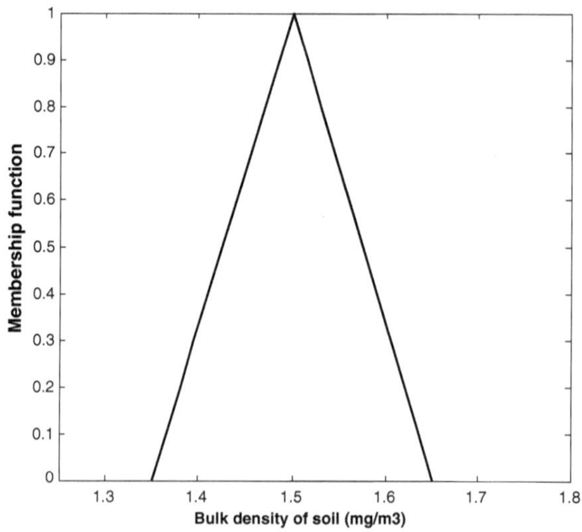

Fig. 5 Triangular fuzzy number of bulk density of soil

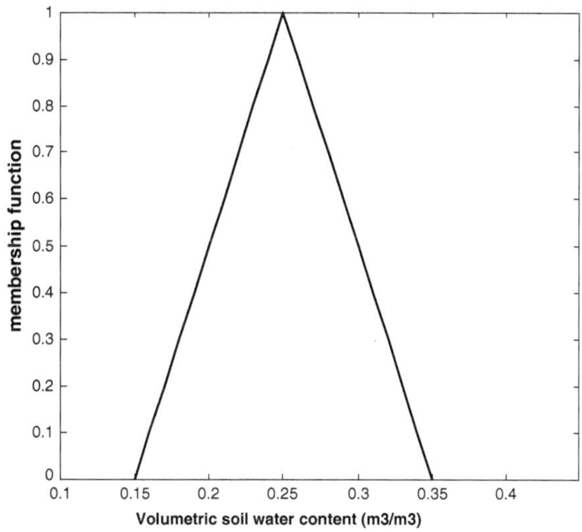

Fig. 6 Triangular fuzzy number of volumetric soil water content

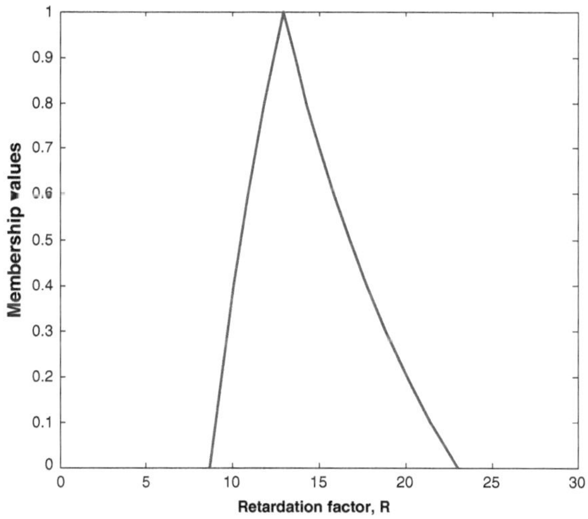

Fig. 7 Triangular fuzzy number of the retardation factor

Uncertainty of the retardation factor using fuzzy set theoretic approach is quantified as per its alpha cut level. In order to compare the degree of uncertainty of retardation factor by this approach and the same by probabilistic approach, we take 0.2 alpha cut representation of the retardation factor as 0.2 alpha cut value which corresponds to the 90 % confidence level as obtained in the probabilistic approach. Figure 7 indicates the 0.2 alpha cut value of the retardation factor as an interval of [10.5, 16.75]. Now we can write the 90 % confidence level for degree of uncertainty using fuzzy set theoretic approach of uncertainty quantification of any model parameter as 0.41 (41 %). It is also obvious that uncertainty estimate of any model using probabilistic approach is lesser than by using fuzzy set theoretic approach.

4 Uncertainty Analysis Using Hybrid Approach

Hybrid approach of uncertainty analysis generally proceeds with the admixture of random and fuzzy parameters of the defined model. Various methodologies of hybrid uncertainty analysis have been found elsewhere in the literature [14, 15]. We have used the concept of fuzzy random variable in formulating the hybrid approach of uncertainty analysis. In order to carry out the uncertainty analysis using hybrid approach we have estimated the fuzzy cumulative distribution function (FCDF) of the retardation factor. Algorithm of our hybrid approach of uncertainty analysis is depicted below.

Step 1: Define the triangular fuzzy membership function for mean and standard deviation of the normally distributed random variable, ρ_b. Define the triangular fuzzy membership function for the lower and upper limits of the uniformly distributed random variable, θ_v.
Step 2: Set α-cut interval as 0.5, i.e., $\alpha = [0 : 0.5 : 1]$. Since the mean and standard deviation of normal distribution and lower and upper limits of uniform distribution are fuzzy numbers, α-cut method is used for discretization purpose. The lower and upper bounds of the intervals for each α-cut for mean and standard deviation $(\tilde{\mu}^\alpha, \tilde{\sigma}^\alpha)$ of the PDF of ρ_b are represented as

$$\tilde{\mu}^\alpha = [\mu_L^\alpha, \mu_U^\alpha], \quad \tilde{\sigma}^\alpha = [\sigma_L^\alpha, \sigma_U^\alpha] \tag{4}$$

The lower and upper bounds of the intervals for each α-cut for lower and upper limits $(L\tilde{B}^\alpha, U\tilde{B}^\alpha)$ of the PDF of θ_v are represented as

$$L\tilde{B} = [LB_L^\alpha, LB_U^\alpha] \text{ and } U\tilde{B} = [UB_L^\alpha, UB_U^\alpha] \tag{5}$$

Step 3: Discretize the random variable domain (R) as R = Rk: k = 1, 2, ..., n, $\forall F(R_n) = 1$.
Step 4: Replace the random number Rk in Eq. (1) with fuzzy normal distributed random number and fuzzy uniform distributed random number using fuzzy arithmetic and inverse transformation method.
Step 5: Calculate the lower and upper bounds of interval of fuzzy function for probability of the retardation factor. All the input variables such as mean and standard deviation of normal distribution and lower limit and upper limit of uniform distribution are greater than zero, it is better to use restricted DSW algorithm [16]. For fuzzy sets [a, b] and [c, d], restricted DSW algorithm states that if a, b, c, d > 0, then

$$\begin{aligned} [a, b] \div [c, d] &= [a \div d, b \div c] \\ [a, b] \times [c, d] &= [a \times c, b \times d] \end{aligned} \tag{6}$$

The restricted DSW algorithm is employed for the calculation of upper and lower bounds of the intervals of the fuzzy function for each α-cut level at $R_k : k = 1 - 11$. The substitution of restricted DSW algorithm makes possible two combinations of

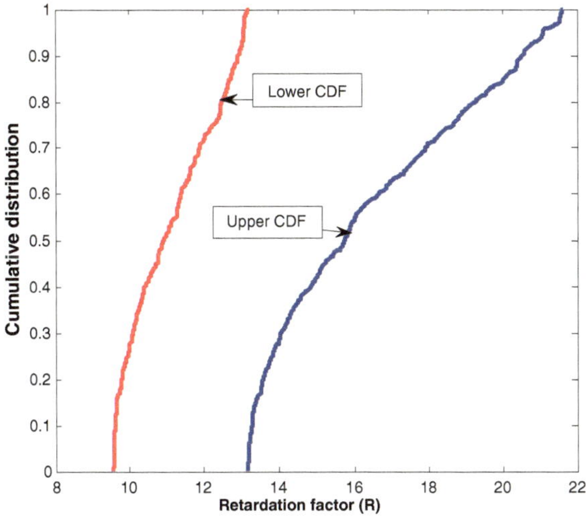

Fig. 8 Lower and upper cumulative probability distributions of retardation factor for alpha cut of 0.2

Table 1 Percentiles of lower and upper CDF of fuzzy random variable, R

Percentiles of retardation factor (R) at α-cut = 0.2	Lower	Upper
2.5	9.58	13.17
5	9.59	13.19
50	10.94	15.75
95	13.07	21.09
97.5	13.10	21.53

lower and upper α-cut values of parameters (mean and standard deviation of normal distribution; lower and upper limits of uniform distribution) to be sufficient to calculate fuzzy CDFs of the retardation factor. Using the above procedure, the fuzzy CDFs of the retardation factor ($F(R)_L^\alpha,\ F(R)_U^\alpha$) are generated for all the lower and upper α-cut levels of the parameters of the distributions. Results of fuzzy CDFs of the retardation factor for α-cut value of 0.2 of the parameters of the respective distributions are shown in Fig. 8. From the lower and upper fuzzy CDFs of the retardation factor we have computed 2.5th, 5th, 50th, 95th, 97.5th percentiles and the results are tabulated in Table 1. Membership function of fuzzy CDF of the retardation factor corresponding to the 50th percentile is as shown in Fig. 9.

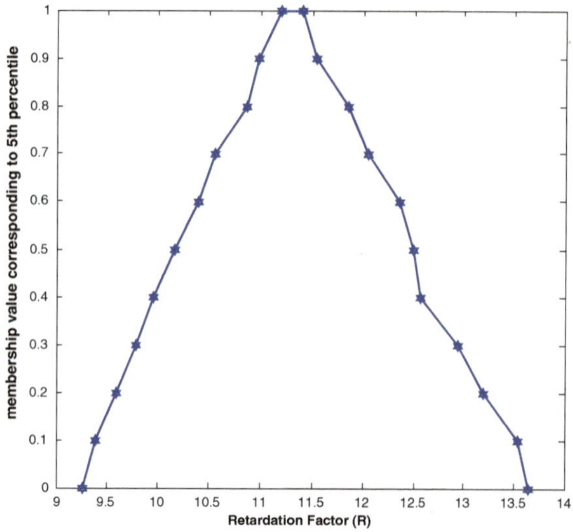

Fig. 9 Triangular membership function of fuzzy retardation factor corresponds to 5th percentile

5 Conclusions

Uncertainty analysis of any model using various methods such as Monte Carlo simulation, Fuzzy set theory and Hybrid approach has been discussed. Utility of fuzzy random variable is presented for demonstrating the hybrid approach of uncertainty modeling. In order to illustrate the various methods of uncertainty we have presented the uncertainty modeling of retardation factor which is one of the most important factors for controlling the transport of contaminant through porous medium such as soil. In the hybrid approach of uncertainty modeling, fuzzy cumulative distribution functions (lower and upper) of the retardation factor at various alpha cut levels are constructed. At a specific percentile from the lower and upper fuzzy CDF, membership function of the retardation factor is constructed and also shown in the paper.

References

1. Vose, D.: Quantitative Risk Analysis—A Guide to Monte Carlo Simulation Modeling. Wiley, New York (1996)
2. Helton, J.C.: Uncertainty and sensitivity analysis techniques for use in performance assessment for radioactive waste disposal. Reliab. Eng. Syst. Saf. **42**, 327–367 (1993)
3. Ross, T.J.: Fuzzy Logic with Engineering Applications, 2nd edn. Wiley, New York (2004)
4. Dubois, D., Prade, H.: Fuzzy Sets and Systems. Academic Press, New York (1980)
5. Dong, W., Shah, H.: Vertex method for computing functions of fuzzy variables. Fuzzy Sets Syst. **24**(1), 65–78 (1987)
6. Kwakernaak, H.: Fuzzy random variables—I, definitions and theorems. Inf. Sci. **15**(1), 1–29 (1978)

7. Kwakernaak, H.: Fuzzy random variables—II, Algorithms and examples for the discrete case. Inf. Sci. **17**(3), 253–278 (1979)
8. Puri, M.L., Ralescu, D.A.: Fuzzy random variables. J. Math. Anal. Appl. **114**, 409–422 (1986)
9. Ferson, S., Tucker, W.T.: Sensitivity analysis using probability bounding. Reliab. Eng. Syst. Saf. **91**, 1435–1442 (2006)
10. Van Genuchten, M.Th., Alves, W.J.: Analytical solutions of the one-dimensional convective dispersive solute transport equation. United States Department of Agriculture, Agricultural Research Service, Technical Bulletin 1661 (1982)
11. Liu, B.: Uncertainty Theory: An Introduction to Its Axiomatic Foundations. Springer, Berlin (2004)
12. Klir, G.J.: The many faces of uncertainty. In: Ayyub, B.M., Gupta, M.M (eds.) Uncertainty Modeling and Analysis: Theory and Applications, pp. 3–19. Elsevier, Amsterdam (1994)
13. Zadeh, L.: Fuzzy Sets. Inf. Control **8**, 338–353 (1965)
14. Baudrit, C., Couso, I., Dubois, D.: Joint propagation of probability and possibility in risk analysis: towards a formal framework. Int. J. Approx. Reason. **45**, 82–105 (2007)
15. Guyonnet, D., Bourgine, B., Dubois, D., Fargier, H., Come, B., Chiles, J.P.: Hybrid approach for addressing uncertainty in risk assessments. J. Environ. Eng. (ASCE) **129**, 68–78 (2003)
16. Givens, J., Tahani, H.: An improved method of performing fuzzy arithmetic for computer vision, In: Proceedings of North American Information Processing Society (NAFIPS), pp. 275–280. Purdue University, West Lafayette (1987)

Part IV
Roll of Uncertainties

Two Person Interaction Detection Using Kinect Sensor

Sriparna Saha, Amit Konar and Ramadoss Janarthanan

Abstract This proposed work explains a noble two-person interaction modelling system using Kinect sensor. Here a pentagon for each person is formed taking the three dimensional co-ordinate information with the help of Microsoft's Kinect sensor. Five Euclidean distances between two pentagon vertices corresponding to two persons are considered as features for each frame. So the body gestures of two persons are analysed employing pentagons. Based on these, eight interactions between two persons are modelled. This system produces the best recognition rate (greater than 90 %) with the virtue of multi-class support vector machine for rotation invariance case and for rotation variance phenomenon, the recognition rate is greater than 80 %.

Keywords Body gesture · Euclidean distance · Kinect sensor · Pentagon

1 Introduction

Full body tracking is an emerging field of human–computer interaction. Human body gesture realization is highly important for surveillance, cybernetics, information gathering from video, gaming purposes and for many more areas. In today's era, human body gesture [1, 2] is modelled using the skeleton structure of persons. The human motion [3, 4] is analysed using different sensors. The rapid development in the field of skeleton detection is possible due to Microsoft's Kinect sensor [5–7].

S. Saha (✉) · A. Konar
Electronics and Telecommunication Engineering Department,
Jadavpur University, Kolkata, India
e-mail: sahasriparna@gmail.com

A. Konar
e-mail: konaramit@yahoo.co.in

R. Janarthanan
Computer Science and Engineering Department, TJS Engineering College, Chennai, India
e-mail: srmjana_73@yahoo.com

M.K. Chakraborty et al. (eds.), *Facets of Uncertainties and Applications*,
Springer Proceedings in Mathematics & Statistics 125,
DOI 10.1007/978-81-322-2301-6_13

Complex body gestures, such as pushing, hugging are successfully tracked using Kinect sensor.

Kinect sensor comprises with RGB camera [8, 9] and 3D depth sensor [10–12]. In [13], two-person interactions are simulated using Kinect sensor. This device is feasible as it is a low-cost device with high efficiency. But the problem of [13] is that the algorithm stated there does not incorporate view invariance interactions. All the two-persons interactions modelled there are parallel to the Kinect sensor. But in our novel work, each interaction is captured in three different angles (0°, 45° and 135°).

In the proposed work, eight two-person interactions are taken into account, namely approaching, departing, kicking, punching, hugging, shaking hands, exchanging and pushing. We have created a pentagon for each person for each frame of a sequence. The recognition of body gesture for each monocular frame [14–16] is the aim of this algorithm. Due to differences in interactions of different individuals based on age, sex and physical built, their gestures for a specific sequence vary greatly from each other. This leads to fuzziness in the input.

As the Kinect sensor models the human body using 20 body joint co-ordinates in three dimension, the pentagon vertices produced by the algorithm also have three-dimensional information. The vertices are configured with the help of average values from head, right hand, left hand, right leg and left leg for a single person. As we are modelling interactions for two persons, for each frame two dissimilar pentagons are configured. The Euclidean distances between similar vertices of two different persons are taken as features for this algorithm. For each interaction, 6 s stream of information of skeleton co-ordinates are taken and five pairs of persons have acted for the preparation of the dataset. For the recognition purpose, multi-class support vector machine (SVM) [17, 18] is utilised. Also a comparative study with multi-layered perceptron [19, 20] and k-nearest neighbours (k-NN) [21, 22] algorithms are performed. Experimentally, it is found that recognition rate for SVM outperforms for both the rotation invariance and rotation variance cases. Total time required for two-person interaction recognition is always less than 3 s using Matlab 2011b.

In this paper, Sect. 2 overviews the fundamental ideas about Kinect sensor, SVM, perceptron and k-NN. Section 3 clearly explains the proposed algorithm, whereas Sects. 4 and 5 illustrates the experimental results and performance analysis. Section 6 concludes with idea about future work.

2 Fundamental Ideas

The subsections below explain the Kinect sensor and multi-class support vector machine algorithm briefly.

Fig. 1 Kinect sensor

2.1 Kinect Sensor

Kinect sensor, consisting of RGB (red, green, blue) camera, infrared (IR) projector, IR camera and microphone, is capable of full human body tracking up to two persons at a time [5, 6]. It looks like a webcam as displayed in Fig. 1. It detects 3D representation of an object using depth sensor [10, 11], which consists of infrared laser [12]. Kinect sensor produces RGB video as the output using 8-bit VGA resolution camera [7]. It tracks the human body using 20 body joint co-ordinates within a finite amount of range 1.23.5 m [8, 9]. Voice gestures can also be recorded with the help of microphone array. A light emitting diode (LED) is present in front of the Kinect sensor to ensure that Kinect is running properly.

Kinect sensor brings forth 3D information about human. Thus 3D co-ordinates of 20 body joints are also received. Figure 2a demonstrates a human standing at 0° angle with Kinect sensor, while Fig. 2b, c clearly explains the scenario when the human is standing at an angle 45° and 135° with respect to the Kinect sensor. In our proposed work, we take account the interaction between two persons with 0° angle with respect to Kinect sensor, as well as the actions take place at two other angles with respect to Kinect sensor. When the two-person interaction encountered with parallel with Kinect sensor (i.e., with 0° angle), then the case is conducted as rotation invariance. With the change of angle, the case leads to rotation variance. In Fig. 2, pentagon which is created for the proposed work is marked with red dotted lines with black star vertices. The formation of pentagon is explained in Sect. 3.1.

2.2 Multi-class Support Vector Machine

Support vector machine (SVM) [17, 18], also known as support vector networks, is a supervised learning algorithm based on concept of dividing the set of inputs by a hyperplane. This algorithm is widely applicable in classification, regression analysis and pattern recognition problems. Here we have used this for classification of multi-class data points.

The simplest form of SVM is linear SVM, which works on the principle of separating two classes of data by constructing a hyper plane. These classes are specified

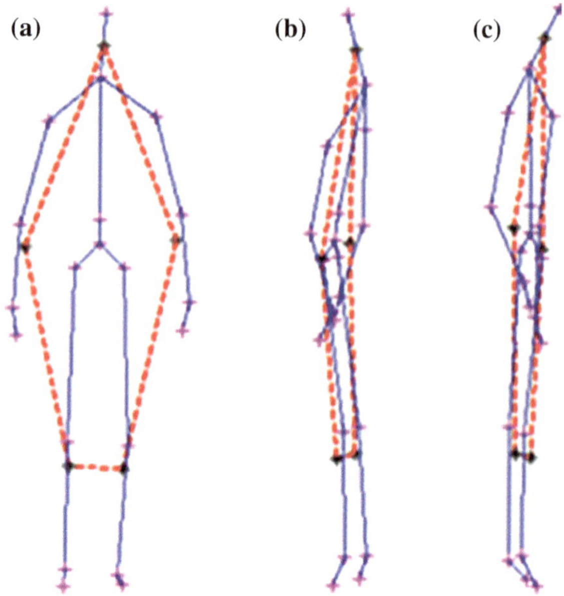

Fig. 2 Human skeleton in three different angles with Kinect sensor **a** 0°, **b** 45°, **c** 135°

by support vectors, within the training data points. The distance margin between the support vectors are taken into account and the aim is to maximise this distance. As linear SVM can be successfully used only where the data are linearly separable, this limitation can be overcome by mapping the data into a larger dimensional space using a kernel function, to make it linearly-separable. A frequently used kernel functions is the radial basis function kernel.

3 Proposed Algorithm

In the proposed algorithm, each person is modelled using a pentagon and the skeleton produced for the total interaction sequence is broke into frames. Thus the body gestures are extracted for each monocular frame [14–16].

3.1 Pentagon Formation

The five vertices of the pentagon are formed using (1)–(5). The Kinect sensor is capable of modelling human body during motion with 20 3D joint co-ordinates. Here we require 18 body joints information, i.e. spine and hip centre joints are neglected for this proposed work. The first vertex is formed by averaging head (H) and shoulder centre (SC) joints. The second vertex is created using mean values of shoulder right (SR), elbow right (ER), wrist right (WR) and hand right (HR). In the same way, mean values of shoulder left (SL), elbow left (EL), wrist left (WL) and hand left (HL) are noted to produce vertex 3. Vertices 4 and 5 are due to the average values of leg co-ordinates for right and left legs. Hip right (HR), knee right (KR), ankle right (AR) and foot right (FR) are taken into account for representation of vertex 4. Similarly, hip left (HL), knee left (KL), ankle left (AL) and foot left (FL) are marked and vertex 5 is obtained using the mean value of those joints.

$$vertex_1 = \frac{H + SC}{2} \tag{1}$$

$$vertex_2 = \frac{SR + ER + WR + HR}{4} \tag{2}$$

$$vertex_3 = \frac{SL + EL + WL + HL}{4} \tag{3}$$

$$vertex_4 = \frac{HR + KR + AR + FR}{4} \tag{4}$$

$$vertex_5 = \frac{HL + KL + AL + FL}{4} \tag{5}$$

As Kinect sensor creates 3D information for each skeleton joint, thus each vertex obtained using the above equations also have three dimensions. In Fig. 2, the calculated pentagon vertices using the above equations are shown using black stars and red dotted lines in the figure picturise the edges of the pentagons.

3.2 Calculation of Five Euclidean Distances

For each person, a definite pentagon is formed at the ith particular frame. Let the left and right persons are represented using L and R respectively. Then the Euclidean distance (ED) between vertex no j (which can be in between 1 and 5) is calculated by (6).

$$ED = \left\| L_{i,j} - R_{i,j} \right\| \tag{6}$$

4 Experimental Results

All the videos are conducted for 6 s duration. As Kinect sensor captures video at 30 frames/second rate, thus total 180 frames are processed for each interaction between two persons when the interaction is performed with 0° angle with respect to camera. We have carried out this experiment also for rotation invariant cases, i.e. with 45° and 135° angles with Kinect sensor. Thus we have for each interaction three different skeleton data. Five different pairs have participated in this proposed work. The starting and ending positions are neutral. Figures 3 and 4 demonstrates the eight instructions for frame no 50, 100 and 150. Twenty body joints for each person are marked using red stars. The skeletons of the persons are represented via blue lines, whereas green dotted lines describe the pentagon formed for each person. Black stars are the vertices of the pentagons. The red lines are the depiction of Euclidean distances between the vertices of the two pentagons, which are the essence of this paper. For all the interactions shown in Figs. 3 and 4, the right person is acting and the left person is reacting to the situation. Total dataset is broken into 4:1 ratio for testing and training purposes, respectively. Table 1 presents the experimental values obtained for five Euclidean distances for frame no 50, 100 and 150.

This system acquires a recognition rate of 93.7, 81.3 and 90.4 % with the virtue of multi-class support vector machine, multi-layered perceptron and k-nearest neighbours algorithm, respectively, when the actions are performed with 0° angle with Kinect sensor. When the angle of interaction varies with respect to kinect sensor, i.e. when the angle is 45° or 135°, then the average performance degrades to 81.3, 69.7 and 80.4 %, respectively, for SVM, perceptron and k-nearest neighbours.

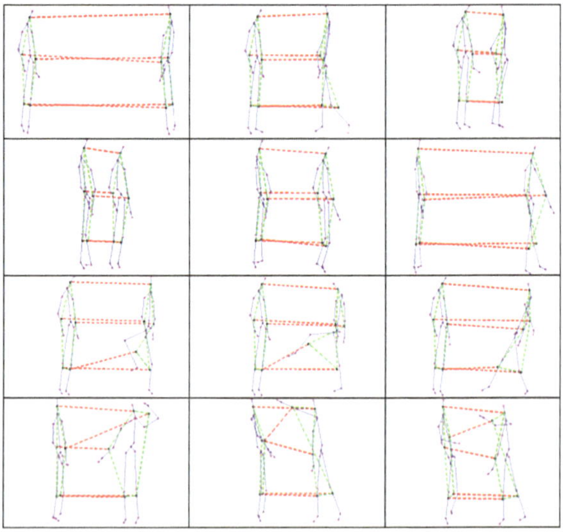

Fig. 3 Two person interaction modelling for approaching, departing, kicking and punching interactions

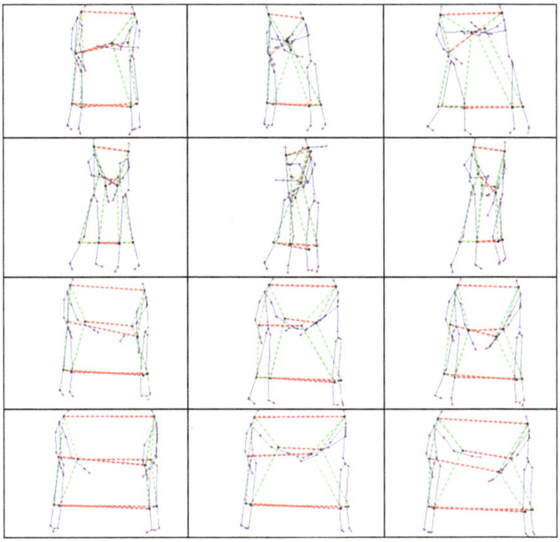

Fig. 4 Two person interaction modelling for hugging, shaking hands, exchanging and pushing interactions

Table 1 The result of five Euclidean distances for frame no 50, 100 and 150

Interaction	Frame no 50	Frame no 100	Frame no 150
Approaching	1.8254 1.7639 1.8385 1.8536 1.8701	1.5966 1.5704 1.6382 1.7058 1.5064	0.4663 0.5595 0.4822 0.4500 0.4758
Departing	0.5025 0.4567 0.5678 0.5227 0.5550	1.2556 1.1610 1.2340 1.1691 1.0693	1.8221 1.7328 1.9395 1.9082 1.7377
Kicking	1.1443 1.0948 1.2887 0.9518 1.2529	1.0961 1.1099 1.3010 0.7541 1.1838	1.2494 1.0866 1.4201 0.7646 1.2355
Punching	1.0687 1.2488 0.8550 1.0523 0.9950	0.8364 0.5797 0.9160 0.9285 0.8306	0.8520 0.6177 0.8544 0.8736 0.8059
Hugging	0.6978 0.6296 0.6832 0.7329 0.8629	0.5584 0.4213 0.5985 0.6992 0.6730	0.9491 0.4275 0.7425 0.7994 0.9934
Shaking Hands	0.4456 0.2101 0.5654 0.4409 0.5648	0.2841 0.1434 0.6302 0.3566 0.4889	0.5084 0.1890 0.6283 0.3732 0.6152
Exchanging	1.1194 0.8327 1.1592 1.1008 1.1855	0.8227 0.4646 0.5806 0.8967 1.0744	0.8275 0.6171 0.7957 0.8966 1.0728
Pushing	1.2713 1.2022 1.3092 1.2965 1.3638	1.2740 0.6237 0.8901 1.3449 1.3964	1.2260 1.0299 1.3138 1.2863 1.3465

Recognition rate comparison is picturised in Fig. 5. Here darker colour bars are for rotation invariance cases and lighter colour bars are for rotation variance cases.

Average computational time for SVM, perceptron and k-NN are 2.573, 2.820 and 2.934 s correspondingly in an Intel Pentium Dual Core processor running Matlab R011b for both the two cases of rotation invariance and rotation variance.

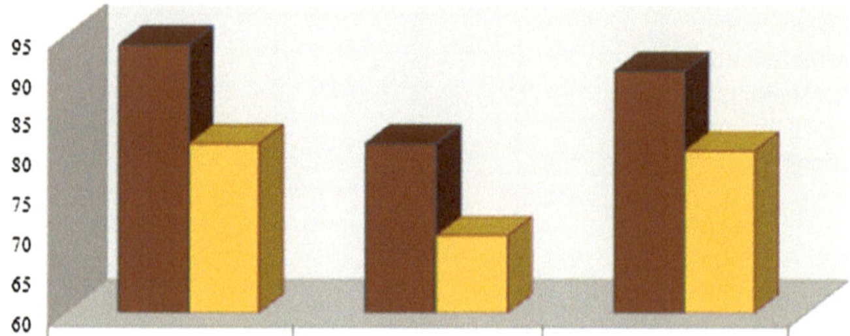

	SVM	Perceptron	k-NN
Rotation invariant	93.7	81.3	90.4
Rotation variant	81.3	69.7	80.4

Fig. 5 Accuracy comparison between SVM, perceptron and k-NN for rotation variance and invariance cases

In [13], two person interactions are modelled using Linear SVMs and multiple instance learning (MILBoost) with achieved maximum recognition of 87.6, 91.1 %. But in this work, the rotation variance is not taken into account. The proposed work in the paper, not only able to produce higher recognition rates than [13] for rotation invariance purpose, but also capable to manage good results, i.e. greater than 80 % for majority of the cases. We have taken 45° and 135° angles for rotation invariance cases. Thus the limitation of [13] is overcome with high efficiency and also with less timing complexity.

5 Performance Analysis

McNemar's Test [23] is used to judge two algorithms. Here, we assume multi-class SVM to be the reference algorithm (A) and compare it with either multi-layered perceptron and k-nearest neighbours (B) at a time using (7). n_{01} is number of samples mapped to a wrong cluster by algorithm A but not by B and n_{10} is number of samples mapped to a wrong cluster by algorithm B but not by A. The critical value of Z for 95 % confidence interval is 3.84 for one degree of freedom. According to Table 2, for both the cases the null-hypothesis is rejected. Hence, the algorithms are not equivalent. This validates our results.

$$Z = \left(|n_{01} - n_{10}| - 1\right)^2 / n_{01} + n_{10} \tag{7}$$

Table 2 Results of statistical test

Classifier used	n_{01}	n_{10}	Z	Comment
Multi-layered perceptron	17	63	0.5625	Reject
k-nearest neighbours	26	79	0.4952	Reject

6 Conclusion and Future Work

The proposed work is to recognise eight interactions between two persons. As the same gesture depicting a particular interaction varies widely across different persons, thus, the input is fuzzy in nature. Hence, multi-class support vector machine is employed.

Till date none of the papers acknowledge rotation invariance interaction modelling. Also Kinect sensor is implemented, so differences in weight, height and body types for different persons do not hamper the results. We have achieved a high accuracy of more than 80 % for all the cases. Hence this proposed work can easily find its place in surveillance purposes.

In future, we will concentrate on much more difficult interactions between two persons.

Acknowledgments We would like to thank University Grant Commission, India and University of Potential Excellence Program (Phase II) in Cognitive Science, Jadavpur University.

References

1. Aggarwal, J.K., Cai, Q.: Human motion analysis: a review. In: IEEE Proceedings of Nonrigid and Articulated Motion Workshop, 1997, pp. 90–102 (1997)
2. Yang, H.-D., Park, A.-Y., Lee, S.-W.: Gesture spotting and recognition for humanrobot interaction. IEEE Trans. Robot. **23**(2), 256–270 (2007)
3. Moeslund, T.B., Granum, E.: A survey of computer vision-based human motion capture. Comput. Vis. Image Underst. **81**(3), 231–268 (2001)
4. Wang, L., Hu, W., Tan, T.: Recent developments in human motion analysis. Pattern Recognit. **36**(3), 585–601 (2003)
5. Leyvand, T., Meekhof, C., Wei, Y.-C., Sun, J., Guo, B.: Kinect identity: technology and experience. Computer **44**(4), 94–96 (2011)
6. Solaro, J.: The kinect digital out-of-box experience. Computer **44**, 97–99 (2011)
7. Clark, J.: Object digitization for everyone. Computer **44**, 81–83 (2011)
8. Zhang, Z.: Microsoft kinect sensor and its effect. IEEE Multimed. **19**(2), 410 (2012)
9. Ryden, F.: Tech to the future: making a "Kinection" with haptic interaction. IEEE Potentials **31**(3), 34–36 (2012)

10. Cho, J.-H., Lee, K.H., Aizawa, K.: Enhancement of depth maps with alpha channel estimation for 3-D video. IEEE J. Sel. Top. Sig. Process. **6**(5), 483–494 (2012)
11. Tong, J., Zhou, J., Liu, L., Pan, Z., Yan, H.: Scanning 3d full human bodies using kinects. IEEE Trans. Vis. Comput. Graph. **18**(4), 643–650 (2012)
12. Herrera, C., Kannala, J.: Joint depth and color camera calibration with distortion correction. IEEE Trans. Pattern Anal. Mach. Intell. **34**(10), 2058–2064 (2012)
13. Yun, K., Honorio, J., Chattopadhyay, D., Berg, T.l., Samaras, D.: Two-person interaction detection using body-pose features and multiple instance learning. In: 2012 IEEE Computer Society Conference on Computer Vision and Pattern Recognition Workshops (CVPRW), pp. 28–35 (2012)
14. Agarwal, A., Triggs, B.: Recovering 3D human pose from monocular images. IEEE Trans. Pattern Anal. Mach. Intell. **28**(1), 4458 (2006)
15. Wachter, S., Nagel, H.-H.: Tracking of persons in monocular image sequences. In: IEEE Proceedings of Nonrigid and Articulated Motion Workshop, 1997, p. 29 (1997)
16. Snidaro, L., Foresti, G.L., Chittaro, L.: Tracking human motion from monocular sequences. Int. J. Image Graph. **8**(03), 455471 (2008)
17. Tax, D.M.J., Duin, R.P.W.: Support vector domain description. Pattern Recognit. Lett. **20**(11), 1191–1199 (1999)
18. Theodoridis, S., Pikrakis, A., Koutroumbas, K., Cavouras, D.: Introduction to Pattern Recognition: A Matlab Approach: A Matlab Approach. Academic Press, New York (2010)
19. Konar, A.: Computational Intelligence: Principles, Techniques and Applications. Springer, The Netherlands (2005)
20. Duin, R.P.W., Juszczak, P., Paclik, P., Pekalska, E., De Ridder, D., Tax, D.M.J., Verzakov, S.: A matlab toolbox for pattern recognition. PRTools version, vol. 3 (2000)
21. Coomans, D., Massart, D.L.: Alternative k-nearest neighbour rules in supervised pattern recognition. Fresenius. Z. Anal. Chem. **313**, 15–27 (1982)
22. Cover, T., Hart, P.: Nearest neighbor pattern classification. IEEE Trans. Inf. Theory **13**(1), 21–27 (1967)
23. Dietterich, T.G.: Approximate statistical tests for comparing supervised classification learning algorithms. Neural Comput. **10**(7), 1895–1923 (1998)

An Improved Genetic Algorithm and Its Application in Constrained Solid TSP in Uncertain Environments

Monoranjan Maiti, Samir Maity and Arindam Roy

Abstract In this paper, we propose an improved genetic algorithm (IGA) to solve Constrained Solid Travelling Salesman Problems (CSTSPs) in crisp, fuzzy, rough, and fuzzy-rough environments. The proposed algorithm is a combination of probabilistic selection, cyclic crossover, and nodes-oriented random mutation. Here, CSTSPs in different uncertain environments have been designed and solved by the proposed algorithm. A CSTSP is usually a travelling salesman problem (TSP) where the salesman visits all cities using any one of the conveyances available at each city under a constraint say, safety constraint. Here a number of conveyances are used for travel from one city to another. In the present problem, there are some risks of travelling between the cities through different conveyances. The salesman desires to maintain certain safety level always to travel from one city to another and a total safety for his entire tour. Costs and safety level factors for travelling between the cities are different. The requirement of minimum safety level is expressed in the form of a constraint. The safety factors are expressed by crisp, fuzzy, rough, and fuzzy-rough numbers. The problems are formulated as minimization problems of total cost subject to crisp, fuzzy, rough, or fuzzy-rough constraints. This problem is numerically illustrated with appropriate data values. Optimum results for the different problems are presented via IGA. Moreover, the problems from the TSPLIB (standard data set) are tested with the proposed algorithm.

M. Maiti (✉)
Department of Applied Mathematics and Oceanology, Vidyasagar University, Medinipur, West Bengal, India
e-mail: mmaiti2005@yahoo.co.in

S. Maity
Department of Computer Science, Vidyasagar University, Medinipur, West Bengal, India
e-mail: maitysamir13@gmail.com

A. Roy
Department of Computer Science, P.K. College, Contai, Purba Mednipur, West Bengal, India
e-mail: royarindamroy@yahoo.com

M.K. Chakraborty et al. (eds.), *Facets of Uncertainties and Applications*,
Springer Proceedings in Mathematics & Statistics 125,
DOI 10.1007/978-81-322-2301-6_14

Keywords Travelling Salesman Problem (TSP) · Solid Travelling Salesman Problem (STSP) · Constraint Solid Travelling Salesman Problem (CSTSP) · Genetic Algorithm (GA) · Probabilistic selection

1 Introduction

Soft Computing (SC) is an association of computing methodologies that includes fuzzy set, fuzzy logic, neuro-computing, evolutionary computing, and probabilistic computing. Soft computing (SC) is a term originally coined by Zadeh to denote the systems that "… exploit the tolerance for imprecision, uncertainty, and partial truth to achieve tractability, robustness, low solution cost, and better rapport with reality" [1]. Traditionally, SC has been comprised of four technical disciplines. The first two, probabilistic reasoning (PR) and fuzzy logic (FL) reasoning systems, are based on knowledge driven reasoning. The other two technical disciplines, neuro computing (NC) and evolutionary computing (EC), are data-driven search and optimization approaches [2]. Here we propose an IGA which is the data-driven searching SC technique.

The Travelling Salesman Problem (TSP) is a well-known NP hard combinatorial optimization problem [3]. In TSP, it is assumed that a salesman travels from one city to another using only one conveyance. But in real-life TSP, a set of conveyances are available at each city. In this case, a salesman has to design his tour for minimum cost maintaining the TSP conditions and using the suitable different conveyances at different cities. This problem is called Solid Travelling Salesman Problem (STSP). Till now, except for one or two researchers [4], none has considered such type of STSP.

In real life, travelling costs from one city to another city depend on the conveyances and its nature can be uncertain, i.e., fuzzy, rough, fuzzy-rough, etc. Due to the conditions of the roads and the vehicles which the salesman chooses, travelling costs are thus normally uncertain in the nonstochastic sense.

Nowadays, travel from one city to another always involves some risks. There are more risks in terrorist influenced areas, hill areas, etc., and also during rainy season. Hence, a salesman will always desire to travel from one city to another using a particular conveyance with minimum safety and for his complete tour also, total minimum safety is ensured. Again, it is difficult to measure the said risk/safety factor precisely in deterministic form. Thus these values are imprecise, i.e., can be fuzzy, rough, fuzzy-rough. Till now, even the usual TSPs have not been formulated and solved using the said imprecise safety constraints.

The present problem under investigation is more complicated for its imprecise costs and constraints on safety level. Fuzziness of the cost and safety level leads to fuzzy total-cost with a fuzzy constraint. As fuzzy objective function of an optimization problem is not well-defined, it is difficult to find optimal paths for the stated problem. Due to this complexity, a fuzzy possibility/necessity based approach is proposed to transfer fuzzy objective into an equivalent crisp objective. Similarly for rough objectives and constraints, trust measure is used following Xu and Zhao [5]. Both expectation and trust measure are applied to convert the fuzzy-rough objective and constraints into corresponding deterministic forms following Xu and Zhao [6].

In the existing literature, there are several soft computing methods such as Nearest Neighborhoods Search (NNS) [7], Simulated Annealing (SA) [8], Tabu Search (TS) [9], Ant Colony System (ACS) [10], and Genetic Algorithm (GA) [11], Particle Swarm Optimization (PSO) [12], etc. In this paper we have developed an improved genetic algorithm (IGA) based on probabilistic selection, cyclic crossover, and node dependent random mutation. Here CSTSPs are formulated in crisp, fuzzy, rough, and fuzzy-rough environments. The travelling costs and safety factors along the different routes are crisp, fuzzy, rough, and fuzzy-rough numbers. The goal of the salesman is to minimize the total travelling cost having a minimum total safety level for the entire tour. Thus the models are formulated as the cost minimization problems with constraint. Here the objectives and constraints are crisp, fuzzy, rough, or fuzzy-rough numbers. The imprecise objectives and constraint are made deterministic using the appropriate technique as mentioned above. The reduced optimization problems are solved by IGA. Some test problems TSPLIB [13] are solved using the proposed IGA and usual GA with Roulette wheel selection, arithmetic crossover, and random mutation. The crisp, fuzzy [14], rough, and fuzzy-rough [16] CSTSPs are illustrated with numerical examples and the optimal results obtained by IGA are presented. Numerical results show the efficiency of our improved algorithm.

2 Mathematical Preliminaries

Let $\tilde{a}$ and $\tilde{b}$ be two fuzzy numbers with membership functions $\mu_{\tilde{a}}(x)$ and $\mu_{\tilde{b}}(x)$ respectively. Then according to [14],

(1) $pos(\tilde{a} * \tilde{b}) = \sup\{\min(\mu_{\tilde{a}}(x), \mu_{\tilde{b}}(y)), x, y \in R, x * y\}$ where the abbreviation pos represents possibility, $*$ is any one of the relations $>, <, =, <=, >=$, and R represents set of real numbers.

(2) $nes(\tilde{a} * \tilde{b}) = 1 - pos(\tilde{a} * \tilde{b})$ where the abbreviation nes represents necessity. If $\tilde{a}, \tilde{b}$ are sub set of R and $\tilde{c} = f(\tilde{a}, \tilde{b})$ where $f : \mathcal{R} \times \mathcal{R} \rightarrow \mathcal{R}$ is a binary operation then membership function $\mu_{\tilde{c}}$ of $\tilde{c}$ is defined as [13].

(3) For each $z \in R$, $\mu_{\tilde{c}} = \sup\{\min(\mu_{\tilde{a}}(x), \mu_{\tilde{b}}(y)), x, y \in R \text{ and } z = f(x, y)\}$

2.1 Triangular Fuzzy Number (TFN)

A TFN $\tilde{a} = (a_1, a_2, a_3)$ (cf. Fig. 1) has three parameters a_1, a_2, a_3 where $a_1 < a_2 < a_3$ and is characterized by the membership function $\mu_{\tilde{a}}$, given as

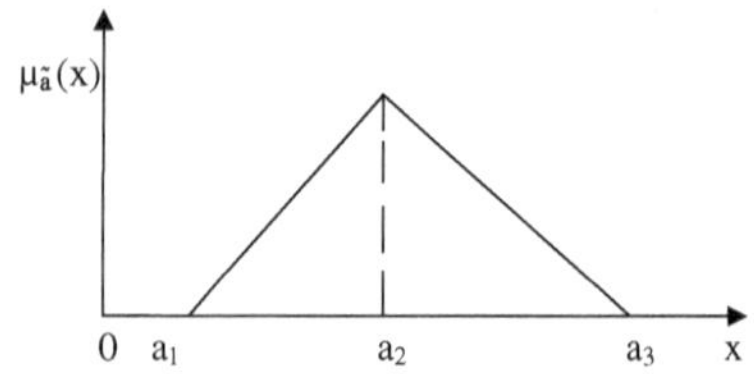

Fig. 1 Triangular fuzzy number $\tilde{a} = (a_1, a_2, a_3)$

$$\mu_{\tilde{a}}(x) = \begin{Bmatrix} \frac{x-a1}{a2-a1} & \text{for } a_1 \le x \le a_2 \\ \frac{a3-x}{a3-a2} & \text{for } a_2 \le x \le a_3 \\ 0 & \text{otherwise} \end{Bmatrix} \quad (1)$$

According to the above definitions the following lemmas can easily be derived.

Lemma 2.1.1 *If* $\tilde{a} = (a_1, a_2, a_3)$ *be a TFN with* $0 < a_1$ *and* b *is a crisp number then* $pos(\tilde{a} < b) \ge \alpha$ iff $\frac{b-a1}{a2-a1} \ge \alpha$.

Lemma 2.1.2 *If* $\tilde{a} = (a_1, a_2, a_3)$ *be a TFN with* $0 < a_1$ *and* b *is a crisp number then* $nes(\tilde{a} < b) \ge \alpha$ iff $\frac{a_3-b}{a_3-a_2} \le 1-\alpha$.

Lemma 2.1.3 *If* $\tilde{a} = (a_1, a_2, a_3)$ *and* $\tilde{b} = (b_1, b_2, b_3)$ *be TFNs with* $0 < a_1$ *and* $0 < b_1$ *then* $pos\,(\tilde{a} > \tilde{b}) \ge \alpha$ iff $\frac{a_3-b_1}{a_3-a_2+b_2-b_1} \ge \alpha$.

Lemma 2.1.4 *If* $\tilde{a} = (a_1, a_2, a_3)$ *and* $\tilde{b} = (b_1, b_2, b_3)$ *be TFNs with* $0 < a_1$ *and* $0 < b_1$ *then* $nes\,(\tilde{a} > \tilde{b}) \ge \alpha$ iff $\frac{b3-a1}{a2-a1+b3-b2} \le 1-\alpha$.

2.2 Rough Set Theory

In this section, we will state some basic concepts and theorems which have been cited on rough theory [5]. These results are crucial for the remainder of this paper. Credibility theory is a branch of Mathematics that studies the behavior of fuzzy phenomena. The fuzzy variable has been defined in many ways.

Definition 2.2.1 Let Θ be a nonempty set, $P(\Theta)$ the power set of Θ, and pos a possibility measure. Then the triplet $(\Theta, P(\Theta), pos)$ is called a possibility space. A fuzzy variable is a function from a possibility space $(\Theta, P(\Theta), pos)$ to the real line R.

Definition 2.2.2 Let Λ be a nonempty set, $\mathcal{A}$ a σ algebra of subsets of Λ, and Δ an element in $\mathcal{A}$, and Π a trust measure. Then $(\Lambda, \Delta, \mathcal{A}, \Pi)$ is called a rough space. A rough variable ξ is a measurable function from the rough space $(\Lambda, \Delta, \mathcal{A}, \Pi)$ to the set of real numbers. That is, for every Borel set B of R, we have $\{\lambda \in \Lambda | \xi(\lambda) \in B\} \in \mathcal{A}$.

The lower and the upper approximations of the rough variable are defined as follows:

$$\underline{\xi}_i = \{\xi(\lambda) | \lambda \in \Delta\}, \qquad \overline{\xi} = \{\xi(\lambda) | \lambda \in \Lambda\}$$

$\underline{\xi}$ is the lower approximation of the rough variable ξ and $\overline{\xi}$ is the upper approximation of the rough variable ξ.

Definition 2.2.3 Let $(\Lambda, \Delta, \mathcal{A}, \Pi)$ be a rough space . The upper trust of an event A is defined as

$$\overline{\text{Tr}}\{A\} = \frac{\Pi\{A\}}{\Pi\{\Lambda\}};$$

The lower trust of the event is defined as

$$\underline{\text{Tr}}\{A\} = \frac{\Pi\{A \cap \Delta\}}{\Pi\{\Delta\}}.$$

The trust of the event is defined as

$$\text{Tr}\{A\} = \frac{1}{2}(\overline{\text{Tr}}\{A\} + \underline{\text{Tr}}\{A\}) \tag{2}$$

When we do not have enough information to determine the measure of Π for a real-life problem, we can assume that all elements in Λ are equally likely to occur. In this case, the measure Π may be viewed as the Lebesgue measure. Then we can get the trust measure of the rough event $\xi\hat{} \geq$ t, $\text{Tr}\{\xi\hat{} \geq t\}$ and its function as presented below where t is a crisp number, ξ is a rough variable given by $\xi\hat{} = ([a, b][c, d])$, $0 \leq c \leq a \leq b \leq d$

$$\text{Tr}\{\tilde{\xi} \geq \delta\} = \begin{cases} 0 & \text{for } d \leq t \\ \frac{(d-t)}{2(d-c)} & \text{for } b \leq t \leq d \\ \frac{1}{2}\left(\frac{(d-t)}{(d-c)} + \frac{b-t}{b-a}\right) & \text{for } a \leq t \leq b \\ \frac{1}{2}\left(\frac{(d-t)}{(d-c)} + 1\right) & \text{for } c \leq t \leq a \\ 1 & \text{for } t \leq c \end{cases} \tag{3}$$

And the rough expectation is $E[\tilde{\xi}] = \frac{1}{4}(a + b + c + d)$.

2.3 *Fuzzy-Rough Variable*

Definition 2.3.1 A fuzzy rough variable is a measurable function from a rough space $(\Lambda, \Delta, \mathcal{A}, \Pi)$ to the set of fuzzy variables such that $\text{pos}\{\xi(\lambda) \in B\}$ is a measurable function of λ for any Borel set B of $\mathbb{R}$. Generally speaking, a fuzzy rough variable is a rough variable taking fuzzy values [6, 16].

Definition 2.3.2 An n-dimensional fuzzy rough vector is a function ξ from a rough space $(\Lambda, \Delta, \mathcal{A}, \Pi)$ to the set of n-dimensional fuzzy vectors such that $\mathrm{Pos}\{\xi(\lambda) \in \mathrm{B}\}$ is a measurable function of λ for any Borel set B of $\mathcal{R}^n$.

Definition 2.3.3 Let $f : \mathcal{R}^n \mapsto \mathcal{R}$ be a function, and $\xi_1, \xi_2, \ldots, \xi_n$ be fuzzy rough variables defined on $(\Lambda, \Delta, \mathcal{A}, \Pi)$, respectively.

Then $\xi = f(\xi_1, \xi_2, \ldots, \xi_n)$ is a fuzzy rough variable defined as:

$$\xi(\lambda) = f(\xi_1(\lambda_1), \xi_2(\lambda_2), \ldots, \xi_n(\lambda_n)), \quad \text{for any } (\lambda_1, \lambda_2, \ldots, \lambda_n) \in \Lambda.$$

Definition 2.3.4 Let $f : \mathcal{R}^n \mapsto \mathcal{R}$ be a function, and ξ_i are fuzzy rough variables defined on $(\Lambda_i, \Delta_i, \mathcal{A}_i, \Pi_i), i = 1, 2, 3, \ldots, n$, respectively. Then $\xi = f(\xi_1, \xi_2, \ldots \xi_n)$ is a fuzzy rough variable defined on the product rough space $(\Lambda, \Delta, \mathcal{A}, \Pi)$ as:

$$\xi(\lambda_1, \lambda_2, \ldots, \lambda_n) = f(\xi_1(\lambda_1), \xi_2(\lambda_2), \ldots, \xi_n(\lambda_n)), \text{ for any } (\lambda_1, \lambda_2, \ldots, \lambda_n) \in \Lambda.$$

Definition 2.3.5 Let $\xi = (\xi_1, \xi_2, \ldots, \xi_n)$ be a fuzzy rough vector on the rough space $(\Lambda, \Delta, \mathcal{A}, \Pi)$, and $g_k : \mathcal{R}^n \to \mathcal{R}$ be continuous functions, $k = 1, 2, \ldots, p$. Then the primitive chance of a fuzzy rough event characterized by $g_k(\xi) \le 0, k = 1, 2, \ldots, p$ is a function from [0, 1] to [0, 1], defined as:

$$\begin{aligned} &\mathrm{Ch}\{g_k(\xi) \le 0, k = 1, 2, \ldots, p\}(\alpha) \\ &\quad = \sup\{\beta / \mathrm{Tr}\{\lambda \in \Lambda / \mathrm{pos}\{g_k(\xi) \le 0, k = 1, 2, \ldots, p\} \ge \beta\} \ge \alpha\}. \end{aligned}$$

Lemma 2.3.5.1 *Assume that ξ is a fuzzy rough vector, i.e., with the n-tuple of fuzzy rough variables $(\xi_1, \xi_2, \ldots, \xi_n)$ and g_k are real-valued continuous functions for $k = 1, 2, \ldots, p$. Then the possibility $\mathrm{pos}\{g_k(\xi(\lambda)) \le 0, k = 1, 2, \ldots, p\}$ is a rough variable.*

Definition 2.3.6 Let $\tilde{\xi}\hat{}$ be a fuzzy rough variable. The expected value of the fuzzy rough variable $\tilde{\xi}\hat{}$ is denoted by $E[\tilde{\xi}\hat{}]$ and defined by

$$\mathbf{E}[\tilde{\xi}\hat{}] = \int_0^{\infty} \boldsymbol{Tr}(\boldsymbol{\lambda} \in \boldsymbol{\Lambda} / \mathbf{E}[\tilde{\boldsymbol{\xi}}(\boldsymbol{\lambda})] \ge \mathbf{r})\boldsymbol{dr} - \int_{-\infty}^{\mathbf{0}} \boldsymbol{Tr}(\boldsymbol{\lambda} \in \boldsymbol{\Lambda} / \mathbf{E}[\tilde{\boldsymbol{\xi}}(\boldsymbol{\lambda})] \le \mathbf{r})\mathbf{dr}$$

Lemma 2.3.6.1 *Let $\tilde{\xi}\hat{} = (\xi\hat{} - L, \xi\hat{}, \xi\hat{} + R)$ be a fuzzy rough variable, where $\xi\hat{} = ([a, b][c, d])$ is a rough variable. The expected value of $\tilde{\xi}\hat{}$ is $\mathbf{E}[\boldsymbol{\xi}\hat{}] = \frac{\mathbf{1}}{\mathbf{4}}[\boldsymbol{a} + \boldsymbol{b} + \boldsymbol{c} + \boldsymbol{d}] + \frac{\boldsymbol{\rho}\mathcal{R} - (\mathbf{1} - \boldsymbol{\rho})\boldsymbol{L}}{\mathbf{2}}$ where $0 \le \rho \le 1$.*

Where L, R are crisp numbers, the lower level and upper level of fuzzy variables.

3 Problem Definition

3.1 General TSP with Safety Constraints

In a classical two-dimensional travelling salesman problem, a salesman has to travel N cities using minimum cost. In his tour salesman starts from a city, visits all the cities exactly once, and comes to the starting city using minimum cost. The salesman should choose a path in which the minimum safety values are ensured. Let c(i, j) be the cost for travelling from ith city to jth city and s(i, j) be the travel comfort in travelling from ith city to jth city. Then the problem can be mathematically formulated as:

$$\left.\begin{aligned} &\text{Minimize } Z = \sum_{i=1}^{N} \sum_{j=1}^{N} t_{ij}\, c(i,j) \\ &\text{Subject to } \sum_{i=1}^{N} t_{ij} = 1 \quad \text{for } j = 1, 2, \ldots, N \\ &\sum_{j=1}^{N} t_{ij} = 1 \quad \text{for } i = 1, 2, \ldots, N \\ &\sum_{i=1}^{N} \sum_{j=1}^{N} t_{ij}\, S(i,j) \geq S_{min} \end{aligned}\right\} \quad (4)$$

where $t_{ij} = 1$ if the salesman travels from city-i to city-j, otherwise $t_{ij} = 0$ and S_{min} is the minimum safety level that should be maintained by the salesman. Let $(x_1, x_2, \ldots, x_N, x_1)$ be a complete tour of a salesman, where x_i €$\{1, 2, \ldots, N\}$ for $i = 1, 2, \ldots, N$ and all x_i are distinct. Then the above problem reduces to

$$\left.\begin{aligned} &\text{Determine a complete tour } (x_1, x_2, \ldots, x_N, x_1) \\ &\text{Minimize } Z = \sum_{i=1}^{N-1} c(x_i, x_{i+1}) + c(x_N, x_1) \\ &\text{Subject to } \sum_{i=1}^{N=1} S(x_i, x_{i+1}) + S(x_N, x_1) \geq S_{min} \end{aligned}\right\} \quad (5)$$

3.2 Solid TSP with Safety Constraints

In a solid travelling salesman problem, a salesman has to travel N cities by choosing any one of the M conveyances available using minimum cost. In his tour the salesman starts from a city, visits all the cities exactly once using suitable conveyances available at the cities, and comes to the starting city using minimum cost. Safety factors in travelling from one city to another using different conveyances are different. The salesman should choose such a path and conveyance that a minimum safety level is maintained. Let c(i, j, k) be the cost for travelling from ith city to jth city using kth type conveyance and s(i, j, k) be the safety level in travelling from ith city to jth using kth type conveyance. Then the salesman has to determine a complete tour $(x_1, x_2, \ldots, x_N, x_1)$ in which a particular or different combinations of conveyance types $(v_1, v_2, \ldots, v_p)$ are to be used for the tour, where x_i €$\{1, 2, \ldots, N\}$ for i =

$1, 2, \ldots, N$, $v_k \in \{1, 2, \ldots, P\}$ for $k = 1, 2, \ldots, P$ and all x_i's are distinct. Then the problem can be mathematically formulated as:

Determine a complete tour $(x_1, x_2, \ldots, x_N, x_1)$ and corresponding conveyance types $(v_1, v_2, \ldots, v_p)$

$$\left.\begin{array}{r}\text{to minimize } Z = \sum_{i=1}^{N-1} c(x_i, x_{i+1}, v_i) + c(x_N, x_1, v_k),\\ \text{Subject to} \sum_{i=1}^{N-1} S(x_i, x_{i+1}, v_i) + S(x_N, x_1, v_k) \geq S_{min,}\end{array}\right\} \quad (6)$$

where $(v_i, v_k) \in \{1, 2, \ldots \text{or } P\}$, S_{min} is the minimum safety level attained by the salesman.

3.3 Constrained Solid TSP with Fuzzy Costs

In the above problem if costs and safety factors are fuzzy numbers, i.e., $\tilde{c}(i, j, k)$ and $\tilde{s}(i, j, k)$, respectively, and safety level limit S_{min} also fuzzy number $\tilde{S}_{min}$, the above problem using article 2.1 reduces to determine a complete tour $(x_1, x_2, \ldots, x_N, x_1)$ and corresponding conveyance $(v_1, v_2, \ldots, v_p)$

$$\left.\begin{array}{l}\text{to minimize } Z = \sum_{i=1}^{N-1} \tilde{c}(x_i, x_{i+1}, v_i) + \tilde{c}(x_N, x_1, v_p), (v_i, v_k) \in \{1, 2, \ldots \text{or } P\}\\ \text{subject to} \sum_{i=1}^{N-1} \tilde{s}(x_i, x_{i+1}, v_i) + \tilde{s}(x_N, x_1, v_p) \geq \tilde{S}_{min}\end{array}\right\}$$

As minimization of fuzzy objective as well as fuzzy constraints are not well-defined the above problem can be rewritten in optimistic sense by article-2, respectively, to determine a complete tour $(x_1, x_2, \ldots, x_N, x_1)$ and corresponding conveyance $(v_1, v_2, \ldots, v_p)$ to minimize F

$$\left.\begin{array}{r}\text{Subject to pos}\left(\sum_{i=1}^{N-1} \tilde{c}(x_i, x_{i+1}, v_i) + \tilde{c}(x_N, x_1, v_k) < F\right) \geq \alpha_1\\ \text{pos}\left(\sum_{i=1}^{N-1} \tilde{s}(x_i, x_{i+1}, v_i) + \tilde{s}(x_N, x_1, v_k) \geq \tilde{S}_{min}\right) \geq \beta_1\end{array}\right\} \quad (7)$$

and in pessimistic sense to determine a complete tour $(x_1, x_2, \ldots, x_N, x_1)$ and corresponding conveyance $(v_1, v_2, \ldots, v_p)$ to minimize F

$$\left.\begin{array}{r}\text{Subject to nes}\left(\sum_{i=1}^{N-1} \tilde{c}(x_i, x_{i+1}, v_i) + \tilde{c}(x_N, x_1, v_i) < F\right) \geq \alpha_2\\ \text{nes}\left(\sum_{i=1}^{N-1} \tilde{s}(x_i, x_{i+1}, v_i) + \tilde{s}(x_N, x_1, v_i) \geq \tilde{S}_{min}\right) \geq \beta_2\end{array}\right\} \quad (8)$$

where $\alpha_1, \beta_1, \alpha_2$ and β_2 are predefined levels of possibility and necessity, respectively, which are entirely determined by the salesman. The meaning of $\alpha_1, \beta_1, \alpha_2$ and β_2 are given in article-2, and F is any crisp parameter. It is clear that minimization of

F implies minimization of fuzzy objective $\tilde{Z}$. For this reason the above approach is used to treat fuzzy objective $\tilde{Z}$.

If we consider the fuzzy numbers as TFNs, $\tilde{c}(i,j,k) = (c(i,j,k)_1, c(i,j,k)_2, c(i,j,k)_3)$, $\tilde{s}(i,j,k) = (s(i,j,k)_1, s(i,j,k)_2, s(i,j,k)_3)$ and $\tilde{S}_{min} = (s_1, s_2, s_3)$, then the above (7) and (8) equations can be written as follows to determine a complete tour $(x_1, x_2, \ldots, x_N, x_1)$ and corresponding conveyance $(v_1, v_2, \ldots, v_p)$ to minimize F

$$\left.\begin{array}{l} \text{Subject to } \frac{F-c1}{c2-c1} \geq \alpha_1 \\ \frac{S3-s1}{S3-S2+s2-s1} \geq \beta_1 \end{array}\right\} \quad (9)$$

$$\text{Where } C_j = \sum_{i=1}^{N-1} c(x_1, x_{i+1}, v_i)_j + c(x_N, x_1, v_i)_j, j = 1, 2, 3$$

$$S_j = \sum_{i=1}^{N-1} s(x_i, x_{i+1}, v_i)_j + s(x_N, x_1, v_k)_j, j = 1, 2, 3$$

To determine a complete tour $(x_1, x_2, \ldots, x_N, x_1)$ and corresponding conveyance $(v_1, v_2, \ldots, v_p)$ to minimize F

$$\left.\begin{array}{l} \text{Subject to } \frac{C3-F}{C3-C2} \geq (1-\alpha_2) \\ \frac{s3-S1}{S2-S1+s3-s2} \leq (1-\beta_2) \end{array}\right\} \quad (10)$$

Thus the condition for the fuzzy case is given to determine a complete tour $(x_1, x_2, \ldots, x_N, x_1)$ and corresponding conveyance $(v_1, v_2, \ldots, v_p)$

$$\left.\begin{array}{l} \text{to minimize } C_1 + \alpha_1(C_2 - C_1) \\ \text{subject to } \frac{S3-s1}{S3-S2+s2-s1} \geq \beta_1 \end{array}\right\} \quad (11)$$

To determine a complete tour $(x_1, x_2, \ldots, x_N, x_1)$ and corresponding conveyance $(v_1, v_2, \ldots, v_p)$

$$\left.\begin{array}{l} \text{to minimize } C_3 - (1-\alpha_2)(C_3 - C_2) \\ \text{subject to } \frac{s3-S1}{S2-S1+s3-s2} \leq (1-\beta_2) \end{array}\right\} \quad (12)$$

If the salesman is most optimistic then he will choose value of α_1, β_1 nearly 0 and in that case minimum possible cost function (C_1) is minimized assuming maximum possible safety level attained of the tour (S_3) reaches the minimum possible safety level requirement (s_1). On the other hand if the salesman is least optimistic then he/she will choose values of α_1, β_1 nearly 1 and in that case most feasible cost function (C_2) is minimized assuming most feasible safety level of the tour (S_2) reaches the most feasible safety level requirement (s_2). The pessimistic salesman will go for the necessity approach. If he is most pessimistic, he will choose values of α_2, β_2 nearly 1 and in that case maximum possible cost function (C_3) is minimized assuming minimum possible safety level of the tour (S_1) reaches the maximum possible safety level requirement (s_3). On the other hand if the salesman is least pessimistic then he/she will choose value of α_2, β_2 nearly 0 and in that case the most feasible cost function (C_2) is minimized assuming the most feasible safety level of the tour (S_2) reaches the most feasible safety level requirement (s_2).

3.4 Constraint Solid TSP with Rough Costs

In CSTSP if costs and safety factors are rough numbers, i.e., $c\hat{}(i, j, k)$ and $s\hat{}(i, j, k)$, respectively, and safety effect limit S_{min} also rough number $\hat{S}_{min}$, then CSTSP reduces to determine a complete tour $(x_1, x_2, \ldots, x_N, x_1)$ and corresponding conveyance $(v_1, v_2, \ldots, v_p)$

$$\text{to minimize } Z\hat{} = \sum_{i=1}^{N-1} c\hat{}(x_i, x_{i+1}, v_i) + c\hat{}(x_N, x_1, v_k),\ (v_i, v_k) \in \{1, 2, \ldots \text{or } P\}$$
$$\text{subject to } \sum_{i=1}^{N-1} s\hat{}(x_i, x_{i+1}, v_i) + s\hat{}(x_N, x_1, v_k) \geq S\hat{}_{min}$$

As minimization of rough objective as well as rough constraints are not well-defined the above problem can be rewritten in trust measure by article-2.6, respectively, to determine a complete tour $(x_1, x_2, \ldots, x_N, x_1)$ and corresponding conveyance $(v_1, v_2, \ldots, v_p)$ to minimize T

$$\text{Subject to Tr}\left(\sum_{i=1}^{N-1} c\hat{}(x_i, x_{i+1}, v_i) + c\hat{}(x_N, x_1, v_k) < T\right) \geq \delta$$
$$\text{Tr}\left(\sum_{i=1}^{N-1} s\hat{}(x_i, x_{i+1}, v_i) + s\hat{}(x_N, x_1, v_k) \geq S\hat{}_{min}\right) \geq \omega$$

where T, δ, ω are any crisp parameters. It is clear that minimization of T implies the minimization of rough objective $Z\hat{}$. For this reason the above maintained approach is used to treat rough objective $Z\hat{}$.

To determine trust measure we generate the rough number from the above problem as

$$c\hat{}(i, j, k) = ([c(i, j, k)_1, c(i, j, k)_2][c(i, j, k)_3, c(i, j, k)_4]),$$
$$s\hat{}(i, j, k) = ([s(i, j, k)_1, s(i, j, k)_2][s(i, j, k)_3, s(i, j, k)_{4}]),$$
$$s\hat{}_{min} = ([s_{min1}, s_{min2}][s_{min3}, s_{min4}]),$$

$$\text{Tr}\{c\hat{} \geq \delta\} = \left\{\begin{array}{ll} 0 & \text{for } c4 \leq \delta \\ \frac{(c4-\delta)}{2(c4-c3)} & \text{for } c2 \leq \delta \leq c_4 \\ \frac{1}{2}\left(\frac{(c4-\delta)}{(c4-c3)} + \frac{(c2-\delta)}{(c2-c1)}\right) & \text{for } c_1 \leq \delta \leq c_2 \\ \frac{1}{2}\left(\frac{(c4-\delta)}{(c4-c3)} + 1\right) & \text{for } c_3 \leq \delta \leq c_1 \\ 1 & \text{for } \delta \leq c_3 \end{array}\right\} \tag{13}$$

where $c_1 = c(i, j, k)_1, c_2 = c(i, j, k)_2, c_3 = c(i, j, k)_3, c_4 = c(i, j, k)_4$, now the expected value of $\hat{c}$ is $E[\hat{c}] = \frac{1}{4}[c_1 + c_2 + c_3 + c_4]$.

Similarly, for safety constraints $\hat{s}_{min}$, we derive it as follows to determine a complete tour $(x_1, x_2, \ldots, x_N, x_1)$ and corresponding conveyance $(v_1, v_2, \ldots, v_p)$ to minimize T.

$$Tr\{\hat{s} \geq \omega\} = \begin{Bmatrix} 0 & \text{for } s_4 \leq \omega \\ \frac{(s4-\omega)}{2(s4-s3)} & \text{for } s_2 \leq \omega \leq s_4 \\ \frac{1}{2}\left(\frac{(s4-\omega)}{(s4-s3)} + \frac{(s2-\omega)}{(s2-s1)}\right) & \text{for } s_1 \leq \omega \leq s_2 \\ \frac{1}{2}\left(\frac{(s4-\omega)}{(s4-s3)} + 1\right) & \text{for } s_3 \leq \omega \leq s_1 \\ 1 & \text{for } \omega \leq s_3 \end{Bmatrix} \tag{14}$$

where $s_1 = s(i, j, k)_1, s_2 = s(i, j, k)_2, s_3 = s(i, j, k)_3, s_4 = s(i, j, k)_4$, now the expected value of $\hat{c}$ is $E[\hat{s}] = \frac{1}{4}[s_1 + s_2 + s_3 + s_4]$.

3.5 Solid TSP with Fuzzy-Rough Costs and Safety Constraints

In CSTSP if costs and safety factors are fuzzy-rough numbers, i.e., $\tilde{c}\hat{}(i, j, k)$ and $\hat{s}(i, j, k)$ respectively and safety effect limit S_{min} also fuzzy-rough number $\hat{S}_{min}$, then CSTSP reduces to determine a complete tour $(x_1, x_2, \ldots, x_N, x_1)$ and corresponding conveyance $(v_1, v_2, \ldots, v_p)$

$$\text{to minimize } \tilde{Z}\hat{} = \sum_{i=1}^{N-1} \tilde{c}\hat{}(x_i, x_i + 1, v_i) + \tilde{c}\hat{}(x_N, x_1, v_k)$$
$$\text{subject to } \sum_{i=1}^{N-1} \tilde{s}\hat{}(x_i, x_i + 1, v_i) + \tilde{s}\hat{}(x_N, x_1, v_k) \geq \tilde{S}\hat{}_{min}$$

As minimization of fuzzy-rough objective as well as fuzzy-rough constraints are not well-defined the above problem can be rewritten in trust measure by article-2.6, respectively, to determine a complete tour $(x_1, x_2, \ldots, x_N, x_1)$ and corresponding conveyance $(v_1, v_2, \ldots, v_p)$ to minimize T

$$\text{Subject to } Tr\left\{Pos\left(\sum_{i=1}^{N-1} \tilde{c}\hat{}(x_i, x_{i+1}, v_i) + \tilde{c}\hat{}(x_N, x_1, v_k) < T\right) \geq \delta 1\right\} \geq \delta$$
$$Tr\left\{Pos\left(\sum_{i=1}^{N-1} \tilde{s}\hat{}(x_i, x_{i+1}, v_i) + \tilde{s}\hat{}(x_N, x_1, v_k) \geq \tilde{S}\hat{}_{min}\right) \geq \omega 1\right\} \geq \omega$$

where $T, \delta, \omega, \delta 1, \omega 1$ are crisp parameter with predetermined confidence levels. It is clear that minimization of T implies the minimization of fuzzy- rough objective $\tilde{Z}\hat{}$.

To determine possibility necessity and trust measure we generate the rough number from the above problem as

$$
\begin{aligned}
\tilde{c}\hat{}(i,j,k) &= (\tilde{c} - L_1, \tilde{c}, \tilde{c} + R_1),\\
&\qquad \text{where } \hat{c}(i,j,k) = ([c(i,j,k)_1, c(i,j,k)_2], [c(i,j,k)_3, c(i,j,k)_4]),\\
\tilde{s}\hat{}(i,j,k) &= (\tilde{s} - L_2, \tilde{s}, \tilde{s} + R_2),\\
&\qquad \text{where } s\hat{}(i,j,k) = ([s(i,j,k)_1, s(i,j,k)_2], [s(i,j,k)_3, s(i,j,k)_4]),\\
\tilde{s}\hat{}_{\min} &= (\tilde{s} - L_3, \tilde{s}, \tilde{s} + R_3), \text{ where } s\hat{}_{\min} = ([s_{\min 1}, s_{\min 2}][s_{\min 3}, s_{\min 4}]),
\end{aligned}
$$

Then the expected values according to article 2.3 are given by

$$
\begin{aligned}
E[\tilde{c}\hat{}] &= \frac{1}{4}[c_1 + c_2 + c_3 + c_4] + \frac{\rho R1 - (1-\rho)L1}{2} \quad \text{where } 0 \le \rho \le 1\\
E[\tilde{s}\hat{}] &= [s_1 + s_2 + s_3 + s_4] + \frac{\rho R2 - (1-\rho)L2}{2} \quad \text{where } 0 \le \rho \le 1\\
E[\tilde{s}\hat{}_{\min}] &= [s_{\min 1} + s_{\min 2} + s_{\min 3} + s_{\min 4}] + \frac{\rho R3 - (1-\rho)L3}{2} \quad \text{where } 0 \le \rho \le 1
\end{aligned}
$$

where $c_1 = c(i,j,k)_1, c_2 = c(i,j,k)_2, c_3 = c(i,j,k)_3, c_4 = c(i,j,k)_4, s_1 = s(i,j,k)_1, s_2 = s(i,j,k)_2, s_3 = s(i,j,k)_3, s_4 = s(i,j,k)_4, L_1, L_2, L_3, R_1, R_2, R_3$ are crisp number.

4 Proposed Improved Genetic Algorithm (IGA)

To solve the above CSTSPs, we developed an Improved Genetic Algorithm (IGA) which is the combination of probabilistic selection, cyclic crossover, and node-oriented random mutation. In the natural genesis, chromosomes are the main carriers of hereditary information from parent to offspring and those genes, which present hereditary factors, are lined up in the chromosomes. At the time of reproduction, crossover and mutation take place among the chromosomes of parents. In this way hereditary factors of parents are mixed-up and carried over to their offspring. Again, the Darwinian principle states that only the fittest can survive in nature. So a pair of fittest parents normally reproduces better offspring. In this context we followed to create a genetic algorithm for an optimization problem. Here potential solutions of the problem are analogous with the chromosomes and chromosome of better offspring with the better solution of the problem. Here we use the probabilistic selection (Boltzaman Probability) and cyclic crossover and node-oriented random mutation happen among a set of potential solutions to get a new set of solutions and it continues until terminating conditions are encountered in IGA. The proposed IGA and its procedures are presented below.

IGA Algorithm

1. Begin
2. Initialize max generation number (s_0), population size (pop size), probability of crossover (p_c) and), probability of mutation (p_m).
3. Randomly generate initial population p(n)
4. Evaluate initial population p(n)
5. While $n \leq s_0$ do

    a. $n = n + 1$.
    b. Select p(n) from $p(n - 1)$.
    c. Alter (crossover and mutate) p(n).
    d. Evaluate p(n).

6. Update
7. End While
8. Print optimum result
9. End

4.1 Proposed IGA for Solid TSPs

4.1.1 Representation

Here a complete tour on N cities represents a solution. So an N-dimensional integer vector $X_i = (x_{i1}, x_{i2}, \ldots, x_{iN})$ is used to represent a solution, where $x_{i1}, x_{i2}, \ldots, x_{iN}$ represent N consecutive cities in a tour. For solid TSP another integer vector $V_k = (v_{k1}, v_{k2}, \ldots, v_{kN})$ is used to represent the conveyances types used travel between different cities. Here v_{kj} represents the conveyance (an integer) used to travel from city x_{ij} to $x_i(j + 1)$ for $j = 1, 2, \ldots, N - 1$ and v_{kN} represents the conveyance type used to travel from city x_{iN} to x_{i1}.

4.1.2 Initialization

Population size number of such solutions $X_i = (x_{i1}, x_{i2}, \ldots, x_{iN})$, $i = 1, 2, \ldots,$ pop size, are randomly generated by random number generator, such that each solution satisfies the constraints of the problem. A separate subfunction check constraint $S(X_i)$ is used for this purpose. For STSP another integer vector $V_k = (v_{k1}, v_{k2}, \ldots, v_{kN})$ is randomly generated corresponding to the solution X_i, to represent the conveyance types used to travel between different cities. So in that case (X_i, V_k) represent a solution.

4.1.3 Evaluation Process

To find fitness of a solution $X_i(X_i, V_k)$ for STSP, the following two steps are used—

- Calculate objective function value OBJ_i for the solution $X_i(X_i, V_k)$ for CSTSP.
- As the problems are minimization type take MVAL-OBJ_i as fitness, FIT_i, of X_i (X_i, V_k) for STSP, where MVAL is a sufficiently large value to make the fitness positive.

4.1.4 Probabilistic Selection

Since we choose the population to solve the TSP, we know TSP always minimizes the problem. It is better to choose that population which in the neighborhood of the minimum solution of the entire solution space tends to convergence very fast. From the initial population choose the best fitted population for TSP; that is choose as most minimum fitness value (say f_{min}). To form the matting pool, we use the Boltzmann-Probability as follows:

$$P = \boldsymbol{e}^{\frac{(f_{min} - f(xi))}{T}}$$

$$\begin{aligned} \text{where}\quad T &= T_0(1-\alpha)^k, \\ k &= \left(1 + 100 * \left(\tfrac{g}{G}\right)\right), \\ T_0 &= [5, 100], \\ g &= \text{Current Generation}, \\ G &= \text{Max Generation}, \\ \alpha &= \text{rand}[0, 1], i = 0, 1, 2, \ldots, N \end{aligned}$$

after finding the above probability, then use it in the following way:

if (P < rand[0, 1])

then select x_i;

else

Select f_{min} in the matting pool

Here T increases and corresponding P also increases as T depends on the current generation.

4.1.5 Cyclic Crossover

(i) Selection for crossover: For each solution of p(n) generate a random number r from the range [0, 1]. If $r < p_c$ then the solution is taken for crossover.
(ii) Crossover process: For simple TSP cyclic crossover process is used. The cyclic crossover focuses on subsets of cities that occupy the same subset of positions in both parents. Then these cities are copied from the first parent to the offspring (at the same positions), and the remaining positions are filled with the cities of the second parent. In this way, the position of each city is inherited from one of the two parents. However, many edges can be broken in the process, because the initial subset of cities is not necessarily located at consecutive positions in the parent tours. To illustrate the process let us consider a TSP consisting of nine cities and consider two parents PR1, PR2 as below:

$$\begin{aligned}&PR1 : 1\ 2\ 3\ 4\ 5\ 6\ 7\ 8\ 9\\&PR2 : 3\ 4\ 5\ 1\ 2\ 9\ 8\ 7\ 6\end{aligned}$$

Let CH1, CH2 be two children born after crossover. The mechanism of birth of CH1, CH2 using cycle crossover is explained with the help of the following steps:

Randomly generate an integer in the range $[1 \ldots 9]$. Let it be 3.

As $PR1[3] = 3$, 3rd element of CH1 is 3, i.e., $CH1[3] = 3$. PR2, is then searched to check for the presence of element 3 and it has been found in the first position. Then first element of CH1 is selected from the first element of PR1, i.e., $CH1[1] = PR1[1] = 1$. PR2, is again searched for the presence of element 1 and it has occurred at the fourth position. Thus fourth element of PR1 has been copied as the fourth element of CH1, i.e., $CH1[4] = PR1[4] = 4$. Similarly, following are obtained $CH1[2] = PR1[2] = 2$, $CH1[5] = PR1[5] = 5$.

This completes one cycle because element 5 is seen to be present at the third position of PR2 and the corresponding third position element of PR1 is element 3, which has already been selected as the starting element of the cycle.

The remaining elements of CH1 are selected directly from PR2 as follows:

$$\begin{aligned}&CH1[6] = PR2[6] = 9, CH1[7] = PR1[7] = 8\\&CH1[8] = PR2[8] = 7, CH1[9] = PR1[9] = 6\end{aligned}$$

Final forms of CH1 and CH2 are as below:

$$\begin{aligned}&CH1 : 1\ 2\ 3\ 4\ 5\ 9\ 8\ 7\ 6\\&CH2 : 3\ 4\ 5\ 1\ 2\ 6\ 7\ 8\ 9\end{aligned}$$

If CH1 satisfies the constraint of the problem then PR1 is replaced by CH1. Similarly, if CH2 satisfies the constraint of the problem then PR2 is replaced by CH2.

For STSP to made crossover on two parents (PR1, V1), (PR2, V2), the same procedure is followed on PR1 and PR2 to obtain CH1 and CH2. To keep randomness in selection of conveyances conveyance sets V1 and V2 remain unchanged, i.e., resultant child after crossover becomes (CH1, V1), (CH2, V2). If (CH1, V1) satisfies the constraint of the problem, then (PR1, V1) is replaced by (CH1, V1). Similarly, if (CH2, V2) satisfies the constraint of the problem then (PR2, V2) is replaced by (CH2, V2).

4.1.6 Random Mutation

(i) Selection for mutation: For each solution of p(n) generate a random number r from the range [0, 1]. If $r < p_m$ then the solution is taken for mutation.
(ii) Mutation process: To mutate a solution $X = (x_1, x_2, \ldots, x_N)$ of TSP with T number of nodes, select T number of nodes randomly from the solution and just replace their places in the solution, i.e., if randomly two nodes x_i, x_j are selected then interchange x_i, x_j to get a child solution. The new solution, if satisfies the constraint of the problem, replaces the parent solution. For CSTSP to mutate a solution (X,V), where $X = (x_1, x_2, \ldots, x_N), V = (v_1, v_2, \ldots, v_p)$ at first an integer is randomly selected in the range [1, 2]. If 1 is selected then another two random integers i, j are selected in the range [1, N]. Then interchange x_i, x_j to get child solution. If 2 is selected then another two random integers i and j are selected in the range [1, N] and [1, P] respectively. The value of v_p is replaced by j to get a child solution. If the child solution satisfies the constraint of the problem then it replaces the parent solution.

5 Numerical Experiments

5.1 Verification of Proposed IGA with Standard TSPLIB Test Problem

We use the proposed IGA in the standard TSP from the TSPLIB and compare with classical GA in number of iteration; the result shows the efficiency of the proposed algorithm (Table 1).

Table 1 Test with standard problems for proposed algorithm

			Solution			
Instance	Problem size	Available best solution	IGA	Iteration	GA	Iteration
bays29	29 × 29	2020	2020	349	2020	571
bayg29	29 × 29	1610	1610	256	1610	480
fri26	26 × 26	937	937	202	937	368
dantzig42	42 × 42	699	699	245	699	986

5.2 Results of CSTSPs

5.2.1 Crisp STSP

Here we take three conveyances and the values are crisp in nature (Tables 2 and 3).

5.2.2 Fuzzy CSTSP

See Tables 4 and 5.

Table 2 Costs for crisp CSTSP

i/j	1	2	3	4	5
1	∞	15, 16, 17	18, 19, 20	12, 13, 14	20, 21, 22
2	27, 28, 29	∞	20, 21, 22	48, 49, 50	35, 36, 37
3	42, 43, 44	28, 29, 30	∞	30, 31, 32	25, 26, 27
4	38, 39, 40	30, 31, 32	8, 9,10	∞	20, 21, 22
5	66, 67, 68	22, 23, 24	35, 36, 37	30, 31, 32	∞

Table 3 Costs of safety matrix for crisp CSTSP

i/j	1	2	3	4	5
1	∞	0.3, 0.4, 0.5	0.5, 0.6, 0.7	0.2, 0.3, 0.4	0.1, 0.2, 0.3
2	0.6, 0.7, 0.8	∞	0.2, 0.3, 0.4	0.5, 0.6, 0.7	0.3, 0.4, 0.5
3	0.2, 0.3, 0.4	0.3, 0.4, 0.5	∞	0.2, 0.3, 0.4	0.1, 0.2, 0.3
4	0.6, 0.7, 0.8	0.4, 0.3, 0.2	0.6, 0.7, 0.8	∞	0.3, 0.4, 0.5
5	0.8, 0.7, 0.6	0.3, 0.2, 0.1	0.6, 0.5, 0.4	0.4, 0.5, 0.6	∞

Table 4 Costs for fuzzy CSTSP

i/j	1	2	3	4	5
1	∞	(14, 15, 16) (15, 16, 17) (16, 17, 18)	(17, 18, 19) (18, 19, 20) (19, 20, 21)	(11, 12, 13) (12, 13, 14) (13, 14, 15)	(19, 20, 21) (20, 21, 22) (21, 22, 23)
2	(26, 27, 28) (27, 28, 29) (28, 29, 30)	∞	(19, 20, 21) (20, 21, 22) (21, 22, 23)	(47, 48, 49) (48, 49, 50) (49, 50, 51)	(34, 35, 36) (35, 36, 37) (36, 37, 38)
3	(41, 42, 43) (42, 43, 44) (43, 44, 45)	(27, 28, 29) (28, 29, 30) (29, 30, 31)	∞	(29, 30, 31) (30, 31, 32) (31, 32, 33)	(24, 25, 26) (25, 26, 27) (26, 27,28)
4	(37, 38, 39) (38, 39, 40) (39, 40, 41)	(29, 30, 31) (30, 31, 32) (31, 32, 33)	(7, 8, 9) (8, 9, 10) (9, 10, 11)	∞	(19, 20, 21) (20, 21, 22) (21, 22, 23)
5	(65, 66, 67) (66, 67, 68) (67, 68, 69)	(21, 22, 23) (22, 23, 24) (23, 24, 25)	(34, 35, 36) (35, 36, 37) (36, 37, 38)	(29, 30, 31) (30, 31, 32) (31, 32, 33)	∞

Table 5 Fuzzy costs of safety matrix for CSTSP

i/j	1	2	3	4	5
1	∞	(0.2, 0.3, 0.4) (0.3, 0.4, 0.5) (0.4, 0.5, 0.6)	(0.4, 0.5, 0.6) (0.5, 0.6, 0.7) (0.6, 0.7, 0.8)	(0.1, 0.2, 0.3) (0.2, 0.3, 0.4) (0.3, 0.4, 0.5)	(0.3, 0.4, 0.5) (0.4, 0.5, 0.6) (0.5, 0.6, 0.7)
2	(0.5, 0.6, 0.7) (0.6, 0.7, 0.8) (0.7, 0.8, 0.9)	∞	(0.1, 0.2, 0.3) (0.2, 0.3, 0.4) (0.3, 0.4, 0.5)	(0.4, 0.5, 0.6) (0.5, 0.6, 0.7) (0.6, 0.7, 0.8)	(0.2, 0.3, 0.4) (0.3, 0.4, 0.5) (0.4, 0.5, 0.6)
3	(0.1, 0.2, 0.3) (0.2, 0.3, 0.4) (0.3, 0.4, 0.5)	(0.2, 0.3, 0.4) (0.3, 0.4, 0.5) (0.4, 0.5, 0.6)	∞	(0.1, 0.2, 0.3) (0.2, 0.3, 0.4) (0.3, 0.4, 0.5)	(0.1, 0.2, 0.3) (0.2, 0.3, 0.4) (0.3, 0.4, 0.5)
4	(0.5, 0.6, 0.7) (0.6, 0.7, 0.8) (0.7, 0.8, 0.9)	(0.3, 0.2, 0.1) (0.4, 0.3, 0.2) (0.5, 0.4, 0.3)	(0.5, 0.6, 0.7) (0.6, 0.7, 0.8) (0.7, 0.8, 0.9)	∞	(0.2, 0.3, 0.4) (0.3, 0.4, 0.5) (0.4, 0.5, 0.6)
5	(0.9, 0.8, 0.7) (0.8, 0.7, 0.6) (0.7, 0.6, 0.5)	(0.4, 0.3, 0.2) (0.3, 0.2, 0.1) (0.2, 0.1, 0.0)	(0.7, 0.6, 0.5) (0.6, 0.5, 0.4) (0.5, 0.4, 0.3)	(0.3, 0.4, 0.5) (0.4, 0.5, 0.6) (0.5, 0.6, 0.7)	∞

5.2.3 Rough CSTSP

See Tables 6 and 7.

Table 6 Costs for rough CSTSP

i/j	1	2	3	4	5
1	∞	([14, 15][13, 16]) ([15, 16][14, 17]) ([16, 17][15, 18])	([17, 18][16, 19]) ([18, 19][17, 20]) ([19, 20][18, 21])	([10, 11][9, 12]) ([11, 12][10, 13]) ([12, 13][11, 14])	([18, 19][17, 20]) ([19, 20][18, 21]) ([20, 21][19, 22])
2	([25, 26][24, 27]) ([26, 27][25, 28]) ([27, 28][26, 29])	∞	([18, 19][17, 20]) ([19, 20][18, 21]) ([20, 21][19, 22])	([46, 47][45, 48]) ([47, 48][46, 49]) ([48, 49][47, 50])	([33, 34][32, 35]) ([34, 35][33, 36]) ([35, 36][34, 37])
3	([40, 41][39, 42]) ([41, 42][40, 43]) ([42, 43][41, 44])	([26, 27][25, 28]) ([27, 28][26, 29]) ([28, 29][27, 30])	∞	([28, 29][27, 30]) ([29, 30][28, 31]) ([30, 31][29, 32])	([23,24][22, 25]) ([24, 25][23, 26]) ([25, 26][24, 27])
4	([36, 37][35, 38]) ([37, 38][36, 39]) ([38, 39][37, 40])	([28, 29][27, 30]) ([29, 30][28, 31]) ([30, 31][29, 32])	([6, 7][5, 8]) ([7, 8][6, 9]) ([8, 9][7, 10])	∞	([18, 19][17, 20]) ([19, 20,][18, 21]) ([20, 21][19, 22])
5	([64, 65][63, 66]) ([65, 66][64, 67]) ([66, 67][65, 68])	([20, 21][19, 22]) ([21, 22][20, 23]) ([22, 23][21, 24])	([33, 34][32, 35]) ([34, 35][33, 36]) ([35, 36][34, 37])	([28, 29][27, 30]) ([29, 30][28, 31]) ([30, 31][29, 32])	∞

Table 7 Rough costs of safety matrix for CSTSP

i/j	1	2	3	4	5
1	∞	([0.1, 0.2][0.01, 0.3]) ([0.2, 0.3][0.1, 0.4]) ([0.3, 0.4][0.2, 0.5])	([0.3, 0.4][0.2, 0.5]) ([0.4, 0.5][0.3, 0.6]) ([0.5, 0.6][0.4, 0.7])	([0.2, 0.3][0.1, 0.4]) ([0.3, 0.4][0.2, 0.5]) ([0.4, 0.5][0.3, 0.6])	([0.5, 0.6][0.4, 0.7]) ([0.6, 0.7][0.5, 0.8]) ([0.7, 0.8][0.6, 0.9])
2	([0.5, 0.6][0.4, 0.7]) ([0.6, 0.7][0.5, 0.8]) ([0.7, 0.8][Z, 0.9])	∞	([0.1, 0.2][0.0, 0.3]) ([0.2, 0.3][0.1, 0.4]) ([0.3, 0.4][0.2, 0.5])	([0.3, 0.4][0.2, 0.5]) ([0.4, 0.5][0.3, 0.6]) ([0.5, 0.6][0.4, 0.7])	([0.2, 0.3][0.1, 0.4]) ([0.3, 0.4][0.2, 0.5]) ([0.4, 0.5][0.3, 0.6])
3	([0.1, 0.2][0.01, 0.3]) ([0.2, 0.3][0.1, 0.4]) ([0.3, 0.4][0.2, 0.5])	([0.2, 0.3][0.1, 0.4]) ([0.3, 0.4][0.2, 0.5]) ([0.4, 0.5][0.3, 0.6])	∞	([0.1, 0.2][0.0, 0.3]) ([0.2, 0.3][0.1, 0.4]) ([0.3, 0.4][0.2, 0.5])	([0.2, 0.3][0.1, 0.4]) ([0.3, 0.4][0.2, 0.5]) ([0.4, 0.5][0.3, 0.6])
4	([0.4, 0.5][0.3, 0.6]) ([0.5, 0.6][0.4, 0.7]) ([0.6, 0.7][0.5, 0.8])	([0.2, 0.3][0.1, 0.4]) ([0.3 0.4][0.2, 0.5]) ([0.4, 0.5][0.3, 0.6])	([0.5, 0.6][0.4, 0.7]) ([0.6, 0.7][0.5, 0.8]) ([0.7, 0.8][0.6, 0.9])	∞	([0.2, 0.3][0.1, 0.4]) ([0.3, 0.4][0.2, 0.5]) ([0.4, 0.5][0.3, 0.6])
5	([0.7, 0.8][0.6, 0.9]) ([0.6, 0.7][0.5, 0.8]) ([0.5, 0.6][0.4, 0.7])	([0.4, 0.5][0.3, 0.6]) ([0.3, 0.4][0.2, 0.5]) ([0.2, 0.3][0.1, 0.4])	([0.6, 0.7][0.5, 0.8]) ([0.5, 0.6][0.4, 0.7]) ([0.4, 0.5][0.3, 0.6])	([0.2, 0.3][0.1, 0.4]) ([0.3, 0.4][0.2, 0.5]) ([0.4, 0.5][0.3, 0.6])	∞

Table 8 Costs for fuzzy-rough CSTSP

i/j	1	2	3	4	5
1	∞	([14, 15][13, 16]) (14.5 − 5, 14.5, 14.5 + 5) ([15, 16][14, 17]) (15.5 − 5, 15.5, 15.5 + 5) ([16, 17][15, 18]) (16.5 − 5, 16.5, 16.5 + 5)	([17, 18][16, 19]) (17.5 − 5, 17.5, 17.5 + 5) ([18, 19][17, 20]) (18.5 − 5, 18.5, 18.5 + 5) ([19, 20][18, 21]) (19.5 − 5, 19.5, 19.5 + 5)	([10, 11][9, 12]) (10.5 − 5, 10.5, 10.5 + 5) ([11, 12][10, 13]) (11.5 − 5, 11.5, 11.5 + 5) ([12, 13][11, 14]) (12.5 − 5, 12.5, 12.5 + 5)	([18, 19,][17, 20]) (18.5 − 5, 18.5, 18.5 + 5) ([19, 20,][18, 21]) (19.5 − 5, 19.5, 19.5 + 5) ([20, 21][19, 22]) (20.5 − 5, 20.5, 20.5 + 5)
2	([25, 26][24, 27]) (250.2 − 5, 25.5, 25.5 + 5) ([26, 27][25, 28]) (26.5 − 5, 26.5, 26.5 + 5) ([27, 28][26, 29]) (27.5 − 5, 27.5, 27.5 + 5)	∞	([18, 19][17, 20]) (18.5 − 5, 18.5, 18.5 + 5) ([19, 20][18, 21]) (19.5 − 5, 19.5, 19.5 + 5) ([20, 21][19, 22]) (20.5 − 5, 20.5, 20.5 + 5)	([46, 47][45, 48]) (46.5 − 5, 46.5, 46.5 + 5) ([47, 48][46, 49]) (47.5 − 5, 47.5, 47.5 + 5) ([48, 49][47, 50]) (48.5 − 5, 48.5, 48.5 + 5)	([33, 34][32, 35]) (33.5 − 5, 33.5, 33.5 + 5) ([34, 35][33, 36]) (34.5 − 5, 34.5, 34.5 + 5) ([35, 36][34, 37]) (35.5 − 5, 35.5, 35.5 + 5)
3	([40, 41][39, 42]) (40.5 − 5, 40.5, 40.5 + 5) ([41, 42][40, 43]) (41.5 − 5, 41.5, 41.5 + 5) ([42, 43][41, 44]) (42.5 − 5, 42.5, 42.5 + 5)	([26, 27][25, 28]) (26.5 − 5, 26.5, 26.5 + 5) ([27, 28][26, 29]) (27.5 − 5, 27.5, 27.5 + 5) ([28, 29][27, 30]) (28.5 − 5, 28.5, 28.5 + 5)	∞	([28, 29][27, 30]) (28.5 − 5, 28.5, 28.5 + 5) ([29, 30][28, 31]) (29.5 − 5, 29.5, 29.5 + 5) ([30, 31][29, 32]) (30.5 − 5, 30.5, 30.5 + 5)	([23,24][22, 25]) (23.5 − 5, 23.5, 23.5 + 5) ([24, 25][23, 26]) (24.5 − 5, 24.5, 24.5 + 5) ([25,26][24, 27]) (25.5 − 5, 25.5, 25.5 + 5)
4	([36, 37][35, 38]) (36.5 − 5, 36.5, 36.5 + 5) ([37, 38][36, 39]) (37.5 − 5, 37.5, 37.5 + 5) ([38, 39][37, 40]) (38.5 − 5, 38.5, 38.5 + 5)	([28, 29][27, 30]) (28.5 − 5, 28.5, 28.5 + 5) ([29, 30][28, 31]) (29.5 − 5, 29.5, 29.5 + 5) ([30, 31][29, 32]) (30.5 − 5, 30.5, 30.5 + 5)	([6, 7][5, 8]) (6.5 − 5, 6.5, 6.5 + 5) ([7, 8][6, 9]) (7.5 − 5, 7.5, 7.5 + 5) ([8, 9][7, 10]) (8.5 − 5, 8.5, 8.5 + 5)	∞	([18, 19][17, 20]) (18.5 − 5, 18.5, 18.5 + 5) ([19, 20][18, 21]) (19.5 − 5, 19.5, 19.5 + 5) ([20, 21][19, 22]) (20.5 − 5, 20.5, 20.5 + 5)
5	([64, 65][63, 66]) (64.5 − 5, 64.5, 64.5 + 5) ([65, 66][64, 67]) (65.5 − 5, 65.5, 65.5 + 5) ([66, 67][65, 68]) (66.5 − 5, 66.5, 66.5 + 5)	([20, 21][19, 22]) (20.5 − 5, 20.5, 20.5 + 5) ([21, 22][20, 23]) (21.5 − 5, 21.5, 21.5 + 5) ([22, 23][21, 24]) (22.5 − 5, 22.5, 22.5 + 5)	([33, 34][32, 35]) (33.5 − 5, 33.5, 33.5 + 5) (34.5 − 5, 34.5, 34.5 + 5) ([35, 36][34, 37]) (35.5 − 5, 35.5, 35.5 + 5)	([28, 29][27, 30]) (28.5 − 5, 28.5, 28.5 + 5) ([29, 30][28, 31]) (29.5 − 5, 29.5, 29.5 + 5) ([30, 31][29, 32]) (30.5 − 5, 30.5, 30.5 + 5)	∞

Table 9 Fuzzy-rough safety matrix for CSTSP

i/j	1	2	3	4	5
1	∞	[0.1, 0.2][0.01, 0.3]) (0.15 − 0.1, 0.15, 0.15 + 0.1) ([0.2, 0.3][0.1, 0.4]) (0.25 − 0.1, 0.25, 0.25 + 0.1) ([0.3, 0.4][0.2, 0.5]) (0.35 − 0.1, 0.35, 0.35 + 0.1)	([0.3, 0.4][0.2, 0.5]) (0.35 − 0.1, 0.35, 0.35 + 0.1) ([0.4, 0.5][0.3, 0.6]) (0.45 − 0.1, 0.45, 0.45 + 0.1) ([0.5, 0.6][0.4, 0.7]) (0.55 − 0.1, 0.55, 0.55 + 0.1)	([0.2, 0.3][0.1, 0.4]) (0.25 − 0.1, 0.25, 0.25 + 0.1) ([0.3, 0.4][0.2, 0.5]) (0.35 − 0.1, 0.35, 0.35 + 0.1) ([0.4, 0.5][0.3, 0.6]) (0.45 − 0.1, 0.45, 0.45 + 0.1)	([0.5, 0.6][0.4, 0.7]) (0.55 − 0.1, 0.55, 0.55 + 0.1) ([0.6, 0.7][0.5, 0.8]) (0.65 − 0.1, 0.65, 0.65 + 0.1) ([0.7, 0.8][0.6, 0.9]) (0.75 − 0.1, 0.75, 0.75 + 0.1)
2	([0.5, 0.6][0.4, 0.7]) (0.55 − 0.1, 0.55, 0.55 + 0.1) ([0.6, 0.7][0.5, 0.8]) (0.65 − 0.1, 0.65, 0.65 + 0.1) ([0.7, 0.8][0.6, 0.9]) (0.75 − 0.1, 0.75, 0.75 + 0.1)	∞	([0.1, 0.2][0.01, 0.3]) (0.15 − 0.1, 0.15, 0.15 + 0.1) ([0.2, 0.3][0.1, 0.4]) (0.25 − 0.1, 0.25, 0.25 + 0.1) ([0.3, 0.4][0.2, 0.5]) (0.35 − 0.1, 0.35, 0.35 + 0.1)	([0.3, 0.4][0.2, 0.5]) (0.35 − 0.1, 0.35, 0.35 + 0.1) ([0.4, 0.5][0.3, 0.6]) (0.45 − 0.1, 0.45, 0.45 + 0.1) ([0.5, 0.6][0.4, 0.7]) (0.55 − 0.1, 0.55, 0.55 + 0.1)	([0.2, 0.3][0.1, 0.4]) (0.25 − 0.1, 0.25, 0.25 + 0.1) ([0.3, 0.4][0.2, 0.5]) (0.35 − 0.1, 0.35, 0.35 + 0.1) ([0.4, 0.5][0.3, 0.6]) (0.45 − 0.1, 0.45, 0.45 + 0.1)
3	([0.3, 0.4][0.2, 0.5]) (0.35 − 0.1, 0.35, 0.35 + 0.1) ([0.4, 0.5][0.3, 0.6]) (0.45 − 0.1, 0.45, 0.45 + 0.1) ([0.5, 0.6][0.4, 0.7]) (0.55 − 0.1, 0.55, 0.55 + 0.1)	([0.2, 0.3][0.1, 0.4]) (0.25 − 0.1, 0.25, 0.25 + 0.1) ([0.3, 0.4][0.2, 0.5]) (0.35 − 0.1, 0.35, 0.35 + 0.1) ([0.4, 0.5][0.3, 0.6]) (0.45 − 0.1, 0.45, 0.45 + 0.1)	∞	([0.1, 0.2][0.01, 0.3]) (0.15 − 0.1, 0.15, 0.15 + 0.1) ([0.2, 0.3][0.1, 0.4]) (0.25 − 0.1, 0.25, 0.25 + 0.1) ([0.3, 0.4][0.2, 0.5]) (0.35 − 0.1, 0.35, 0.35 + 0.1)	([0.2, 0.3][0.1, 0.4]) (0.25 − 0.1, 0.25, 0.25 + 0.1) ([0.3, 0.4][0.2, 0.5]) (0.35 − 0.1, 0.35, 0.35 + 0.1) ([0.4, 0.5][0.3, 0.6]) (0.45 − 0.1, 0.45, 0.45 + 0.1)
4	([0.4, 0.5][0.3, 0.6]) (0.45 − 0.1, 0.45, 0.45 + 0.1) ([0.5, 0.6][0.4, 0.7]) (0.55 − 0.1, 0.55, 0.55 + 0.1) ([0.6, 0.7][0.5, 0.8]) (0.65 − 0.1, 0.65, 0.65 + 0.1)	([0.2, 0.3][0.1, 0.4]) (0.25 − 0.1, 0.25, 0.25 + 0.1) ([0.3, 0.4][0.2, 0.5]) (0.35 − 0.1, 0.35, 0.35 + 0.1) ([0.4, 0.5][0.3, 0.6]) (0.45 − 0.1, 0.45, 0.45 + 0.1)	([0.5, 0.6][0.4, 0.7]) (0.55 − 0.1, 0.55, 0.55 + 0.1) ([0.6, 0.7][0.5, 0.8]) (0.65 − 0.1, 0.65, 0.65 + 0.1) ([0.7, 0.8][0.6, 0.9]) (0.75 − 0.1, 0.75, 0.75 + 0.1)	∞	([0.2, 0.3][0.1, 0.4]) (0.25 − 0.1, 0.25, 0.25 + 0.1) ([0.3, 0.4][0.2, 0.5]) (0.35 − 0.1, 0.35, 0.35 + 0.1) ([0.4, 0.5][0.3, 0.6]) (0.45 − 0.1, 0.45, 0.45 + 0.1)
5	([0.7, 0.8][0.6, 0.9]) (0.75 − 0.1, 0.75, 0.75 + 0.1) ([0.6, 0.7][0.5, 0.8]) (0.65 − 0.1, 0.65, 0.65 + 0.1) ([0.5, 0.6][0.4, 0.7]) (0.55 − 0.1, 0.55, 0.55 + 0.1)	([0.4, 0.5][0.3, 0.6]) (0.45 − 0.1, 0.45, 0.45 + 0.1) ([0.3, 0.4][0.2, 0.5]) (0.35 − 0.1, 0.35, 0.35 + 0.1) ([0.2, 0.3][0.1, 0.4]) (0.25 − 0.1, 0.25, 0.25 + 0.1)	([0.6, 0.7][0.5, 0.8]) (0.65 − 0.1, 0.65, 0.65 + 0.1) ([0.5, 0.6][0.4, 0.7]) (0.55 − 0.1, 0.55, 0.55 + 0.1) ([0.4, 0.5][0.3, 0.6]) (0.45 − 0.1, 0.45, 0.45 + 0.1)	([0.2, 0.3][0.1, 0.4]) (0.25 − 0.1, 0.25, 0.25 + 0.1) ([0.3, 0.4][0.2, 0.5]) (0.35 − 0.1, 0.35, 0.35 + 0.1) ([0.4, 0.5][0.3, 0.6]) (0.45 − 0.1, 0.45, 0.45 + 0.1)	∞

Table 10 Results of different models by proposed IGA

Model	Path	Cost	S_{min}
Crisp CSTSP	(1,1)(4,2)(3,2)(5,3)(2,3)	100	2
Fuzzy CSTSP	(1,2)(4,1)(3,1)(5,1)(2,3)	110, 114, 118	1.9, 2.2, 2.5
Rough CSTSP	(1,1)(4,1)(3,2)(5,1)(2,2)	[85, 90][80, 95]	[2.0, 2.5][1.5, 3.1]
Fuzzy-rough CSTSP	(1,3)(4,2)(3,3)(5,3)(2,1)	[91, 96][86, 101] 68.5, 93.5,118.5	[2.3, 2.8][1.8, 3.3] 2.05, 2.55, 3.05

5.2.4 Costs for Fuzzy-Rough CSTSP

Our proposed algorithm is compared with another soft computing technique GA for standard TSPLIB [13] problems which are shown in Table 1. In every test problem, our algorithm gives the better result compared to the number of iterations. In Table 2 CSTSP with three conveyances, and 5×5 crisp cost matrix is presented. In Table 3, we have given the individual safety of the corresponding conveyances. In Table 4 the fuzzy cost values of CSTSP are given, in Table 5 the fuzzy values of the safety factors are declared. In this fuzzy environment, the results obtained via different conveyances are shown in Table 10. And similarly we construct the rough CSTSP with rough costs and safety factors which as shown in Tables 6 and 7 respectively. For the fuzzy-rough environment, we describe the cost and safety factors as fuzzy-rough variables which are presented in Tables 8 and 9 respectively. The final Table 10 gives the results of rough and fuzzy-rough CSTSPs. All these results are obtained using IGA.

6 Conclusion

The results show the efficiency of this developed algorithm, IGA. Here for the first time CSTSP is modeled with safety constraints. Since the travelling costs are completely uncertain, we describe these as fuzzy, rough, or fuzzy rough costs. To solve the real-life CSTSP in crisp, fuzzy, rough, fuzzy-rough environments, here the proposed IGA is most suitable. In future the costs may be taken as other uncertain variables, such as random, fuzzy- random, random-fuzzy, and accordingly several uncertain CSTSPs can be formulated and solved.

References

1. Zadeh, L.A.: Fuzzy logic and soft computing: issues, contentions and perspectives. In: Proceedings of IIZUKA'94: Third International Conference on Fuzzy Logic, Neural Nets and Soft Computing, pp. 1–2. Iizuka, Japan (1994)
2. Zadeh, L.A.: Some reflection on soft computing, granular computing and their roles in the conception, design and utilization of information/intelligent systems. Soft Comput. Fusion Found. Methodol. Appl. **2**(1), 23–25 (1998)
3. Garey, M.R., Johson, D.S.: Computers and Intractability: A Guide to the Theory of NP—Completeness. W.H Freeman, New York (1979)
4. Changdar, C., Maiti, M.K., Maiti, M.: A constrained solid TSP in fuzzy environment: two heuristic approaches. Iran. J. Fuzzy Syst. **7**, 15–29 (2011)
5. Xu, J., Zhao, L.: A class of fuzzy rough expected value multi-objective decision making model and its application to inventory problems. Comput. Math. Appl. **56**, 2107–2119 (2008)
6. Xu, J., Zhao, L.: A Multi-objective decision-making model with fuzzy rough coefficients and application to the inventory problem. Inf. Sci. **180**, 679–696 (2010)
7. Chang, T., Wan, Y., Ooi, W.T.: A stochastic dynamic travelling salesman problem with hard time windows. Eur. J. Oper. Res. **198**(3), 748–759 (2009)
8. Chiang, W.C., Russell, R.A.: Simulated annealing met heuristics for the vehicle routine problem with time windows. Ann. Operat. Res. **63**, 3–27 (1996)
9. Knox, J.: The application of Tabu search to the symmetric traveling salesman problem, Ph.D. Dissertation, University of Colorado (1989)
10. Vescan, A., Pintea, C.-M.: Ant colony component-based system for travelling salesman problem. Appl. Math. Sci. **1**(28), 1347–1357 (2007)
11. Goldberg, D.E.: Genetic Algorithms in Search, Optimization and Machine Learning. Addsion Wesley, Reading (1989)
12. Eberhart, R.C., Kennedy, J.: Particle swarm optimization. In: Proceedings of the IEEE Conference on Neural Networks, pp. 1942–1948 (1995)
13. TSPLIB. www.comopt.ifi.uni-heidelberg.de/software/TSPLIB95/
14. Dubosis, D., Prade, H.: Possibility Theory: Analysis of Fuzzy Information **2**, 3–39 (1988)
15. Liu, B.: Theory and Practice of Uncertain Programming. Physica-Verlag, Heidelberg (2002)
16. Mondal, M., Maity, A.K., Maiti, M.K., Maiti, M.: A production—repairing inventory model with fuzzy rough coefficients under inflation and time value of money. Appl. Math. Model. **37**(5), 3200–3215 (2013)

A Novel Soft Theoretic AHP Model for Project Management in Multi-criteria Decision Making Problem

Tuli Bakshi, T. Som and B. Sarkar

Abstract The present paper introduces a model of decision-making problem in multicriteria optimization domain for project management. The model is built by combining the fuzzy soft set theory and analytical hierarchical model. The later is the well-known method of ranking the alternatives for multicriteria decision-making problem. The ultimate project selection is the best of many decisions associated with project management. Here we illustrate the hybrid method by means of an application of the new mathematical model of soft set theory.

Keywords Soft set · Analytic Hierarchy Process · Reduction · Decision Support System · Multi-criteria Optimization

1 Introduction

In the real world, we have to deal with many complex computational problems pertaining to the areas of engineering, medical sciences, environmental sciences, economics, social sciences etc. which involve data that are not always crisp and precise. Therefore, most of our traditional models for formal reasoning and computing the crisp, deterministic, and precise data fail. We cannot use the well-known classical methods because of various inherent uncertainties present in those problems. There are theories such as theory of probability, theory of fuzzy sets [1], and theory of intuitionistic fuzzy sets [2, 3], vague sets [4], and rough sets [5] which can also be taken into consideration for mathematical model formulation and for dealing with uncertainties. But all the above theories have their inherent difficulties, including

T. Bakshi
Jadavpur University, Kolkata 700032, India

T. Som (✉)
Department of Mathematical Sciences, Indian Institute of Technology (BHU), Varanasi, India
e-mail: tsom.apm@iitbhu.ac.in

B. Sarkar
Department of Production Engineering, Jadavpur University, Kolkata 700032, India

M.K. Chakraborty et al. (eds.), *Facets of Uncertainties and Applications*,
Springer Proceedings in Mathematics & Statistics 125,
DOI 10.1007/978-81-322-2301-6_15

lack of parameterization of tools due to which they are not capable of successfully solving such complicated problems. Reason for these difficulties being the inadequacy of the parameterization tool of these theories, Molodtsov [6] introduced the concept of soft set theory as a mathematical tool for dealing with such uncertainties. We know that Pawlak [7] first used and defined the term "soft set". However, that was a different concept.

According to Hwang and Yoon [8], multicriteria decision making (MCDM) is applied preferably for decisions among available classified alternatives with multiple attributes. So MCDM is one of the most widely used decision methodology in project selection problems [9]. The MCDM is a method that follows the analysis of several criteria, simultaneously. In this method, economic, environmental, social, and technological factors are considered for the selection of the project and for making the choice sustainable. Several frameworks have been proposed for solving MCDM problems, namely Analytical Hierarchy Process [AHP] [10] and Analytical Network Process [ANP] [11], which deal with decisions in absence of knowledge of the independence of higher level elements from lower level elements and about the independence of the elements within a level. Other framework available are data envelopment analysis (DEA), technique for order performance by similarity to ideal solution (TOPSIS) [12], VIKOR [13], COPRAS [14], with gray number [15, 16], Simple Additive weighting (SAW) LINMAP [17] etc. With these techniques, alternative ratings are measured and the weight of the criteria is expressed in precise number. The projects' life cycle assessment is to be determined and the impact of all elements is to be measured. There are some mandatory axioms that the criteria describing feasible alternatives are dimensions which have relative importance to determine the performance.

There are several decision-support system models available in the domain of multicriteria optimization. These have been successfully used for making decisions in multiobjective constraint satisfaction problem.

An application of AHP [18] to the project selection problem is not now in the art. Satty's Analytical Network Process [19] is assumed suitable for project evaluation process. On the other hand, in the attempt to integrate the cardinal and ordinal preferences using ANP/AHP for project selection, the decisions failed to give stable models [20–22].

Bakshi et al. [23] have successfully established Fuzzy AHP-QFD model for software project selection. Fuzzy methods are applied to the multiattributes decision model [24, 25]. Sevkli [26] has proposed a method for project selection hybridized with fuzzy linear programming and AHP. The weights of the project selection criteria are measured using AHP model. Several types of integrated QFD techniques [27] have been proposed for ranking the candidate supplier.

Pawlak [5] proposed the application of soft sets in decision-making problem with the help of rough sets. Earlier Lin [28] and Yao [29] have presented the rough set model for decision-making problem. Maji et al. [30] has given a soft set theoretic model for decision-making problem and Som [31] has given a fuzzy soft matrix model for decision-making method. However, it is noted that so far no researcher has developed soft set theoretic hybrid model combining AHP. So it can be assumed to be an introductory attempt of making soft set theory a tool for deriving a mathematical model for decision support system.

2 Theoretical Preliminaries

In this section, we present the concept of decision support system and decision-making problem briefly. Thereafter, we introduce multicriteria decision-making problem. In the last phase, we shall give a short introduction of fuzzy soft set and Analytical Hierarchical Process Model.

(a) **Multicriteria Decision Making Problem**:

Decision making is the core area of administrative activities. By decision making, we mean a specific type of human activities aimed to choose the best among available alternatives [32]. This definition includes the three necessary objectives in decision-making process. The problem to be solved can be stated as "A the decision to be taken by a person or collective body out of several alternatives with the objective of having the best choice among the alternatives."

The decision-making procedure is essentially maintained by the contents, scale, and time interval in which the problem is required to be solved. We can formulate a decision-making problem in a logical statement of the form [33]:

$$\text{"Given : } V\text{; required: } W\text{;"} \tag{2.1}$$

where V is the specified condition and W is the objective to be fulfilled. As a first approximation, the specified condition V includes V^{s}—the set of probable states of some objectives. V^{p}—set of operators transferring the object from one state to another. Obviously, there can be a set of mappings of the subset V^{s} into V^{p}. The objective W determines the state of objects.

Since the problem-solving procedure mainly depends on the statement and structure of the problem itself, we consider the general formal structure given in [33]:

$$\text{"Given: } Y, Z, D, S, U\text{; required: } W\text{" } (Y, Z, D, S, U), \tag{2.2}$$

where Y is the set of input factors which is under control, Z is the set of unrestricted input factors, S is the set of outcomes or final results, D is the set of operators d from $Y \times Z$ to S, W is the objective of choosing subset $S*$ from S (where $S*$ can consist of a single element of criterion from U), U is the set of criteria for evaluation of elements of S and selection of $S*$.

The real-world decision problem necessitates the development of a model to construct the set of admissible alternatives, from a criteria space, orders the alternatives by aspects, and obtains the estimates under the chosen criteria. The methods for solving the estimation problems are based on the use of expert opinion.

The expert evaluation is applied with the idea of feedback systems when experts obtain the result of processing their estimates by a specific algorithm. A quantitative composition of the expert team is important in the expert evaluation.

(b) **Brief introduction towards multicriterion problem of selection and optimization**:

In the problems of selecting, the set of alternatives is known and the principle of optimality is generally unknown. In the classical mathematical programming of optimization problem, we consider a possibility to use the theory of choice and the theory of optimization, where set C is referred as the set of controls and the mapping $\phi : C \rightarrow E_m$ are specified. The vector $\phi(c) \in E_m$ is interpreted as the outcome from C, where E_m is the decision-making environment.

In general, we formulate the multicriteria problem as follows:
Find all or some $c^* \in C$ such that $\phi(c^*) \subseteq C_{optimal}(\phi)$, specific types of problems can be obtained by specifying the principle of optimality, the type of the set C and the mapping ϕ.

The multicriteria decision-making problem can be represented as the triplet

$$\langle C, \phi, R \rangle \tag{2.3}$$

where C is the set of control variables, ϕ is the mapping from C in E_m, R is the binary relation on E_m by which the alternative outcomes are compared.

The method of comparison of alternative outcomes should be equipped with numerical evaluation of alternative utility and preference relations. Here, we summarize the main notion of utility theory as follows [33, 34].

The set of alternatives A together with the preference relation $\prec$ specified on it is called the structure or the preference space. The utility function is called the real-valued strictly isotone function on A; if there exists function c such that

$$x \prec y \Rightarrow c(x) \prec c(y) \tag{2.4}$$

From (2.4), we can say that the relation $\prec$ is acyclic. This condition is normally supplemented with the constancy condition for function c on equivalence classes:

$$x \approx y \Rightarrow c(x) = c(y) \tag{2.5}$$

If the preference relation is not acyclic, it can no longer be represented in any sense by the ordinary utility function. Nevertheless, any relation $\prec$ can be represented by some function c defined on $A \times A$ in the sense that

$$x \succ y \Leftrightarrow u(x, y) > 0 \tag{2.6}$$

where u is the comparative utility function.

(c) **Decision Making under incomplete Information**:

The quality of decision-making process depends directly on the extent to which all the control factors essential to making decisions and to decisions effects are allowed for. The decision authority often has to perform under uncertainties where it has a smaller amount of information than the requirement for reasonable actions during decision making. The uncertainty can be partially minimized by the information available or the information additionally received by the decision-making authority.

Uncertainty in decision making is characterized by the insufficient reliability and the amount of information on the basis of which the decision-making authority (DM) chooses a decision. We summarize the various kind of uncertainties commonly occurring during the decision-making process as follows:

(i) Uncertainty in principle.
(ii) Uncertainty due to lack of information.
(iii) Uncertainty generated by decision authority.
(iv) Uncertainty involving constraint in decision making process.
(v) Uncertainty caused by behavior of environment.

Another important class of uncertain situation is based on Zadeh's notion of fuzzy set. These tools are adequate to the description of situations having a clear cut boundary.

(d) **Fuzzy Decision Making Environment**:

The uncertain information situation characterizes the case where the control authority (c) has a "Fuzzy" knowledge of states of environment (E_m). We assume that the control authority C has an exact knowledge of the complete set Δ of probable state ∂_j of environment, the set ϕ has its decision ϕ_k and the value of evaluation functional

$$F = \{f_{jk}\}_{j,k=1}^{n,m}$$

Based on the concepts of theory of fuzzy sets, we model the "behavior" of uncertainties and define the decision situations as triplet $\{\phi, R_{\theta,} F\}$, where R_θ are the fuzzy sets or fuzzy random event determined by membership function μ_R and the probability distribution p in the states of environment E_m.
We list the main operations of fuzzy sets as follows:

(i) Equivalence $A \sim B \Leftrightarrow \mu_A(x) \equiv \mu_B(x)$.
(ii) Inclusion $A \subset B \Leftrightarrow \mu_A(x) \leq \mu_B(x)$.
(iii) Complement $\overline{A} \Leftrightarrow \mu_{\overline{A}}(x) = 1 - \mu_A(x)$.
(iv) Union $A \cup B \Leftrightarrow \mu_{A\cup B}(x) = \max\{\mu_A(x), \mu_B(x)\}$.
(v) Intersection $A \cap B \Leftrightarrow \mu_{A\cap B}(x) = \min\{\mu_A(x), \mu_B(x)\}$.
(vi) Product $A \cdot B \Leftrightarrow \mu_{A\cdot B}(x) = \mu_A(x) \cdot \mu_B(x)$.
(vii) Sum $A + B \Leftrightarrow \mu_{A+B}(x) = \mu_A(x) + \mu_B(x) - \mu_A(x).\mu_B(x)$.
(viii) Multiplication of A by $\alpha \in [0, 1] : \alpha \Leftrightarrow \mu_{\alpha A}(x) = \alpha\mu_A(x)$.
(ix) Exponentiation of A to $\alpha > 0 : A^\alpha \Leftrightarrow \mu_{A^\alpha}(x) = (\mu_A(x))^\alpha$.

(x) Concentration CON (A) = A^k, $k \geq 2$.
(xi) Dilation $DIL(A) = A^{0,5}$.

The notion of fuzzy sets and relations defines the construction of various models for fuzzy specification of the "behavior" of environment E_m as applied to formal scheme of definition of decision situations $\{\phi, \Delta, F\}$, which have been discussed above previously.

(e) **Introduction to Soft Set Theory**:

In this subsection, we try to give a precised introduction of soft set theory and its competence in decision making.

Let U be the initial Universal set and let Q be the set of parameters.

Definition 1 ([30]) A pair (F, Q) is called a soft set over U if and only if F is a mapping of Q into the set of all subsets of the set U i.e., $F : Q \to P(U)$, where $P(U)$ is the power set of U.

Soft set is a parameterized family of subsets of the set U. Every set $F(\varepsilon)$, $for \varepsilon \in Q$ from this family may be considered as the set of ε-elements of the soft set (F, Q) or as the set of ε-approximate elements of the soft set.

According to Zadeh, Fuzzy sets can be considered as a special case of soft set. Let A be a fuzzy set of U with membership value, μ_A , i.e., μ_A is a mapping of U into [0,1]. Let us consider the family of α-level sets for the function μ_A given by

$$F(\alpha) = \{x \in U; \mu_A(x) \geq \alpha\}, \quad \alpha \in [0, 1].$$

If we know the family F, we can find the functions $\mu_A(x)$ by means of the following formulae:

$$\mu_A(x) = \sup_{\alpha \in [0,1].x \in F(\alpha)} (\alpha)$$

Thus every fuzzy set A may be considered as the soft set (F[0,1]).

3 Proposed Fuzzy Soft AHP Hybrid Model of Decision Making

We construct a hybrid model of decision-making problem as follows:

First, we represent the problem and convert it into equivalent binary tabular representation. Then reduce the table of binary information into reduced soft set.

In the next step, we construct an algorithm to select the best project using choice criteria.

In the third step, we construct the weighted table of the proposed soft set problem. For finding the weight, fuzzy AHP method has been used. Finally, the best alternatives are chosen.

Schematic Diagram of Fuzzy-AHP

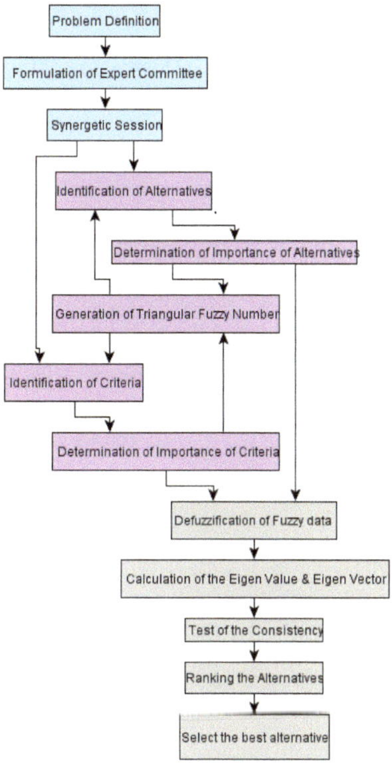

Proposed Soft theoretic AHP Model

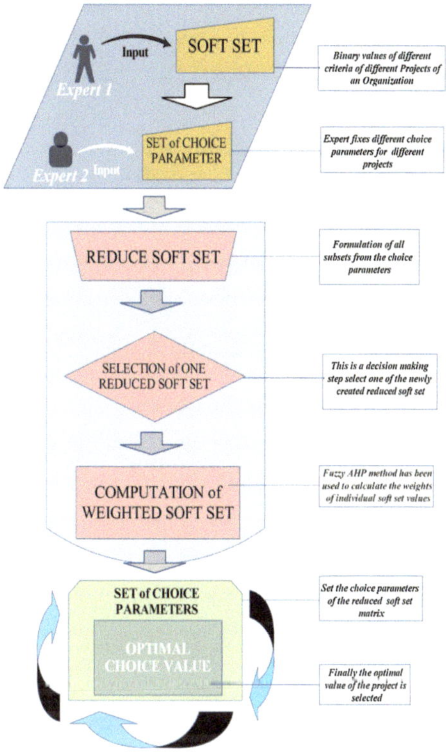

Case Study:

Let us consider four projects: P_1, P_2, P_3 and P_4.

Let E = {NPV; ROR; PB; PR; Highly Beneficial; Beneficial; Average; Poor} be a set of parameters.

Consider the Soft Set (F, E) which describes the "Profit of the Organization" given by

(F, E) = {Max NPV $\{P_4, P_2\}$, Max ROR $\{P_3, P_2\}$, Max PB $\{P_1, P_2, P_3, P_4\}$, Min PR $\{P_1, P_2\}$, Highly Beneficial $\{P_4, P_3, P_1, P_2\}$, Beneficial $\{P_3, P_2\}$, Average $\{P_4, P_1\}$, Poor $\{P_3, P_1, P_2\}$}.

Suppose that an organization is interested to take the project on the basis of its choice of parameters: "Maximum ROR," "Maximum Payback Period," "Beneficial," and "Minimum Project Risk" etc. which consider the subset

P = {Max ROR; Max PB; Beneficial; Min Project Risk, Max NPV} of the set E.

This means, out of available projects in U, organization would select that the project which qualifies with all (or with max number of) parameters of the soft set P.

Tabular representation of a soft set (F, P) above on the basis of the set P of the choice parameters of the organization A. We can represent this soft set in a tabular form as shown below. This style of representation will be useful for storing a soft set in a computer memory. If $h_i \in F(\varepsilon)$ then $h_{ij} = 1$, otherwise $h_{ij} = 0$ where h_{ij} are the entries in Table 1.

Reduce—Table of a Soft Set:

From the table we see that $\{e_1, e_2, e_4, e_5\}$, $\{e_2, e_3, e_4, e_5\}$ are the two reduces of P = $\{e_1, e_2, e_3, e_4, e_5\}$.

Choose any one say Q = $\{e_1, e_2, e_4, e_5\}$.

Incorporating the choice values, the reduced soft set can be represented in Table 2.

Now, having the reduced table by Fuzzy AHP method, use the revised algorithm for the selection of the best project using Fuzzy AHP Method, i.e.,

1. Input the Soft Set (F, E).
2. Input the set P of choice parameters of the organization is a subset of E.
3. Compute all reduce soft sets of (F, P).
4. Choose one reduce soft set say (F, Q) of (F, P).
5. Compute weighted table of the soft set (F, Q) according to the weights computed by Fuzzy AHP Method.
6. Find k, for which $C_k = \max C_i$. Then h_k is the optimal choice object.

Algorithm of Fuzzy AHP Method:

The fuzzy AHP technique can be viewed as an advanced analytical method developed from the traditional AHP. According to the method of Chang's (1992) [35] extent analysis, each criterion is taken into account and the extent analysis for each criterion g_i is performed on the set of criteria, respectively. Therefore, m extent analysis values for each criterion can be obtained by using following notation:

Table 1 Initial Soft set Formation

	Max ROR	Max PB	Beneficial	Min PR	Max NPV
U	e_1	e_2	e_3	e_4	e_5
P_1	0	1	0	1	0
P_2	1	1	1	1	1
P_3	1	1	1	0	0
P_4	0	1	0	0	1

Table 2 Reduce Soft set Formation

U	e_1	e_2	e_4	e_5	Choice value
P_1	0	1	1	0	$C_1 = 2$
P_2	1	1	1	1	$C_2 = 4$
P_3	1	1	0	0	$C_3 = 2$
P_4	0	1	0	1	$C_4 = 2$

$M_{g_i}^1, M_{g_i}^2, M_{g_i}^3, M_{g_i}^4, M_{g_i}^5, \ldots, M_{g_i}^m$, where g_i is the goal set ($i = 1, 2, 3, 4, \ldots, n$) and all $M_{g_i}^j$ ($j = 1, 2, 3, 4, \ldots m$) are Triangular Fuzzy Numbers (Tfns). The steps of the analysis can be given as follows:

Step 1: The fuzzy synthetic extent value (S_i) with respect to the ith criterion is defined in Eq. (3.1):

$$S_i = \sum_{j=1}^{m} M_{g_i}^j \otimes \left(1/\left[\sum_{i=1}^{n}\sum_{j=1}^{m} M_{g_i}^j\right]\right) \tag{3.1}$$

(Operation $\otimes$ is defined as the one to one multiplication)

To obtain (2.2) as

$$\sum_{j=1}^{m} M_{g_i}^j \tag{3.2}$$

Perform the fuzzy addition operation of m extent analysis values for a particular matrix given in Eq. (3.3) below, at the end step of calculation, new ($l, m,$ and u) set is obtained and used for the next:

$$\sum_{j=1}^{m} M_{g_i}^j = \left(\sum_{j=1}^{m} lj, \sum_{j=1}^{m} mj, \sum_{j=1}^{m} uj\right) \tag{3.3}$$

where l is the lower limit value, m is the most promising value, and u is the upper limit value and to obtain (3.4):

$$\left(1/\left[\sum_{i=1}^{n}\sum_{j=1}^{m} M_{g_i}^j\right]\right) \tag{3.4}$$

Perform the fuzzy addition operation of $M_{g_i}^j$ ($j = 1, 2, 3, \ldots, m$) values given as Eq. (3.5):

$$\sum_{i=1}^{n}\sum_{j=1}^{m} M_{g_i}^j = \left(\sum_{i=1}^{n} li, \sum_{i=1}^{n} mi, \sum_{i=1}^{n} ui\right) \tag{3.5}$$

Then compute the inverse of the vector in the Eq. (3.5) and obtain the inverse Eq. (3.6) as:

$$\left(1/\left[\sum_{i=1}^{n}\sum_{j=1}^{m} M_{g_i}^j\right]\right) = \left[\frac{1}{\sum_{i=1}^{n} u_i}, \frac{1}{\sum_{i=1}^{n} m_i}, \frac{1}{\sum_{i=1}^{n} l_i}\right] \tag{3.6}$$

Step 2: The degree of possibility of $M_2 = (l_2, m_2, u_2) \geq M_1 = (l_1, m_1, u_1)$ is defined as Eq. (3.7):

$$V(M_2 \geq M_1) = \sup_{y \geq x}[\min(\mu_{M1}(x), \mu_{M2}(y))] \tag{3.7}$$

In addition, x and y are the values on the axis of membership function of each criterion. This equation can be written as:

$$\begin{aligned} V(M_2 \geq M_1) &= 1, \quad \text{if } m_2 \geq m_1 \\ &= 0, \quad \text{if } l_1 \geq u_2 \\ &= \frac{l_1 - u_2}{(m_2 - u_2) - (m_1 - l_1)}, \quad \text{otherwise} \end{aligned} \tag{3.8}$$

Step 3: The degree possibility for a convex fuzzy number to be greater than k convex fuzzy numbers $M_i (i = 1, 2, 3, 4, 5 \ldots k)$ can be defined by

$$V(M \geq M_1, M_2, M_3 \ldots, M_k) = \min V(M \geq M_i),$$

$i = 1, 2, \ldots, k$. Assume that Eq. (3.9) is

$$d^*(A_i) = \min V(S_i \geq S_k) \tag{3.9}$$

For $k = 1, 2, 3, \ldots, n; k \neq i$. Then the weight vector is given by Eq. (3.10):

$$W^* = (d^*(A_1), d^*(A_2), \ldots d^*(A_n))^T \tag{3.10}$$

where $A_i (i = 1, 2, 3 \ldots n)$ are n elements.
Step 4: Via normalization, the normalized weight vectors are given in Eq. (3.11) as

$$W = (d(A_1), d(A_2), d(A_3) \ldots d(A_n))^T \tag{3.11}$$

where W has nonfuzzy numbers.

Case Study of Fuzzy AHP Method:

According to expert's decision, the following matrix is formed and then by using triangular fuzzy number the Fuzzy evaluation matrix is formed [36].

Now calculating all the values by applying Chang's [35] theory, the following results are obtained:

$$\begin{cases} S_{NPV} = (3.50,\ 5.00,\ 6.50) \otimes (0.04, 0.057, 0.078) = (0.14, 0.28, 0.51) \\ S_{ROR} = (4.13,\ 6.00,\ 9.33) \otimes (0.04, 0.057, 0.078) = (0.17, 0.34, 0.73) \\ S_{PB} = (3.13,\ 3.83,\ 5.33) \otimes (0.04, 0.057, 0.078) = (0.13, 0.22, 0.42) \\ S_{PR} = (2.08,\ 2.75,\ 3.75) \otimes (0.04, 0.057, 0.078) = (0.08, 0.16, 0.29) \end{cases}$$

and

$$\begin{cases} V(S_{NPV} \geq S_{ROR}) = 0.85, \; V(S_{NPV} \geq S_{PB}) = 1, \\ V(S_{NPV} \geq S_{PR}) = 1, \\ V(S_{ROR} \geq S_{NPV}) = \; 1, \; V(S_{ROR} \geq S_{PB}) = 1, \\ V(S_{ROR} \geq S_{PR}) = 1, \\ V(S_{PB} \geq S_{NPV}) = 0.82 \; V(S_{PB} \geq S_{ROR}) = 0.67, \\ V(S_{PB} \geq S_{PR}) = 1, \\ V(S_{PR} \geq S_{NPV}) = 0.55, \; V(S_{PR} \geq S_{ROR}) = 0.4, \\ V(S_{PR} \geq S_{PB}) = 0.73 \end{cases}$$

Minimum of all values (0.85, 1, 0.67, and 0.4)
The weight $W = (0.29, 0.34, 0.23, 0.14)$
So in our case study, e_1 denotes the Max ROR and its weight $w_1 = 0.34$
e_2 denotes the Max PB and its weight $w_2 = 0.23$
e_4 denotes the Min PR and its weight $w_4 = 0.14$
e_5 denotes the Max NPV and its weight $w_5 = 0.29$.
Using these weighted values, Table 3 is constructed:
So $P_2 \to P_3 \to P_4 \to P_1$ i.e., P_2 is the best project (Tables 4 and 5).

Table 3 Evaluation matrix

Criteria	NPV	ROR	PB	PR
NPV	1	1	2	1
ROR	1	1	2	2
PB	0.5	1	1	1.33
PR	0.5	0.5	0.75	1

Table 4 Fuzzy evaluation matrix

Criteria	NPV	ROR	PB	PR
NPV	(1, 1, 1)	(0.75, 1, 1.25)	(1, 2, 3)	(0.75, 1, 1.25)
ROR	(0.8, 1, 1.33)	(1, 1, 1)	(1, 2, 3)	(1.33, 2, 4)
PB	(0.33, 0.5, 1)	(0.8, 1, 1.33)	(1, 1, 1)	(1, 1.33, 2)
PR	(0.25, 0.5, 0.75)	(0.33, 0.5, 1)	(0.5, 0.75, 1)	(1, 1, 1)

Table 5 Final Matrix with weighted Value

Weight	0.34	0.23	0.14	0.29	Choice value
U	$e_1 \cdot w_1$	$e_2 \cdot w_2$	$e_4 \cdot w_4$	$e_5 \cdot w_5$	
P_1	0	1	1	0	$C_1 = 0.37$
P_2	1	1	1	1	$C_2 = 1$
P_3	1	1	0	0	$C_3 = 0.57$
P_4	0	1	0	1	$C_4 = 0.52$

4 Conclusion

In the present paper, we modeled an application of fuzzy soft theory in decision support system. In this context, we have introduced the soft theoretic model of analytic hierarchical process (AHP) to have a better decision. This proposed decision support strategy for an intended project manager helped to take decision in the perspective environment. The dataset used in this paper is collected from experts' opinion. The algorithm, evolved from the resultant soft set theoretic AHP; it lead us to maximize the proper choice in the environment of imprecise information. The main advantage of this method, when compared to others, is that this hybrid method is very simple in terms of calculation and the computational complexity of the proposed algorithm is very low.

References

1. Zadeh, L.A.: Fuzzy sets. Inf. Control **8**, 338–353 (1965)
2. Atanassov, K.: Intutionistic fuzzy sets. Fuzzy Sets Syst. **20**, 87–96 (1986)
3. Atanassov, K.: Operators over interval valued intutionistic fuzzy sets. Fuzzy Sets Syst. **64**, 159–174 (1994)
4. Gau, W.L., Buehrer, D.J.: Vague sets. IEEE Trans. Syst. Man Cyber. **23**(2), 610–614 (1993)
5. Pawlak, Z.: Rough sets. Int. J. Inf. Comput. Sci. **11**, 341–356 (1982)
6. Molodtsov, D.: Soft set theory-first results. Comput. Math Appl. **37**(4/5), 19–31 (1999)
7. Pawlak, Z.: Hard set and soft sets. ICS research report, Institute of Computer Science, Poland (1994)
8. Hwang, C.L., Yoon, K.P.: Multiple Attribute Decision Making and Introduction, p. p2. Sage Publication, London (1995)
9. Bakshi, T., Sarkar, B.: MCA based performance evaluation of project selection. Int. J. Softw. Eng. Appl. (IJSEA) **2**(2), 14–22 (2011)
10. Saaty, T.L.: The Analytic Hierarchy Process: Planning, Priority Setting, Resource Allocation. Mcgraw-Hill, New York (1980)
11. Satty, T.L.: The Analytic Network Process. RWS Publications, Pittsburgh (1996)
12. Hwang, C.L., Yoon, K.: Multiple Attribute Decision Making: Methods and Applications. A state of the Art Survey., 1st edn, p. 259. Springer, Berlin (1981)
13. Wang, T.C., Lee, H.-D.: Developing a fuzzy TOPSIS approach based on subjective weights and objective weights. Expert Syst. Appl. **36**(5), 8980–8985 (2009)

14. Datta, S., Beriha, G.S., Patnaik, B., Mahapatra, S.S.: Use of compromise ranking method for supervisor selection: a multi-criteria decision making (MCDM) approach. Int. J. Vocat. Tech. Educ. **1**(1), 007–013 (2009)
15. Zavadskas, E.K., Turskis, Z., Tamosaitiene, J., Marina, V.: Multicriteria selection of project managers by applying grey criteria. Tech. Econ. Dev. Econ. **14**(4), 462–477 (2008)
16. Zavadskas, E.K., Kaklauskas, A., Turskis, Z., Tamosaitiene, J.: Multi-attribute decision-making model by applying grey numbers. Informatica **20**(2), 305–320 (2009)
17. Srinivasan, V., Shocker, A.D.: Linear programming techniques for multidimensional analysis of privileged. Psychometrika **38**, 337–369 (1973)
18. Bayazit, O., Karpak, B.: An AHP application in vendor selection. In: ISAHP 2005, Honolulu, Hawaii, July 8–10, pp. 1–24 (2005)
19. Sarkis, J., Talluri, S.: A model for strategic supplier selection. J. Supply Chain Manag.: A Glob. Rev. Purch. Supply (Winter), pp. 18–28 (2002)
20. Bayazit, O.: Use of analytic network process in vendor selection decisions. Benchmarking: Int. J. **13**(5), 566–579 (2006)
21. Chan, F.T.S.: Interactive selection model for supplier selection process: an analytical hierarchy process approach. Int. J. Prod. Res. **41**(15), 3549–3579 (2003)
22. Kahraman, C., Cebeci, U., Ulukan, Z.: Multi-criteria supplier selection using fuzzy AHP. Logist. Inf. Manag. **16**(6), 382–394 (2003)
23. Bakshi T., Sanyal S.K.: A soft-computing approach for software project selection. In: IEEE International Conference on Recent Trends in Information Systems (RETIS-11), Kolkata, pp. 18–23, 978- 1–4577-0792-6/11. IEEE, 21–23 December 2011
24. Razmi, J., Rafiei, H., Hashemi, M.: Designing a decision support system to evaluate and select suppliers using fuzzy analytic network process. Comput. Ind. Eng. **57**(4), 1282–1290 (2009)
25. Sreekumar, Mahapatra, S.S.: A fuzzy multi-criteria decision making approach for supplier selection in supply chain management. Afr. J. Bus. Manag. **3**(4), 168–177 (2009)
26. Sevkli, M., Koh, S.C.L., Zaim, S., Demirbag, M., Tatoglu, E.: Hybrid analytical hierarchy process model for supplier selection. Ind. Manag. Data Syst. **108**(1), 122–142 (2008)
27. Ju, J.-M., Hwang, S.-G.: A study on supplier evaluation and selection method based on dependence. J. Adv. Comput. Intell. Intell. Inform. **8**(4), 415–420 (2004)
28. Lin, T.Y., Skoworn, A., Polkowski, L.: Granular computing on binary relations II: rough set representations and belief functions. In: Skoworn, A., Polkowski, L. (eds.) Rough Sets in Knowledge Discovery, pp. 121–140. Springer, Berlin (1998)
29. Yao, Y.Y.: Relational interpretations of neighbourhood operators and rough set approximation operators. Inf. Sci. **111**(1–4), 239–259 (1998)
30. Maji, P.K., Biswas, R., Roy, A.R.: Soft set theory. Comput. Math. Appl. **45**, 555–562 (2003)
31. Som, T.: The theory of fuzzy soft sets, fuzzy soft matrix and their application in decision making problem. In: Proceedings of National Conference on Mathematics of Soft Computing, National Institute of Technology Calicut, 5–7 July 2012
32. Trukhayev, R.I.: Decision Models Under Uncertainty. Nauka, Moscow (1981)
33. Kolbin, V.V.: Systems Optimization Methodology. World Scientific Publishing Company, Singapore (2000–2001)
34. Kolbin, V.V., Suvorova, M.A: Decision Making under Incomplete Information, Saint-Petersburg State University, p. 81 (2002)
35. Chang, D.Y.: Theory and methodology applications of the extent analysis method on fuzzy AHP. Eur. J. Oper. Res. **95**, 649–655 (1996)
36. Bakshi T., Sarkar B., Sanyal S.K.: A new soft-computing based framework for project management using game theory. In: International Conference on Communications, Devices and Intelligent Systems (CODIS-2012), 978-1-4673-4700-6/12/$31.00_c, pp. 282–285. IEEE (2012)

An Application of Weighted Neutrosophic Soft Sets in a Decision-Making Problem

Pabitra Kumar Maji

Abstract Decision-making problems in an imprecise environment are of paramount importance in recent years. Here we consider an object recognition problem in an imprecise environment. In this paper we study the concept of weighted neutrosophic soft sets. A multiobserver decision-making problem has been considered here as an application of weighted neutrosophic soft sets. We have considered here a recognition strategy based on multiobserver input parameter data set.

Keywords Soft sets · Neutrosophic set · Neutrosophic soft set · Weighted neutrosophic soft set

1 Introduction

Works on soft set theory is growing rapidly since its initiation [1]. The novel concept of soft set theory plays an important role as a mathematical tool for dealing with uncertainties. The basic properties of the theory may be found in [2]. Ali et al. [3] presented some new algebraic operations on soft sets. Chen et al. [4] presented a new definition of soft set parameterization reduction and compared this definition with the related concept of knowledge reduction in the rough set. Feng et al. [5] introduced the concept of semirings and we can also find the concept of soft groups in [6]. Xu et al. [7] introduced vague soft sets which is a combination of soft sets and vague sets. Some applications of soft sets may be found in [4, 8–10].

The problem of object recognition has received paramount importance in recent years. The recognition problem may be viewed as a multiobserver decision-making problem, where the final identification of the object is based on set of inputs from different observers who provide the overall object characterization in terms of diverse sets of choice parameters. In this paper we present the concept of weighted neutrosophic soft sets. A multiobserver decision-making problem has been considered

P.K. Maji (✉)
Department of Mathematics, Bidhan Chandra College, Asansol, West Bengal, India
e-mail: pabitra_maji@yahoo.com

M.K. Chakraborty et al. (eds.), *Facets of Uncertainties and Applications*,
Springer Proceedings in Mathematics & Statistics 125,
DOI 10.1007/978-81-322-2301-6_16

here as an application of weighted neutrosophic soft set theoretic approach toward the solution of the above decision-making problem.

In Sect. 2 of this paper we briefly recall some relevant preliminaries centered around our problem. Some basic definitions on weighted neutrosophic soft set relevant to this work are available in Sect. 3. A decision-making problem has been discussed and solved in Sect. 4. Conclusions are drawn in the concluding Sect. 5.

2 Preliminaries

In most real-life problems in the fields of medical sciences, economics, engineering, etc., the data involved are imprecise in nature. The classical mathematical tools are not capable to handle such problems. The novel concept 'soft set theory' initiated by Molodtsov [1] is a new mathematical tool to deal with such problems. For better understanding we now recapitulate some preliminaries relevant to the work.

Definition 2.1 [1] Let U be an initial universe set and E be a set of parameters. Let P(U) denote the power set of U. Consider a nonempty set A, A $\subset$ E.

A pair (F, A) is called a soft set over U, where F is a mapping given by F: A $\rightarrow$ P (U).

A soft set over U is a parameterized family of subsets of the universe U. For $\epsilon \in$ A, F(ϵ) may be considered as the set of ϵ- approximate elements of the soft set (F, A).

Definition 2.2 [2] For two soft sets (F, A) and (G, B) over a common universe U, we say that (F, A) is a soft subset of (G, B) if

(i) A $\subset$ B, and
(ii) $\forall\ \epsilon \in$ A, F(ϵ) and G(ϵ) are identical approximations.

We write (F, A) $\tilde{\subset}$ (G, B).

(F, A) is said to be a soft superset of (G, B), if (G, B) is a soft subset of (F, A). We denote it by (F, A) $\tilde{\supset}$ (G, B).

Let A and B be two subsets of E, the set of parameters. Then A $\times$ B $\subset$ E $\times$ E. Now we are in the position to define 'AND', 'OR' operations on two soft sets.

Definition 2.3 [2] If (F, A) and (G, B) be two soft sets then "(F, A) AND (G, B)" denoted by (F, A) $\wedge$ (G, B) is defined by (F, A) $\wedge$ (G, B) = (H, A $\times$ B), where H(α, β) = F(α) $\bigcap$ G(β), $\forall$ (α, β) $\in$ A $\times$ B.

Definition 2.4 [2] If (F, A) and (G, B) are two soft sets then "(F, A) OR (G, B)" denoted by (F,A) $\vee$ (G, B) is defined by (F, A) $\vee$ (G, B) = (O, A $\times$ B), where, O(α, β) = F(α) $\bigcup$ G(β), $\forall$ (α, β) $\in$ A $\times$ B.

In 1960 Abraham Robinson developed the nonstandard analysis, a formalization of analysis and a branch of mathematical logic that rigorously defines the infinitesimals. Informally, an infinitesimal is an infinitely small number. Formally, x is said to be infinitesimal if and only if for all positive integers n one has $|x| < \frac{1}{n}$.

Let $\epsilon > 0$ be a such infinitesimal number. Let us consider the nonstandard finite numbers $1^+ = 1 + \epsilon$, where '1' is its standard part and 'ϵ' its nonstandard part, and $^-0 = 0 - \epsilon$, where '0' is its standard part and 'ϵ' its nonstandard part.

Definition 2.5 [11] A neutrosophic set A on the universe of discourse X is defined as A= $\{< x, T_A(x), I_A(x), F_A(x) >, x \in X\}$, where $T_A, I_A, F_A : X \to]^-0, 1^+[$ and $^-0 \leq T_A(x) + I_A(x) + F_A(x) \leq 3^+$. Here T_A, I_A, F_A are, respectively, the true membership, indeterministic membership, and false membership function of an object $x \in X$.

From a philosophical point of view, the neutrosophic set takes the value from real standard or nonstandard subsets of $]^-0, 1^+[$. But in real-life application in scientific and engineering problems it is difficult to use neutrosophic set with value from real standard or nonstandard subset of $]^-0, 1^+[$. Hence we consider the neutrosophic soft set which takes the value from the subset of [0, 1].

Definition 2.6 [12] Let U be an initial universe set and E be a set of parameters. Consider $A \subset$ E. Let P(U) denote the set of all neutrosophic sets of U.

The collection (F, A) is termed to be the neutrosophic soft set (N S S) over U, where F is a mapping given by F: A $\to$ P (U).

For illustration we consider the following example.

Example 1 Let U be the set of houses under consideration and E the set of parameters. Each parameter is a neutrosophic word or sentence involving neutrosophic words. Consider E = {beautiful, large, very large, small, average, costly, cheap, brick build}. In this case, to define a neutrosophic soft set means to point out beautiful houses, large houses, very large houses, and so on. Suppose there are five houses in the universe U given by, U = $\{h_1, h_2, h_3, h_4, h_5\}$ and the set of parameters A = $\{e_1, e_2, e_3, e_4\}$, where e_1 stands for the parameter 'large', e_2 stands for the parameter 'very large', e_3 stands for the parameter 'small', and e_4 stands for the parameter 'average'. Suppose,

F(large) = $\{< h_1, 0.6, 0.4, 0.7 >, < h_2, 0.5, 0.6, 0.8 >, < h_3, 0.8, 0.7, 0.7 >, < h_4, 0.6, 0.4, 0.8 >, < h_5, 0.8, 0.6, 0.7 >\}$,

F(very large) = $\{< h_1, 0.5, 0.3, 0.6 >, < h_2, 0.8, 0.5, 0.7 >, < h_3, 0.9, 0.7, 0.8 >, < h_4, 0.7, 0.6, 0.7 >, < h_5, 0.6, 0.7, 0.9 >\}$,

F(small) = $\{< h_1, 0.3, 0.8, 0.9 >, < h_2, 0.4, 0.6, 0.8 >, < h_3, 0.6, 0.8, 0.4 >, < h_4, 0.7, 0.7, 0.6 >, < h_5, 0.6, 0.7, 0.9 >\}$,

F(average) = $\{< h_1, 0.8, 0.3, 0.4 >, < h_2, 0.9, 0.6, 0.8 >, < h_3, 0.8, 0.7, 0.8 >, < h_4, 0.6, 0.7, 0.5 >, < h_5, 0.7, 0.6, 0.8 >\}$.

For the purpose of storing a neutrosophic soft set in a computer, we could represent it in the form of a table as shown below (corresponding to the neutrosophic soft set in the above example). In this table, the entries are c_{ij} corresponding to the house h_i and the parameter e_j, where c_{ij} = (true-membership value of h_i, indeterminacy-membership value of h_i, falsity-membership value of h_i) in F(e_j). The tabular representation of the neutrosophic soft set (F, A) is as follows (Table 1).

Table 1 Tabular form of the NSS (F, A)

U	e_1 = large	e_2 = very large	e_3 = small	e_4 = average
h_1	(0.6, 0.4, 0.7)	(0.5, 0.3, 0.6)	(0.3, 0.8, 0.9)	(0.8, 0.3, 0.4)
h_2	(0.5, 0.6, 0.8)	(0.8, 0.5, 0.7)	(0.4, 0.6, 0.8)	(0.9, 0.6, 0.8)
h_3	(0.8, 0.7, 0.7)	(0.9, 0.7, 0.8)	(0.6, 0.8, 0.4)	(0.8, 0.7, 0.8)
h_4	(0.6, 0.4, 0.8)	(0.7, 0.6, 0.7)	(0.7, 0.7, 0.6)	(0.6, 0.7, 0.5)
h_5	(0.8, 0.6, 0.7)	(0.6, 0.7, 0.9)	(0.6, 0.7, 0.9)	(0.7, 0.6, 0.8)

Definition 2.7 [12] Let (F, A) and (G, B) be two neutrosophic soft sets over the common universe U. (F, A) is said to be a neutrosophic soft subset of (G, B) if $A \subset B$, and $T_{F(e)}(x) \leq T_{G(e)}(x)$, $I_{F(e)}(x) \leq I_{G(e)}(x)$, $F_{F(e)}(x) \geq F_{G(e)}(x)$, $\forall e \in A$.

We denote it by (F, A) $\subseteq$ (G, B). (F, A) is said to be neutrosophic soft superset of (G, B) if (G, B) is a neutrosophic soft subset of (F, A). We denote it by (F, A) $\supseteq$ (G, B).

Definition 2.8 [12] AND operation on two neutrosophic soft sets. Let (H, A) and (G, B) be two NSSs over the same universe U. Then the 'AND' operation on them is denoted by '(H, A) $\bigwedge$ (G, B)' and is defined by (H, A) $\bigwedge$(G, B) = (K, A $\times$ B), where the truth-membership, indeterminacy-membership, and falsity-membership of (K, A$\times$B) are as follows: $T_{K(\alpha,\beta)}(m) = min(T_{H(\alpha)}(m), T_{G(\beta)}(m))$, $I_{K(\alpha,\beta)}(m) = \frac{I_{H(\alpha)}(m)+I_{G(\beta)}(m)}{2}$, $F_{K(\alpha,\beta)}(m) = max(F_{H(\alpha)}(m), F_{G(\beta)}(m))$, $\forall \alpha \in A, \forall \beta \in B$.

An application of NSS may be foundin [13]. The decision maker may not have equal choice for all the parameters. He may impose some conditions to choose the parameters for which the decision will be taken. The conditions may be imposed in terms of weights (Positive real numbers less than 1). This condition motivates us to define weighted neutrosophic soft set.

3 Weighted Neutrosophic Soft Sets

Definition 3.1 A neutrosophic soft set is termed to be a weighted neutrosophic soft set if the weights (w_i, a real positive number) are imposed on its parameters. The entries of the weighted neutrosophic soft set, $d_{ij} = w_i \times c_{ij}$, where c_{ij} is the ijth entry in the table of neutrosophic soft set.
For illustration we consider the following example.

Example 2 Consider the Example 1. Suppose the decision maker has no equal preference for each of the parameters. He may impose the weights of preference for the parameters 'e_1 = large' as '$w_1 = 0.8$', 'e_2 = very large' as '$w_2 = 0.4$', 'e_3 = small' as '$w_3 = 0.5$', 'e_4 = average large' as '$w_4 = 0.6$'. Then the weighed neutrosophic soft set obtained from (F, A) be (H, A) and its tabular representation are as in Table 2.

Table 2 Tabular form of the weighted NSS (H, A)

U	$e_1, w_1 = 0.8$	$e_2, w_2 = 0.4$	$e_3, w_3 = 0.5$	$e_4, w_4 = 0.6$
h_1	(0.48, 0.32, 0.56)	(0.20, 0.12, 0.24)	(0.15, 0.40, 0.45)	(0.48, 0.18, 0.24)
h_2	(0.40, 0.48, 0.64)	(0.32, 0.20, 0.28)	(0.20, 0.30, 0.40)	(0.54, 0.36, 0.48)
h_3	(0.64, 0.56, 0.56)	(0.36, 0.28, 0.32)	(0.30, 0.40, 0.20)	(0.48, 0.42, 0.48)
h_4	(0.48, 0.32, 0.64)	(0.28, 0.24, 0.28)	(0.35, 0.35, 0.30)	(0.36, 0.42, 0.30)
h_5	(0.64, 0.48, 0.56)	(0.24, 0.28, 0.36)	(0.30, 0.35, 0.45)	(0.42, 0.36, 0.48)

Definition 3.2 Weighted Comparison Matrix It is a matrix whose rows are labeled by the object names $h_1, h_2, \ldots, h_n$ and the columns are labeled by the weighted parameters $p_1, p_2, \ldots, p_m$, where $p_i = e_i \times w_i$, for i = 1, …, m. The entries c_{ij} of the weighted comparison matrix are evaluated by c_{ij} = a + b − c, where 'a' is the positive integer calculated as 'how many times $T_{h_i}(p_j)$ exceeds or is equal to $T_{h_k}(p_j)$', for $i \neq k$, $\forall k = 1, 2, \ldots, n$, 'b' is the integer calculated as 'how many times $I_{h_i}(p_j)$ exceeds or is equal to $I_{h_k}(p_j)$', $i \neq k$, $\forall k = 1, 2, \ldots, n$ and 'c' is the integer 'how many times $F_{h_i}(p_j)$ exceeds or is equal to $F_{h_k}(p_j)$', $i \neq k$, $\forall k = 1, 2, \ldots, n$.

Definition 3.3 Weighted Choice Value Score of an Object The weighted choice value score of an object h_i is S_i and is calculated as $S_i = \sum_j c_{ij}$.

Here we consider a problem to choose an object from a set of given objects with respect to a set of choice parameters P. We follow an algorithm to identify an object based on multiobserver (considered here three observers with their own choices) input data characterized by colors (F, A), size (G, B) and surface textures (H, C) features. The algorithm to choose an appropriate object depending on the choice parameters is given below

3.1 Algorithm

1. Input the neutrosophic soft sets (H, A), (G, B) and (H, C) (for three observers)
2. Input the parameter set P as preferred by the decision maker
3. Compute the corresponding NSS (S, P) from the NSSs (H, A), (G, B) and (H, C) and place in tabular form
4. Compute the NSS (S, Q) which is the weighted NSS obtained from he NSS (S, P), the weights ($w_i > 0$) depend on the decision maker
5. Compute the weighted comparison matrix of the NSS (S, Q)
6. Compute the score S_i of $h_i, \forall i$
7. The decision is h_k if $S_k = max_i S_i$
8. If k has more than one value then any one of h_i may be chosen.

As indeterministic part of data involved in a decision-making problem plays an important role we cannot use the so-called classical methods. Neutrosophic soft set approach considers the membership, nonmembership, and indeterministic part of data to make a decision. So, this approach is better than any other traditional methods.

4 Application in a Decision-Making Problem

Let U = $\{h_1, h_2, h_3, h_4, h_5\}$, be the set of objects characterized by different sizes, texture, and colors. Consider the parameter set, E = {blackish, dark brown, yellowish, reddish, large, small, very small, average, very large, coarse, moderately coarse, fine, extra fine}. Also consider A ($\subset$ E) to represent the size of the objects and B ($\subset$ E) represents the texture granularity while C ($\subset$ E) represents different colors of the objects. Let A = {average, small, large}, B = {coarse, fine, moderately coarse, extra fine} and C = {whitish, reddish, blackish} be three subsets of the set of parameters E. Now, suppose the NSS (F, A) describes the 'objects having size', the NSS (G, B) describes the 'surface texture of the objects,' and the NSS (H, C) describes the 'objects having color space.' The problem is to identify an unknown object from the multiobservers neutrosophic data, specified by different observers (we consider here three observers), in terms of NSSs (F, A), (G, B), and (H, C) as described above. These NSSs as computed by the three observers Mr. X, Mr. Y, and Mr. Z, respectively, given below in their respective tabular forms (Tables 3, 4, and 5).

Consider the above two NSSs (F, A) and (G, B) given above if we evaluate '(F, A) AND (G, B)' then we will have $3 \times 4 = 12$ parameters of the form e_{ij}, where $e_{ij} = a_i \wedge b_j$, for i = 1, 2, 3 and j = 1, 2, 3, 4. If we consider the NSS for the parameters R = $\{e_{13}, e_{23}, e_{31}, e_{24}, e_{33}\}$, (depending on the choice of the decision maker) then the resultant NSS obtained from the NSSs (F, A) and (G, B) is (K, R). Computing '(F, A) AND (G, B)' for the choice parameters R, we have the tabular representation of the resultant NSS (K, R) as in Table 6.

Table 3 Tabular form of the NSS (F, A)

U	a_1 = average	a_2 = small	a_3 = large
h_1	(0.5, 0.4, 0.6)	(0.7, 0.8, 0.7)	(0.6, 0.4, 0.4)
h_2	(0.6, 0.7, 0.8)	(0.8, 0.7, 0.6)	(0.7, 0.6, 0.8)
h_3	(0.8, 0.2, 0.9)	(0.5, 0.6, 0.8)	(0.8, 0.8, 0.9)
h_4	(0.7, 0.3, 0.6)	(0.4, 0.3, 0.6)	(0.6, 0.7, 0.4)
h_5	(0.8, 0.7, 0.8)	(0.8, 0.2, 0.3)	(0.7, 0.8, 0.9)

Table 4 Tabular form of the NSS (G, B)

U	b_1 = coarse	b_2 = fine	b_3 = moderately coarse	b_4 = extra fine
h_1	(0.7, 0.6, 0.8)	(0.8, 0.6, 0.7)	(0.6, 0.8, 0.5)	(0.6, 0.7, 0.8)
h_2	(0.8, 0.5, 0.6)	(0.9, 0.6, 0.8)	(0.7, 0.6, 0.8)	(0.4, 0.3, 0.5)
h_3	(0.7, 0.6, 0.8)	(0.8, 0.4, 0.6)	(0.6, 0.8, 0.8)	(0.7, 0.4, 0.8)
h_4	(0.6, 0.8, 0.4)	(0.7, 0.5, 0.8)	(0.7, 0.6, 0.7)	(0.8, 0.5, 0.6)
h_5	(0.7, 0.6, 0.8)	(0.6, 0.8, 0.5)	(0.6, 0.8, 0.8)	(0.7, 0.6, 0.8)

Table 5 Tabular form of the NSS (H, C)

U	c_1 = whitish	c_2 = reddish	c_3 = blackish
h_1	(0.8, 0.6, 0.6)	(0.6, 0.7, 0.9)	(0.8, 0.9, 0.5)
h_2	(0.7, 0.5, 0.8)	(0.8, 0.6, 0.7)	(0.6, 0.3, 0.8)
h_3	(0.6, 0.4, 0.9)	(0.5, 0.6, 0.8)	(0.6, 0.6, 0.8)
h_4	(0.9, 0.6, 0.5)	(0.6, 0.8, 0.7)	(0.7, 0.8, 0.9)
h_5	(0.8, 0.5, 0.8)	(0.7, 0.8, 0.5)	(0.9, 0.6, 0.8)

Table 6 Tabular form of the NSS (K, R)

U	e_{13}	e_{23}	e_{31}	e_{24}	e_{33}
h_1	(0.5, 0.6, 0.6)	(0.6, 0.8, 0.7)	(0.6, 0.5, 0.8)	(0.6, 0.75, 0.8)	(0.6, 0.6, 0.5)
h_2	(0.6, 0.65, 0.8)	(0.7, 0.65, 0.8)	(0.7, 0.55, 0.8)	(0.4, 0.50, 0.6)	(0.7, 0.6, 0.8)
h_3	(0.6, 0.5, 0.9)	(0.5, 0.7, 0.8)	(0.7, 0.7, 0.9)	(0.5, 0.5, 0.8)	(0.6, 0.8, 0.9)
h_4	(0.7, 0.45, 0.7)	(0.4, 0.45, 0.7)	(0.6, 0.75, 0.4)	(0.4, 0.4, 0.6)	(0.6, 0.65, 0.7)
h_5	(0.6, 0.75, 0.8)	(0.6, 0.5, 0.8)	(0.7, 0.7, 0.9)	(0.7, 0.4, 0.8)	(0.6, 0.8, 0.9)

We now compute the NSS (S, P) from the NSSs (K, R) and (H, C) for the specified parameters P = $\{e_{13} \wedge c_2, e_{23} \wedge c_2, e_{31} \wedge c_1, e_{33} \wedge c_3\}$. Then the tabular form of the NSS (S, P) is as in Table 7.

Suppose the decision maker imposes the weights $w_1 = 0.8, w_2 = 0.6, w_3 = 0.3, w_4 = 0.7$ for the parameters $p_1 = e_{13} \wedge c_2$, $p_2 = e_{23} \wedge c_2$, $p_3 = e_{31} \wedge c_1$ and $p_4 = e_{33} \wedge c_3$ respectively. Then the tabular form of the weighted NSS (S, Q) is as in Table 8.

Then the tabular form of the comparison matrix for the weighted NSS (S, Q) is as in Table 9.

Computing the score for each of the objects we have the scores as below.

U	Score
h_1	13
h_2	11
h_3	0
h_4	16
h_5	14

Clearly, the maximum score is 16 and is scored by the object h_4. The selection will be in favor of h_4. In case the decision maker does not choose it then his next choice will go for the object having the next score, i.e., 14. So his next choice will be h_5.

Table 7 Tabular form of the NSS (S, P)

U	$e_{13} \wedge c_2$	$e_{23} \wedge c_2$	$e_{31} \wedge c_1$	$e_{33} \wedge c_3$
h_1	(0.5, 0.65, 0.9)	(0.6, 0.75, 0.9)	(0.6, 0.55, 0.8)	(0.6, 0.75, 0.5)
h_2	(0.6, 0.625, 0.8)	(0.7, 0.625, 0.8)	(0.7, 0.525, 0.8)	(0.6, 0.45, 0.8)
h_3	(0.5, 0.55, 0.9)	(0.5, 0.65, 0.8)	(0.6, 0.55, 0.9)	(0.6, 0.7, 0.9)
h_4	(0.6, 0.625, 0.7)	(0.4, 0.625, 0.7)	(0.6, 0.675, 0.5)	(0.6, 0.725, 0.9)
h_5	(0.6, 0.725, 0.8)	(0.6, 0.65, 0.8)	(0.7, 0.6, 0.9)	(0.6, 0.7, 0.9)

Table 8 Tabular form of the weighted NSS (S, Q)

U	$p_1, w_1 = 0.8$	$p_2, w_2 = 0.6$	$p_3, w_3 = 0.3$	$p_4, w_4 = 0.7$
h_1	(0.4, 0.52, 0.72)	(0.36, 0.45, 0.54)	(0.18, 0.165, 0.24)	(0.42, 0.525, 0.35)
h_2	(0.48, 0.5, 0.64)	(0.42, 0.375, 0.48)	(0.21, 0.158, 0.24)	(0.42, 0.315, 0.56)
h_3	(0.40, 0.44, 0.72)	(0.30, 0.39, 0.48)	(0.18, 0.165, 0.27)	(0.42, 0.49, 0.63)
h_4	(0.48, 0.5, 0.56)	(0.24, 0.375, 0.42)	(0.18, 0.202, 0.15)	(0.42, 0.508, 0.63)
h_5	(0.48, 0.62, 0.64)	(0.36, 0.39, 0.48)	(0.21, 0.18, 0.27)	(0.42, 0.49, 0.63)

Table 9 Tabular form of the comparison matrix of the weighted NSS (S, Q)

U	$p_1, w_1 = 0.8$	$p_2, w_2 = 0.6$	$p_3, w_3 = 0.3$	$p_4, w_4 = 0.7$
h_1	0	3	2	8
h_2	4	2	2	3
h_3	−3	1	0	2
h_4	6	1	6	3
h_5	6	3	3	2

5 Conclusion

Since its introduction the soft set theory has played an important role as a mathematical tool for dealing with problems involving uncertain, vague data. In this paper we present an application of weighted neutrosophic soft set in object recognition problem. The recognition strategy is based on multiobserver input data set. We introduce an algorithm to choose an appropriate object from a set of objects depending on some specified parameters.

Acknowledgments The author is grateful to the anonymous referee and the editor for their valuable suggestions to improve the presentation of this paper.

References

1. Molodtsov, D.: Soft set theory-first results. Comput. Math. Appl. **37**, 19–31 (1999)
2. Maji, P.K., Biswas, R., Roy, A.R.: Soft set theory. Comput. Math. Appl. **45**, 555–562 (2003)
3. Ali, M.I., Feng, F., Liu, X., Min, W.K., Shabir, M.: On some new operations in soft set theory. Comput. Math. Appl. **57**(9), 1547–1553 (2009)
4. Chen, D., Tsang, E.C.C., Yeung, D.S., Wang, X.: The parameterization reduction of soft sets and its applications. Comput. Math. Appl. **49**(5–6), 757–763 (2005)
5. Feng, F., Jun, Y.B., Zhao, X.: Soft semirings. Comput. Math. Appl. **56**(10), 2621–2628 (2008)
6. Aktas, H., Cagman, N.: Soft sets and soft groups. Inform. Sci. **177**(13), 2726–2735 (2007)
7. Xu, W., Ma, J., Wang, S., Hao, G.: Vague soft sets and their properties. Comput. Math. Appl. **59**, 787–794 (2010)
8. Maji, P.K., Biswas, R., Roy, A.R.: An application of soft set in a decision making problem. Comput. Math. Appl. **44**, 1077–1083 (2002)
9. Roy, A.R., Maji, P.K.: A fuzzy soft set theoretic approach application of soft set in a decision making problem. Comput. Appl. Math. **203**, 412–418 (2007)
10. Jun, Y.B., Park, C.H.: Applications of soft sets in ideal theory of BCK/BCI algebras. Inform. Sci. **178**(11), 2466–2475 (2008)
11. Smarandache, F.: Neutrosophic set, a generalisation of the intuitionistic fuzzy sets. Int. J. Pure Appl. Math. **24**, 287–297 (2005)
12. Maji, P.K.: Neutrosophic soft set. Ann. Fuzzy Math. Inform. **5**(1), 157–168 (2013)
13. Maji, P.K.: A neutrosophic soft set approach to a decision making problem. Ann. Fuzzy Math. Inform. **3**(2), 313–319 (2012)

Further Studies:

14. Pawlak, Z.: Rough Sets: Theoretical Aspects of Reasoning About Data. Kluwer Academic Publishers, Boston (1991)
15. Zimmerman, H.I.: Fuzzy Set Theory and its Applications. Kluwer Academic Publishers, Boston (1996)

Approximate Reasoning in Management of Hypertension

Banibrata Mondal and Swapan Raha

Abstract In this paper, we propose a concrete application of similarity-based approximate reasoning (SAR) to the management of hypertension. It is one of the silent killer diseases that threatens the lives of millions of people in developed and developing nations. The need to optimize the management of hypertension using SAR may improve the medicine diagnostic support system. It diagnoses the possibility of the disease and its severity. Systolic blood pressure (SBP), diastolic blood pressure (DBP), age and body mass index (BMI) are taken as input parameters of the fuzzy expert system and hypertension risk is the output parameter. SAR is the inference mechanism. Based on the result obtained, fuzzy diagnosis resembles human decision making with its ability to work using similarity-based approximate reasoning and ultimately find a precise solution.

Keywords Similarity-based reasoning · Hypertension management

1 Introduction

Hypertension (HTN) or high blood pressure, sometimes called arterial hypertension, is a chronic medical condition in which the blood pressure in the arteries is elevated. This requires the heart to work harder than in normal conditions to circulate blood through the blood vessels. Blood pressure is summarized by two measurements, systolic and diastolic, which depend on whether the heart muscle is contracting (systole) or relaxed between beats (diastole). Normal blood pressure at rest is within the

---

This research has been partially supported by the UGC SAP (DRS) Phase-II programme of the Department of Mathematics, Visva-Bharati. Authors appreciate UGC's financial support.

B. Mondal (✉) · S. Raha
Department of Mathematics, Visva-Bharati, Santiniketan 731235, West Bengal, India
e-mail: mbanibrata@gmail.com

S. Raha
e-mail: swapan.raha@visva-bharati.ac.in

M.K. Chakraborty et al. (eds.), *Facets of Uncertainties and Applications*,
Springer Proceedings in Mathematics & Statistics 125,
DOI 10.1007/978-81-322-2301-6_17

range of 100–140 mmHg systolic (top reading) and 60–90 mmHg diastolic (bottom reading). High blood pressure is said to be present if it is persistently at or above 140/90 mmHg. Hypertension is rarely accompanied by any symptoms and its identification is usually through screening or when seeking health care for an unrelated problem. In people aged 18 years or older, hypertension is defined as a systolic and/or a diastolic blood pressure measurement consistently higher than an accepted normal value (currently 139 mmHg systolic, 89 mmHg diastolic). Lower thresholds are used (135 mmHg systolic or 85 mmHg diastolic) if measurements are derived from 24-h ambulatory or home monitoring. Overweight and obese individuals are at increased risk for many diseases and health conditions, including hypertension. BMI provides a simple numeric measure of person's thickness or thinness. A BMI of 18.5–25 may indicate optimal weight; a BMI lower than 18.5 suggests a person underweight while a number above 25 may indicate the person is overweight. A number above 30 suggests the person is obese. It is calculated as given by

$$\text{BMI} = \frac{\text{mass (kg)}}{[\text{height (m)}]^2}.$$

These recommended distinctions along the linear scale may vary from age to age, time to time, and country to country. Recent international hypertension guidelines have also created categories below the hypertensive range to indicate a continuum of risk with higher blood pressures in the normal range. Complication arising from hypertension could lead to stroke or heart failure. Such complications may be caused by improper diagnosis and/or improper management of the disease. The latter may be due to inaccessibility to proper medical care at the time of need. One way to deal with this problem is to build an intelligent decision support system which can mimic the reasoning of medical experienced doctors in diagnosis of hypertension. Fuzzy logic technology provides a simple way to arrive at a definite conclusion from vague, ambiguous, imprecise, or noisy data (as found in medical data) using linguistic variables that are not necessarily precise [1, 2]. Till date, many well-known expert systems in medicine such as MYCIN—a computer-based medical consultation [3], INTERNIST 1—a computer-based diagnostic consultant for general internal medicine [4], CADIAG-2—Computer-Assisted Medical Diagnosis Using Fuzzy Subsets [5], LDDS—A fuzzy rule-based lung diseases diagnostic system combining positive and negative knowledge [6], MEDDIAG—a Medical Diagnostic Support System for the Management of Hypertension [7], etc., have already been built and applied in clinical applications. However, none can apply similarity in diagnostic systems to make a conclusion.

In this paper, we take the knowledge base in terms of fuzzy IF-THEN rules in MEDDIAG [7] and use the concept of similarity [8, 9] in approximate reasoning (call it, SAR) as inference mechanism along with fuzzification and defuzzification processes.

The study is aimed to design a medical diagnostic support system for the management of hypertension. Its uses can assist medical experts in the tedious and complicated task of diagnosing hypertension and the designed system can provide scheme that will assist medical personnel especially in rural areas, where there is shortage of doctors, in the process of offering primary health care to the patients.

2 Fuzzy Logic and Hypertension

Fuzzy logic allows nonlinear input/output relationships to be expressed by a set of qualitative IF-THEN rules. Behavior of many nonlinear processes may be expressed in the form of a set of fuzzy rules. Such systems are mostly hand-crafted by human experts to capture some desired input/output relationship that the expert has in mind. However, often an expert cannot express his or her knowledge explicitly; and, for many applications, an expert may not even exist. Hence, there is considerable interest in being able to automatically extract fuzzy rules from experimental input/output data. The key motivation for capturing data behavior in the form of fuzzy rules instead of polynomials is that the fuzzy rules are easy to understand, verify, and extend. A system designer can check the automatically extracted rules against intuition, check the rules for completeness, and easily fine-tune or extend the system by editing the rule base.

A fuzzy expert system is a form of artificial intelligence that uses a collection of membership functions and rules to reason about data. Systolic blood pressure (SBP), diastolic blood pressure (DBP), age, and body mass index (BMI) are taken as input parameters and "hypertension risk" is the output parameter to the fuzzy expert system.

The linguistic values *Mild, Moderate, Severe* and *Very Severe* were used for SBP and DBP, the linguistic values *Young, Middle Age, Old* and *Very Old* for AGE, and *Low, Normal, High* and *Very High* for BMI.

The knowledge was elicited from the expert through interview and literature search. The knowledge was represented in the system using the rule-based approach. The fuzzy expert system underwent three transformational stages such as Fuzzification, Rule base, and Defuzzification processes.

Membership functions of Input and output Parameters:

The various membership functions for both input and output parameters with their linguistic values are shown in Figs. 1a, b and 2a, b (Fig. 3).

Fuzzification:

Fuzzification is a process that determines the degree of membership to the fuzzy set based on fuzzy membership function. Given a fuzzy equivalence relation and a crisp point "*a*" in the domain, we can define (generate) a fuzzy set about the point "*a*." This is called fuzzification and plays an important role in the design of fuzzy systems. We show this fuzzification by the following algorithm.

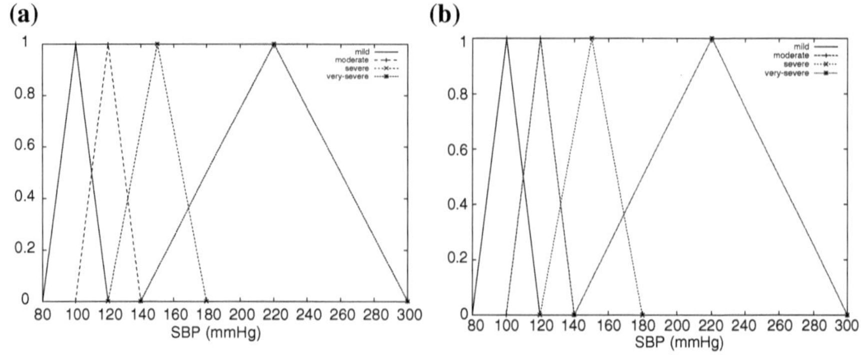

Fig. 1 Membership function for SBP and DBP. **a** Membership function for SBP. **b** Membership function for DBP

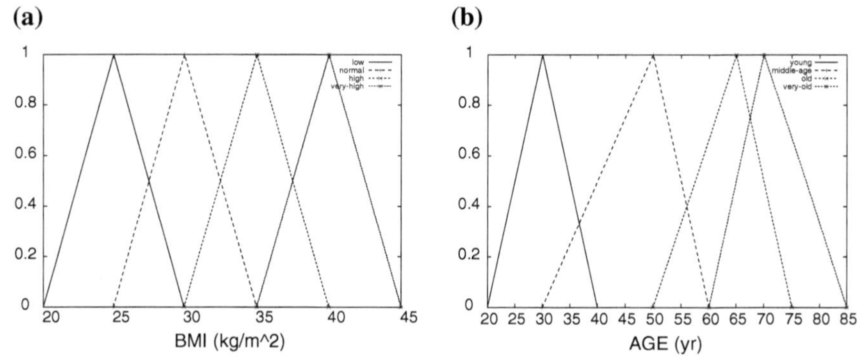

Fig. 2 Membership function for BMI and AGE. **a** Membership function for BMI. **b** Membership function for AGE

Fig. 3 Membership function for hypertension risk

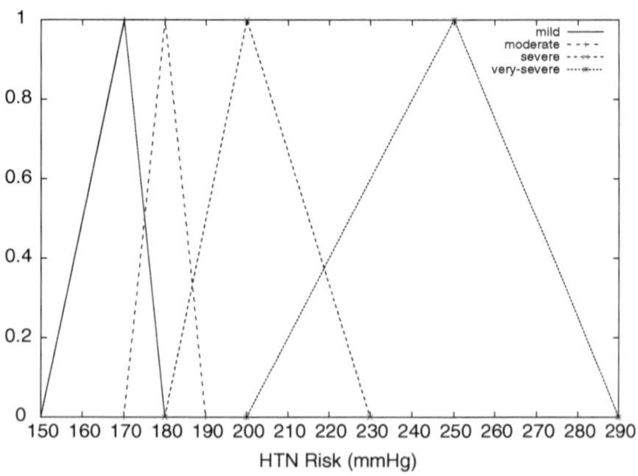

ALGORITHM: FUZZ-Fuzzification

Step 1. Given n fuzzy sets $A_1, A_2, \ldots, A_n$ defined over some universe of discourse U and for $u \in U$; $\mu_{A_i}(u),\ i = 1, \ldots, n$ are the corresponding membership degrees of given fuzzy sets A_i.
Step 2. Construct a fuzzy equivalence relation $E(u_1, u_2)$ from $A_1, A_2, \ldots, A_n$ on U using

$$E(u_1, u_2) = 1 - \sqrt{\frac{\Sigma_i(\mu_i(u_1) - \mu_i(u_2))^2}{n}}, \tag{1}$$

Step 3. Set $\delta > 0$. Define a fuzzy set about the point $a \in U$ from the fuzzy equivalence relation $E(u_1, u_2)$ by

$$\begin{aligned}\mu_a(u_1) &= E(u_1, a),\ a - \delta \leq u_1 \leq a + \delta,\ \delta > 0\\ &= 0, \qquad \text{otherwise.}\end{aligned}$$

Sample data collected from patients by some physical medical test shown in Table 1 are fuzzified into the fuzzy value range by the domain expert given in Table 2.

Rule Base:

Rule Base is the nucleus of the fuzzy logic expert system. Rules are predefined and evaluated by combining degrees of membership to form output strengths. The Rule Base consists of a set of fuzzy propositions and is derived from knowledge base of medical experts. A fuzzy proposition or a statement establishes a relationship between different input and output fuzzy sets. Fuzzy Logic offers possibility to update the knowledge base continuously and by this improves previous diagnosis. The fuzzy

Table 1 Sample data collected from 5 patients

Sl	SBP (mmHg)	DBP (mmHg)	BMI (kg/m^2)	AGE
1	70	60	35	28
2	78	30	30	55
3	68	110	41	36
4	90	47	30	57
5	66	80	47	65

Table 2 Fuzzification of sample patient's data

Sl	SBP	DBP	BMI	AGE
1	Mild	Mild	High	Young
2	Moderate	Mild	Normal	Middle
3	Mild	Severe	High	Young
4	Moderate	Moderate	Normal	Middle
5	Mild	Severe	Very high	Old

Table 3 Sample Rules for the proposed fuzzy expert system

Rule no.	IF				THEN
	SBP	DBP	AGE	BMI	Hypertension risk
1	Mild	Severe	Young	Low	Mild
2	Moderate	Moderate	Middle age	Normal	Mild
3	Severe	Mild	Old	High	Moderate
4	Severe	Mild	Very old	Very high	Moderate
5	Mild	Mild	Very old	Normal	Mild
6	Moderate	Moderate	Old	Very high	Moderate
7	Mild	Severe	Middle age	High	Severe
8	Moderate	Mild	Young	Low	Severe
9	Severe	Mild	Middle age	Low	Mild
10	Moderate	Moderate	Young	Normal	Moderate
11	Mild	Severe	Old	High	Moderate
12	Mild	Severe	Very old	Very high	Mild
13	Moderate	Moderate	Young	Low	Moderate
14	Severe	Mild	Middle age	Normal	Severe
15	Severe	Mild	Old	Normal	Severe
16	Moderate	Moderate	Very old	Low	Mild
17	Mild	Severe	Old	Normal	Mild
18	Mild	Moderate	Young	High	Severe
19	Severe	Mild	Middle age	Very high	Severe
20	Moderate	Moderate	Very old	Normal	Mild

rules for this research were developed with the assistance of domain experts (five medical doctors) in the field of internal medicine, which was proposed in [7, 10]. Sample rules base for hypertension diagnosis is shown in Table 3.

3 Implementation and System Performance

The process of drawing conclusion from exiting data is called inference. The fuzzy inference mechanism employed in this research is the Similarity-based approximate reasoning (SAR) proposed by us in [11]. The fuzzy inference engine uses the rules in the knowledge base and derives conclusion base on the rules. A mathematical formulation of the above information on medical diagnosis is provided as in the following:

Let us consider a generalized model as presented in Table 4. This form of reasoning is used in many rule-based fuzzy systems. Let there be mn-linguistic variables associated with another linguistic variable Y according to the following m-fuzzy

Table 4 Applicable form of approximate reasoning

IF	X_{11} is A_{11}	and	X_{12} is A_{12}	$\cdots$	X_{1n} is A_{1n}	THEN	Y is B_1
ELSE IF	X_{21} is A_{21}	and	X_{22} is A_{22}	$\cdots$	X_{2n} is A_{2n}	THEN	Y is B_2
			$\vdots$		$\vdots$		
ELSE IF	X_{m1} is A_{m1}	and	X_{m2} is A_{m2}	$\cdots$	X_{mn} is A_{mn}	THEN	Y is B_m
	X_1 is A_1	and	X_2 is A_2	$\cdots$	X_n is A_n		
Conclusion							Y is B

rules. The problem is to find the linguistic value of the variable Y as suggested by the rules, when the values of some of the mn-variables are given.

The scheme proposed here, for computing the final conclusion, is based on a measure of similarity. The method is based on rule-selection and then rule-execution. In both cases, we use the concept of similarity between fuzzy sets as a basis of the task. Let $X_1, X_2, \ldots, X_n$ corresponds to $X_{i1}, X_{i2}, \ldots, X_{in}$ of the ith fuzzy rule. Let there be m such rules in the rule base satisfying the same. Now, the similarity between the antecedent fuzzy set and the corresponding observed fuzzy set is computed, i.e., we compute $S(A_{ij}, A_i)$; $i = 1, 2, \ldots, m$ and perform the same operation for different $j = 1, 2, \ldots, n$. Let s_{ij} denote the different similarity values. Next, we compute the overall rule matching index from the above data as

$$s^i = \min_j s_{ij} \tag{2}$$

From among the m distinct rules, we choose those rules for which $s^i > \epsilon$. This ϵ can be interpreted as a threshold in our case. Then, we apply algorithm **SAR-HTN** to generate a conclusion from each rule conformal for firing. Here, fewer rules are fired and the output of each rule is significant.

ALGORITHM : SAR-HTN

Step 1. Compute $s_{ij} = S(A_{ij}, A_i)$ for $i = 1, 2, \ldots, m$; $j = 1, 2, \ldots, n$ and then s^i according to Eq. (2).
Step 2. Set ϵ and find the rules conformal for firing.
Step 3. If $s^i > \epsilon$ go to **Step 4**.
Step 4. Translate ith rule, as obtained in **Step 3** and compute the relation R_i using any T-norm operator.
Step 5. Modify R_i to obtain R_i' by $R_i' = s_i \rightarrow R_i$ where $\rightarrow$ is an implication operator.
Step 6. Obtain B_i' by $B_i' = proj R_i'$.
Step 7. Compute $B = \bigcup_i B_i'$.

Defuzzification:

The defuzzifier translates the output from the inference engine into crisp output. This is due to the fact that the output from the inference engine is usually a fuzzy set, while

Table 5 Hypertension risk of sample patients

Sl	Hypertension risk	Crisp output %
1	Moderate	58
2	Mild	27
3	Moderate	63
4	Severe	70
5	Very severe	97

for most real life applications, crisp values are often required. The defuzzification technique employed in this research is a specificity-based defuzzification method [11]. B'_k ,$(1 \leq k \leq m)$ may be the output fuzzy sets on the universe V. The specificity measure [12] of a fuzzy set B'_k denoted by sp_k is defined as

$$sp_k = \int_0^{\bar{\alpha}} \frac{1}{|(B'_k)_\alpha|} d\alpha \tag{3}$$

where $\bar{\alpha} = \max_{v \in V} \mu_{B'_k(v)}$; $(B'_k)_\alpha = \{v \in V | \mu_{B'_k(v)} \geq \alpha\}$ and $|.|$ denote the cardinality.

Let there be l clipped fuzzy sets $B_k^{(p)}$, $p = 1, \ldots, l$ and let $\{sp_k^{(p)}, pk^{(p)}, p = 1, \ldots, l\}$ be the specificity associated with $B_k^{(p)}$ as well as the peak of $B_k^{(p)}$. Then defuzzified value will be given by

$$v^* = \frac{\sum\limits_{p=1}^{l} pk^{(p)}.sp_k^{(p)}}{\sum\limits_{p=1}^{l} sp_k^{(p)}}. \tag{4}$$

Result and Discussion:

Based on the Algorithm SAR-HTN, the hypertension risks of the sample of five patients shown in Table 2 are calculated. The result is given in the following Table 5.

The computed value for hypertension risk for the first sample of patient's data is 58 %. The crisp output of 58 % shows that the patient has a moderate risk of hypertension. Thus, the patient needs close monitoring and possible indication for treatment of it.

4 Conclusion

Our result, based on real patient data, confirms that the fuzzy logic expert system can represent the expert's thinking in a satisfactory manner in handling complex tradeoffs. We have to work on more real data and have to update the rule base. Comparison with other diagnostic system is yet to be done. In the end, we hope to formulate a physical diagnostic support system to help the medical practitioner.

References

1. Straszecka, E.: Defining membership functions. In: Szczepaniak, P.S., Lisboa, P.J., Kacprzyk, J. (eds.) Fuzzy Systems in Medicine. Studies in Fuzziness And Soft Computing, vol. 41, pp. 32–47. Springer, Berlin (2000)
2. Lavanya, K., Durai, M., Iyengar, N.N.: Fuzzy rule based inference system for detection and diagnosis of lung cancer. Int. J. Latest Trends Comput. **2**(1), 165–171 (2011)
3. Shortliffe, E.H.: Computer Based Medical Consultations: MYCIN. Elsevier, New York (1976)
4. Miller, R.A., Pople, H.E., Myers, J.D.: INTERNIST-1: an experimental computer based diagnostic consultant for general internal medicine. New Engl. J. Med. **307**(8), 468–476 (1982)
5. Adlassnig, K.P., Kolarz, G.: CADIAG-2—computer-assisted medical diagnosis using fuzzy subsets. In: Gupta, M.M., Sanchez, E. (eds.) Approximate Reasoning in Decision Analysis, pp. 219–247. North-Holland, Amsterdam (1982)
6. Phuong, N.H., Nghi, L.H., Cho, H., Kwak, Y.S.: LDDS—a fuzzy rule based lung diseases diagnostic system combining positive and negative knowledge. J. Korean Med. Inform. **9**(2), (2003)
7. Djam, X.Y., Kimbi, Y.H.: Fuzzy expert system for the management of hypertension. Pac. J. Sci. Technol. **12**, 390–402 (2011)
8. Mondal, B., Raha, S.: Similarity-based inverse approximate reasoning. IEEE Trans. Fuzzy Syst. **19**(6), 1058–1071 (2011)
9. Mondal, B.: Similarity based inverse approximate reasoning. Ph.D. thesis, Visva-Bharati University, Santiniketan, India (2012)
10. Djam, X.Y., Kimbi, Y.H.: A medical diagnostic support system for the management of hypertension (MED DIAG). J. Sci. Multidiscip. Res. **3**, 16–30 (2011). December
11. Mondal, B., Mazumdar, D., Raha, S.: Similarity in approximate reasoning. Int. J. Comput. Cogn. **4**(3), 46–56 (2006)
12. Yager, R.R.: Measuring tranquility and anxiety in decision making: an application of fuzzy sets. Int. J. Gen. Syst. **8**, 139–146 (1982)

The Hesitant Fuzzy Soft Set and Its Application in Decision-Making

Sujit Das and Samarjit Kar

Abstract This article introduces the concept of hesitant fuzzy soft set (HFSS) by combining Torra's (2010) hesitant fuzzy set and Molodtsov's (1999) soft set theory. In order to handle uncertain and imprecise situation especially in medical diagnosis hesitant fuzzy soft sets are found to be more useful. This article investigates a couple of distance measurements procedures and aggregation operators applicable for HFSS. An algorithmic approach is proposed to solve multiple attribute decision-making (MADM) problems using HFSS with the help of aggregation operators and hesitant fuzzy soft matrix (HFSM). Finally, an illustrative example is presented to analyze the proposed approach.

Keywords Hesitant fuzzy set · Soft set · Hesitant fuzzy soft set · Multiple attribute decision making

1 Introduction

Zadeh initiated fuzzy set [1] in 1965 which started a new dimension to handle imprecise and uncertain information. Based on Zadeh's fuzzy set several extensions and generalizations were developed such as intuitionistic fuzzy set [2], type-2 fuzzy set [3, 4], type-n fuzzy set [3], fuzzy multiset [5, 6], and hesitant fuzzy set [7, 8]. Many researchers have been contributed on the first four types of fuzzy sets. But only a few contributions can be found in hesitant fuzzy sets (HFSs). Hesitant fuzzy sets as generalization of fuzzy set were first introduced by Torra [7] and Torra and Narukawa [8] which permits the membership having a set of possible values.

S. Das (✉)
Department of CSE, Dr. B.C. Roy Engineering College,
Durgapur 713206, India
e-mail: sujit_cse@yahoo.com

S. Kar
Department of Mathematics, National Institute of Technology,
Durgapur 713209, India
e-mail: kar_s_k@yahoo.com

M.K. Chakraborty et al. (eds.), *Facets of Uncertainties and Applications*,
Springer Proceedings in Mathematics & Statistics 125,
DOI 10.1007/978-81-322-2301-6_18

They discussed the relationship between hesitant fuzzy set and intuitionistic fuzzy set and showed that the envelope of hesitant fuzzy set is an intuitionistic fuzzy set. They introduced the extension principle to apply it in decision-making and uncertain situations. Since then, hesitant fuzzy set theory has been applied to several practical problems, primarily in the area of decision-making. Xia and Xu [9] developed a series of aggregation operators for hesitant fuzzy information and applied to multi-criteria decision-making (MCDM) problems with anonymity. Some ordered aggregation operators and induced ordered aggregation operators based on Quasi arithmetic were discussed by Xia et al. [10]. Authors applied these operators in group decision-making. Xu and Xia [11, 12] proposed a variety of distance measures for hesitant fuzzy sets and discussed their properties and relations as their parameters change. Wei et al. [13] developed two hesitant fuzzy Choquet integral aggregation operators: hesitant fuzzy choquet ordered averaging (HFCOA) operator and hesitant fuzzy choquet ordered geometric (HFCOG) operator and applied these operators to multiple attribute decision-making (MADM) with hesitant fuzzy information. Wei [14] developed some prioritized aggregation operators for aggregating hesitant fuzzy information and applied them for hesitant fuzzy multiple attribute decision-making problems in which the attributes were in different priority level. Zhu et al. [15] defined the hesitant fuzzy geometric Bonferroni mean (HFGBM). Gu et al. [16] investigated the evaluation model for risk investment with hesitant fuzzy information and utilized the hesitant fuzzy weighted averaging (HFWA) operator to aggregate the hesitant fuzzy information.

In real world, decision-making with uncertain and vague information has been found to be very much complex task as decision-makers experience and knowledge might not be sufficient to deal with. Often decision-makers may hesitate to put their opinions for some attributes with respect to the alternatives as they might suffer from lack of skills or information in that domain. For example, when a board of medical experts diagnose a patient for better treatment, this might be the case that opinions of some experts are missing in some set of symptoms. Also opinions of different experts may be found to be different for different symptoms. Some set of experts may think a subset of symptoms is very much important for diagnosing the patient, while others may be silent on those symptoms as they think other subset of symptoms to be crucial for diagnosing purpose. Often decision-makers are keen to investigate a subset of attributes rather than the entire set for evaluating the alternatives. This kind of situation has led us to consider a variation of hesitant fuzzy set, i.e., hesitant fuzzy soft set. Soft set theory was originally proposed by Molodtsov [17] as a general mathematical tool for dealing with uncertainty in 1999. This theory has proven useful in many different fields such as decision-making [18–22], data analysis [23], forecasting [24], and simulation [25].

In this paper we extend the concept of hesitant fuzzy set with soft set to introduce hesitant fuzzy soft set. Hesitant fuzzy soft sets are specifically useful in complex and uncertain situations where one hesitate to forward his opinion due to lack of experience or knowledge. HFSS is more flexible to represent a group of decision-makers' judgment in multiple attribute decision-making problems maintaining anonymity using membership values. HFSS allows decision-makers to select a subset of attributes as per their own intuition in an unbiased manner. We have developed hesitant fuzzy soft matrix (HFSM) to represent a hesitant fuzzy soft set. Motivated by the distance measurements approach for hesitant fuzzy sets proposed by Xu and Xia [11], we have investigated HFSS-based Hamming distance and Euclidean distance measurements with necessary examples. This paper presents mainly two types of aggregation operators such as Hesitant Fuzzy Soft Weighted Averaging (HFSWA) Operator and Hesitant Fuzzy Soft Weighted Geometric (HFSWG) Operator which we have used in the proposed multiple attribute decision-making approach. The proposed approach simulates the membership values obtained from different decision-makers on various attributes over a set of alternatives to yield a ranking among the alternatives.

This paper is organized in five sections. Section 2 provides the preliminaries and useful definitions which sets the background for presenting the concepts. Section 3 introduces the concept of hesitant fuzzy soft set. This section also investigates basic distance measurements techniques, score functions, and aggregation operators based on HFSS. Section 4 presents the proposed approach to solve a multiple attribute decision-making problem with an illustrative example. Finally, Sect. 5 concludes the paper.

2 Preliminaries

This section recalls some preliminaries that are used throughout this work. Here we briefly describe some basic concepts and operational laws related to hesitant fuzzy sets and soft sets.

Definition 1 ([17]) Let U refers to an initial universe set, E is a set of parameters, $P(U)$ is the power set of U, and $A \subset E$. A pair (F, A) is called a soft set over U, where F is a mapping given by $F : A \rightarrow P(U)$. In other words, a soft set over U is a mapping from parameters to $P(U)$, and it is not a set, but a parameterized family of subsets of U. For any parameter $e \in A$, $F(e)$ may be considered as the set of e-approximate elements of the soft set (F, A).

Definition 2 ([21]) Suppose U and E are same as Definition 1. Let $\widetilde{P}(U)$ denotes the set of all fuzzy subsets of U, A pair $\left(\widetilde{F}, A\right)$ is called a fuzzy soft set (FSS) over U, where $\widetilde{F}$ is a mapping given by $\widetilde{F} : A \rightarrow \widetilde{P}(U)$.

Definition 3 ([7, 8]) Let X be a fixed set, a HFS on X is defined in terms of a function $h_M(x)$ that returns a subset of [0, 1] when it is applied to X. This can be represented using the following mathematical expression: $M = \{< x, h_M(x) > | x \in X\}$ where $h_M(x)$ is a set of values in [0, 1], denoting the possible membership degrees of the element $x \in X$ to the set M. For convenience Xia and Xu [9] called $h = h_M(x)$ a hesitant fuzzy element (HFE) and H is the set of all HFEs.

It is noted that the number of values in different HFEs may be different. A HFE $h_M(x)$ with k number of values can be defined as $h_M^k(x)$ where k is a positive integer.

Here the membership degree of an element for a set can be represented by several possible values between 0 and 1. HFSs are mainly used in situations where one has hesitation in giving his/her preferences over objects in a decision-making process.

Torra [7, 8] defined some operations on HFEs h_1 and h_2 which can be described as

$$(1)\ h^c = \bigcup_{\gamma \in h} \{1-\gamma\},\ (2)\ h_1 \cup h_2 = \bigcup_{\gamma_1 \in h_1, \gamma_2 \in h_2} \max\{\gamma_1, \gamma_2\},$$
$$(3)\ h_1 \cap h_2 = \bigcup_{\gamma_1 \in h_1, \gamma_2 \in h_2} \min\{\gamma_1, \gamma_2\}$$

Xia and Xu [9] defined some operations on the HFEs h, h_1 and h_2.

$$(1)\ h^\lambda = \cup_{\gamma \in h}\{\gamma^\lambda\},\ \lambda > 0,\ (2)\ \lambda h = \cup_{\gamma \in h}\{1-(1-\gamma)^\lambda\},\ \lambda > 0,$$
$$(3)\ h_1 \oplus h_2 = \cup_{\gamma_1 \in h_1, \gamma_2 \in h_2}\{\gamma_1 + \gamma_2 - \gamma_1\gamma_2\},\ (4)\ h_1 \otimes h_2 = \cup_{\gamma_1 \in h_1, \gamma_2 \in h_2}\{\gamma_1\gamma_2\}.$$

Remark 1 A hesitant fuzzy element of dimension 1 is a Zadeh fuzzy set and a hesitant fuzzy element of dimension 2 with $h^1_M(x) + h^2_M(x) \le 1$ is an Atanassov intuitionistic fuzzy set.

Remark 2 If $\sum_{i=1}^{k} h^i_M(x) \le 1, \forall x \in X$, then the HFE of dimension k is called a normalized HFE.

Remark 3 If for the hesitant membership values $\{h^1_M(x), h^2_M(x), \ldots, h^k_M(x)\}$, $\sum_{i=1}^{k} h^i_M(x) = l > 1, \forall x \in X$, then the membership values can be normalized as $\frac{1}{l}\{(h^1_M(x), h^2_M(x), \ldots, h^k_M(x)\}$.

Example 1 Let $X = \{x_1, x_2, x_3\}$ be a fixed set. $h_M(x_1) = \{0.2, 0.4, 0.5\}$, $h_M(x_2) = \{0.3, 0.4\}$, and $h_M(x_3) = \{0.3, 0.2, 0.5, 0.6\}$ be the HFEs of $x_i (i = 1, 2, 3)$ to the set M. Then the hesitant fuzzy set M is given by $M = \{(x_1, \{0.2, 0.4, 0.5\}), (x_2, \{0.3, 0.4\}), (x_3, \{0.3, 0.2, 0.5, 0.6\})\}$ and the corresponding normalized hesitant fuzzy set is given by $M^n = \{(x_1, \{0.18, 0.36, 0.46\}), (x_2, \{0.3, 0.4\}), (x_3, \{0.19, 0.12, 0.32, 0.37\})\}$

Remark 4 It is noticed that the number of values in different HFEs may be different. Suppose that k be the number of elements in $h^k_M(x), x \in X$. To have a correct comparison, two HFEs should have same length. If fewer elements are in $h^{k_1}_{M_1}(x)$ than in $h^{k_2}_{M_2}(x)$ where $k_1 < k_2$, then an extension of $h^{k_1}_{M_1}(x)$ should be considered optimistically by repeating its final element until it has the same length with $h^{k_2}_{M_2}(x)$, i.e., $k_1 = k_2$.

Definition 4 ([9]) For a HFE h, $s(h) = \frac{1}{\#h}\sum_{\gamma \in h} \gamma$ is called the score function of h, where $\#h$ is the number of elements in h. For two HFEs h_1 and h_2 if $s(h_1) > s(h_2)$, then $h_1 > h_2$; if $s(h_1) = s(h_2)$ then $h_1 = h_2$.

Definition 5 ([11]) The score function of HFS $M = \{< x, h_M(x) > | x \in X\}$ is given by

$$Score(M) = \frac{1}{n}\sum_{i=1}^{n}\left(\frac{1}{k}\sum_{k} h_M^k(x_i)\right)$$

where n be the number of HFEs and k be the dimension of ith HFE in the HFS M.

Definition 6 ([9]) Let $h_j (j = 1, 2, \ldots, n)$ be a collection of HFEs. A hesitant fuzzy weighted averaging (HFWA) operator is a mapping $H^n \to H$ such that

$$HFWA(h_1, h_2, \ldots, h_n) = \sum_{j=1}^{n} w_j h_j = \cup_{\gamma_1 \in h_1, \gamma_2 \in h_2, \ldots, \gamma_n \in h_n}\left\{1 - \prod_{j=1}^{n}(1-\gamma_j)^{w_j}\right\}$$

where $w = (w_1, w_2, \ldots, w_n)^T$ is the weight vector of $h_j (j = 1, 2, \ldots, n)$ with $w_j \in [0, 1]$ and $\sum_{j=1}^{n} w_j = 1$.

Definition 7 ([9]) Let $h_j (j = 1, 2, \ldots, n)$ be a collection of HFEs. A hesitant fuzzy weighted geometric (HFWG) operator is a mapping $H^n \to H$ such that

$$HFWG(h_1, h_2, \ldots, h_n) = \prod_{j=1}^{n} h_j^{w_j} = \cup_{\gamma_1 \in h_1, \gamma_2 \in h_2, \ldots, \gamma_n \in h_n}\left\{\prod_{j-1}^{n}(\gamma_j)^{w_j}\right\}$$

where $w = (w_1, w_2, \ldots, w_n)^T$ is the weight vector of $h_j (j = 1, 2, \ldots, n)$ with $w_j \in [0, 1]$ and $\sum_{j=1}^{n} w_j = 1$.

3 Hesitant Fuzzy Soft Set with Distance Measurements, Score Functions, and Aggregation Operators

This section describes hesitant fuzzy soft set, score functions, a couple of distance measurement techniques and aggregation operators on hesitant fuzzy soft sets.

3.1 Hesitant Fuzzy Soft Set

Definition 8 Let $\widetilde{H}(X)$ be the set of all hesitant fuzzy sets of X where X is a fixed set. Let E be a set of parameters and $A \subseteq E$. A hesitant fuzzy soft set over X is defined by the set of ordered pairs $(\widetilde{F_A}, E)$ where $\widetilde{F_A}$ is a mapping given by, $\widetilde{F_A} : E \to \widetilde{H}(X)$. For any parameter $e \in A$, $\widetilde{F}(e)$ is a hesitant fuzzy subset of X and may be considered as e-elements or e-approximate elements in the HFSS. Clearly, $\widetilde{F}(e)$ can be written

Table 1 Tabular representation of HFSM

	e_1	e_2	$\cdots$	e_n
x_1	$h_M^k(x_1, e_1)$	$h_M^k(x_1, e_2)$	$\cdots$	$h_M^k(x_1, e_n)$
x_2	$h_M^k(x_2, e_1)$	$h_M^k(x_2, e_2)$	$\cdots$	$h_M^k(x_2, e_n)$
$\vdots$	$\vdots$	$\vdots$	$\ddots$	$\vdots$
x_m	$h_M^k(x_m, e_1)$	$h_M^k(x_m, e_2)$	$\cdots$	$h_M^k(x_m, e_n)$

as a hesitant fuzzy set such that $\widetilde{F}(e) = \{< x, h_M(x) > | x \in X\}$ where $h_M(x)$ is a set of values in [0, 1].

Definition 9 HFSS set can be well represented by HFSM. If $X = \{x_1, x_2, \ldots, x_m\}$ and $E = \{e_1, e_2, \ldots, e_n\}$, then HFSM could be defined as $F = (f_{ij})_{m \times n}$, where $f_{ij} = h_M^k(x_i, e_j), i = 1, 2, \ldots, m, j = 1, 2, \ldots, n$. Here $k > 0$ be the number of values which may be different for different HFE. Table 1 presents a HFSM.

Example 2 Let X be a set of three shirts, i.e., $X = \{x_1, x_2, x_3\}$. Let $E = \{e_1, e_2, e_3\}$, where e_1 = bright, e_2 = colorful, and e_3 = light. Let $\omega = (\widetilde{F}_A, E)$ be the HFSS over X defined as follows:

$$\widetilde{F}_A(x_1, e_1) = (0.6, 0.3, 0.7), \widetilde{F}_A(x_1, e_2) = (0.8, 0.7), \quad \widetilde{F}_A(x_1, e_3) = (0.3, 0.5, 0.7),$$
$$\widetilde{F}_A(x_2, e_1) = (0.7, 0.8), \quad \widetilde{F}_A(x_2, e_2) = (0.3, 0.7, 0.5), \widetilde{F}_A(x_2, e_3) = (0.9, 0.3),$$
$$\widetilde{F}_A(x_3, e_1) = (0.5, 0.6, 0.3), \widetilde{F}_A(x_3, e_2) = (0.4, 0.3), \quad \widetilde{F}_A(x_3, e_3) = (0.1, 0.4).$$

Thus as per the Definition 9 we get the HFSM as follows:

$$[a_{ij}] = \begin{bmatrix} (0.6, 0.3, 0.7) & (0.8, 0.7) & (0.3, 0.5, 0.7) \\ (0.7, 0.8) & (0.3, 0.7, 0.5) & (0.9, 0.3) \\ (0.5, 0.6, 0.3) & (0.4, 0.3) & (0.1, 0.4) \end{bmatrix}$$

Then with the above representation, the HFSS $(\widetilde{F}_A, E)$ is represented by the matrix $[a_{ij}]_{m \times n}$ and we can write $(\widetilde{F}_A, E) = [a_{ij}]_{m \times n}$. Let $(\widetilde{F}_A, E) = [a_{ij}]_{m \times n}$ and $(\widetilde{G}_A, E) = [b_{ij}]_{m \times n}$ be two hesitant fuzzy soft sets. Clearly $(\widetilde{F}_A, E) = (\widetilde{G}_A, E)$ if and only if $[a_{ij}]_{m \times n} = [b_{ij}]_{m \times n}$.

3.2 Distance Measure Between Hesitant Fuzzy Soft Sets

Let $\omega = (\widetilde{F}_A, E)$ and $\varpi = (\widetilde{G}_A, E)$ be two hesitant fuzzy soft sets over X, i.e., $\omega, \varpi \in HFSS(X)$. Let 'd' be a mapping $d : HFSS(X) \times HFSS(X) \to R^+ \cup \{0\}$ (where $R^+ \cup \{0\}$ denotes a set of nonnegative real numbers). It can be easily verified that distance $d(\omega, \varpi)$ between hesitant fuzzy soft sets ω and ϖ satisfies the following properties ((P1)–(P3)):

$$(P1)\, d(\omega, \varpi) \geq 0;\ (P2)\, d(\omega, \varpi) = d(\varpi, \omega);\ (P3)\, d(\omega, \varpi) = 0 \text{ iff } (\omega = \varpi).$$

Definition 10 Let $\omega = (\widetilde{F}_A, E) = [a_{ij}]_{m\times n}$ and $\varpi = (\widetilde{G}_A, E) = [b_{ij}]_{m\times n}$ be two HFSSs over X. Hamming distance $d^H_{HFSS}(\omega, \varpi)$ and Euclidean distance $d^E_{HFSS}(\omega, \varpi)$ between ω and ϖ are defined as follows:

$$d^H_{HFSS}(\omega, \varpi) = \sum_{i=1}^{m}\sum_{j=1}^{n}\sum_{k} \frac{|h^k_{\widetilde{F}_A}(x_i, e_j) - h^k_{\widetilde{G}_A}(x_i, e_j)|}{k}$$

$$d^E_{HFSS}(\omega, \varpi) = \left(\sum_{i=1}^{m}\sum_{j=1}^{n}\sum_{k} \frac{\left\{h^k_{\widetilde{F}_A}(x_i, e_j) - h^k_{\widetilde{G}_A}(x_i, e_j)\right\}^2}{k}\right)^{\frac{1}{2}}.$$

Definition 11 Normalized Hamming distance $d^{nH}_{HFSS}(\omega, \varpi)$ and normalized Euclidean distance $d^{nE}_{HFSS}(\omega, \varpi)$ between ω and ϖ can be defined as follows:

$$d^{nH}_{HFSS}(\omega, \varpi) = \frac{d^H_{HFSS}(\omega, \varpi)}{mn},\ d^{nE}_{HFSS}(\omega, \varpi) = \frac{d^E_{HFSS}(\omega, \varpi)}{(mn)^{\frac{1}{2}}}$$

Here it is easy to verify that the aforementioned distance satisfies the properties of distance (i.e., (P1)–(P3)) and also the following properties:

$$0 \leq d^H_{HFSS}(\omega, \varpi) \leq mn, 0 \leq d^{nH}_{HFSS}(\omega, \varpi) \leq 1, 0 \leq d^E_{HFSS}(\omega, \varpi) \leq \sqrt{mn},$$
$$0 \leq d^{nE}_{HFSS}(\omega, \varpi) \leq 1.$$

Example 3 Assume that a hesitant fuzzy soft set $\omega = (\widetilde{F}_A, E) = [a_{ij}]_{3\times 3}$ is given in Example 2. Another hesitant fuzzy soft set $\varpi = (\widetilde{G}_A, E) = [b_{ij}]_{3\times 3}$ is given as follows:

$$[b_{ij}] = \begin{bmatrix} (0.7, 0.1) & (0.4, 0.5, 0.4) & (0.9, 0.1) \\ (0.5, 0.4, 0.2) & (0.5, 0.6) & (0.4, 0.2, 0.1) \\ (0.8, 0.1) & (0.2, 0.3, 0.6) & (0.3, 0.5, 0.1) \end{bmatrix}$$

Then the various distance measurement results are given below.

$$d^H_{HFSS}(\omega, \varpi) = 2.6,\ d^{nH}_{HFSS}(\omega, \varpi) = 0.29,\ d^E_{HFSS}(\omega, \varpi) = 1.06,$$
$$d^{nE}_{HFSS}(\omega, \varpi) = 0.35.$$

3.3 Score of HFSS

Definition 12 Score of a hesitant fuzzy soft set $\omega = (\widetilde{F}_A, E)$ can be defined by:

$$Score(\omega) = \frac{1}{mn}\sum_{i=1}^{m}\sum_{j=1}^{n}\left(\frac{1}{k}\sum_{k} h_{\omega}^{k}(x_i, e_j)\right), e \in E, x \in X.$$

Score of a parameter $e \in E$ can be defined by:

$$Score_{\{e\}}(\omega) = \frac{1}{m}\sum_{i=1}^{m}\left(\frac{1}{k}\sum_{k} h_{\omega}^{k}(x_i, e)\right), x \in X.$$

Example 4 For the HFSM $[b_{ij}]$ (Example 3) score of the parameters $E = \{e_1, e_2, e_3\}$ are defined by

$$\begin{aligned} Score_{\{e_1\}}(\varpi) &= \frac{1}{3}\sum_{i=1}^{3}\left(\frac{1}{k}\sum_{k} h_{\varpi}^{k}(x_i, e_1)\right) \\ &= \frac{1}{3}\left(\frac{(0.7+0.1)}{2} + \frac{(0.5+0.4+0.2)}{3} + \frac{(0.8+0.1)}{2}\right) = 0.41, \end{aligned}$$

$$\begin{aligned} Score_{\{e_2\}}(\varpi) &= \frac{1}{3}\sum_{i=1}^{3}\left(\frac{1}{k}\sum_{k} h_{\varpi}^{k}(x_i, e_2)\right) \\ &= \frac{1}{3}\left(\frac{(0.4+0.5+0.4)}{3} + \frac{(0.5+0.6)}{2} + \frac{(0.2+0.3+0.6)}{3}\right) = 0.45, \end{aligned}$$

$$\begin{aligned} Score_{\{e_3\}}(\varpi) &= \frac{1}{3}\sum_{i=1}^{3}\left(\frac{1}{k}\sum_{k} h_{\varpi}^{k}(x_i, e_3)\right) \\ &= \frac{1}{3}\left(\frac{(0.9+0.1)}{2} + \frac{(0.4+0.2+0.1)}{3} + \frac{(0.3+0.5+0.1)}{3}\right) = 0.34. \end{aligned}$$

3.4 Aggregation Operators in HFSS

Definition 13 Let $h_j (j = 1, 2, \ldots, n)$ be a collection of HFEs for attribute $e \in E$ and alternative $x \in X$. HFSWA operator is a mapping $H^n \to H$ such that

$$HFSWA(h_1, h_2, \ldots, h_n)(x_i, e_j)$$
$$= \sum_{j=1}^{n} w_j h_j(x_i, e_j) = \cup_{\gamma_1 \in h_1, \gamma_2 \in h_2, \ldots, \gamma_n \in h_n}(x_i, e_j) \left\{ 1 - \prod_{j=1}^{n} (1 - \gamma_j)^{w_j} \right\}, \forall i.$$

where k is the dimension of h_j, $w = (w_1, w_2, \ldots, w_n)^T$ is the weight vector of $h_j (j = 1, 2, \ldots, n)$ with $w_j \in [0, 1]$ and $\sum_{j=1}^{n} w_j = 1$.

If $w = (1/n, 1/n, \ldots, 1/n)^T$ then the HFSWA operator reduces to hesitant fuzzy soft averaging (HFSA) operator:

$$HFSA(h_1, h_2, \ldots, h_n)(x_i, e_j)$$
$$= \sum_{j=1}^{n} \left(\frac{1}{n} h_j \right)(x_i, e_j) = \cup_{\gamma_1 \in h_1, \gamma_2 \in h_2, \ldots, \gamma_n \in h_n}(x_i, e_j) \left\{ 1 - \prod_{j=1}^{n} (1 - \gamma_j)^{1/n} \right\}, \forall i.$$

Definition 14 Let $h_j (j = 1, 2, \ldots, n)$ be a collection of HFEs for attribute $e \in E$ and alternative $x \in X$. HFSWG operator is a mapping $H^n \rightarrow H$ such that

$$HFSWG(h_1, h_2, \ldots, h_n)(x_i, e_j) = \prod_{j=1}^{n} h_j^{w_j}(x_i, e_j)$$
$$= \cup_{\gamma_1 \in h_1, \gamma_2 \in h_2, \ldots, \gamma_n \in h_n}(x_i, e_j) \left\{ \prod_{j=1}^{n} (\gamma_j)^{w_j} \right\}, \forall i.$$

where k is the dimension of h_j, $w = (w_1, w_2, \ldots, w_n)^T$ is the weight vector of $h_j (j = 1, 2, \ldots, n)$ with $w_j \in [0, 1]$ and $\sum_{j=1}^{n} w_j = 1$.

In the case where $w = (1/n, 1/n, \ldots, 1/n)^T$ then the HFSWG operator reduces to hesitant fuzzy soft geometric (HFSG) operator:

$$HFSG(h_1, h_2, \ldots, h_n)(x_i, e_j) = \prod_{j=1}^{n} h_j^{1/n}(x_i, e_j)$$
$$= \cup_{\gamma_1 \in h_1, \gamma_2 \in h_2, \ldots, \gamma_n \in h_n}(x_i, e_j) \left\{ \prod_{j=1}^{n} (\gamma_j)^{1/n} \right\}, \forall i.$$

4 An Approach to Multiple Attribute Decision-Making with Hesitant Fuzzy Soft Set

In some real-life problems, for example, to properly diagnose a patient, anonymity is required in order to protect the decision-makers' privacy or avoid influencing each other. In this section, we utilize the hesitant soft aggregation operators to MADM with HFSS. Let $A = \{A_1, A_2, \ldots, A_m\}$ be a discrete set of alternatives, $G = \{G_1, G_2, \ldots, G_n\}$ be a set of attributes, and $w = \{w_1, w_2, \ldots, w_n\}^T$ is the weighting vector of the attribute $G_j\{j = 1, 2, \ldots, n\}$, where $w_j \in [0, 1]$, $\sum_{j=1}^{n} w_j = 1$.

Step 1. The decision-makers provide their evaluations about the alternative A_i under the attribute G_j denoted by the hesitant fuzzy elements $h_{ij}(i = 1, 2, \ldots, m; j = 1, 2, \ldots, n)$ in terms of HFSM.
Step 2. Aggregate all hesitant fuzzy values $h_i(i = 1, 2, \ldots, m)$ by using the proposed soft aggregation operators for the alternatives $A = \{A_1, A_2, \ldots, A_m\}$.

$$HFSWA(h_1, h_2, \ldots, h_n)(x_i, e_j) = \sum_{j=1}^{n} w_j h_j(x_i, e_j) = \cup_{\gamma_1 \in h_1, \gamma_2 \in h_2, \ldots, \gamma_n \in h_n}(x_i, e_j)\left\{1 - \prod_{j=1}^{n}(1-\gamma_j)^{w_j}\right\}, \forall i.$$

$$HFSWG(h_1, h_2, \ldots, h_n)(x_i, e_j) = \prod_{j=1}^{n} h_j^{w_j}(x_i, e_j) = \cup_{\gamma_1 \in h_1, \gamma_2 \in h_2, \ldots, \gamma_n \in h_n}(x_i, e_j)\left\{\prod_{j=1}^{n}(\gamma_j)^{w_j}\right\}, \forall i.$$

Step 3. Calculate the score values $s(h_i)\{i = 1, 2, \ldots, m\}$ of hesitant fuzzy values $h_i\{i = 1, 2, \ldots, m\}$ for the alternatives $A_i\{i = 1, 2, \ldots, m\}$.
Step 4. Rank all the alternatives $A = \{A_1, A_2, \ldots, A_m\}$ in accordance with $s(h_i)\{i = 1, 2, \ldots, m\}$ to find out the priorities of those alternatives.

Example 5 A group of medical experts consisting of five members is empanelled to diagnose a patient. Suppose a set of four possible diseases (alternatives) $D = \{D_1, D_2, D_3, D_4\}$ are to be evaluated against a set of four common symptoms (attributes) $S = \{S_1, S_2, S_3, S_4\}$. Opinions of all experts against each symptoms and diseases are recorded using HFSS which is represented by HFSM (Table 2). As opinions of each expert might differ, one might ignore the importance of one or more symptoms for a particular disease. As a result opinions of some experts may be missing for some symptoms as they are not interested or expertise in those symptoms. In this example missing opinions of experts are marked by '--'. For example, if an HFE is (--, 0.3, 0.7, --, --), then opinions of expert 1, 4, and 5 are said to be missing and expert 2 gives his opinion as 0.3, expert 3 gives the same as 0.7. This type of uncertain situations can be well expressed in the framework of HFSS. In this example, Tables 3 and 4 show the assigned weight vector, score values, and the corresponding ranking of alternatives obtained, respectively, by applying HFSWA and HFSWG operator on

Table 2 Hesitant fuzzy soft matrix as input

Diseases/ symptoms	S_1	S_2	S_3	S_4
D_1	(0.3, 0.7, 0.5, --, --)	(0.3, 0.1, 0.2, 0.5, --)	(0.7, 0.6, 0.4, --, --)	(0.5, 0.6, 0.7, --, --)
D_2	(0.4,0.3,0.5, --, --)	(0.1, 0.7, -- ,--, --)	(0.3, 0.3, 0.4, 0.8, 0.3)	(0.3, 0.7, 0.8, --, --)
D_3	(0.7, 0.9, 0.3, 0.2, 0.1)	(0.3, 0.2, 0.1, --, --)	(0.8, 0.7, --, --, --)	(0.6, 0.7, 0.6, --, --)
D_4	(0.5, 0.4, 0.5, --, --)	(0.5, 0.3, 0.5, 0.7, 0.6)	(0.7, 0.6, 0.7, --, --)	(0.4, 0.3, --, --, --)

Table 3 Score values obtained by HFSWA operator and the rankings of alternatives

	Weight vector	D_1	D_2	D_3	D_4	Ranking
HFSWA	$\{0.2, 0.3, 0.15, 0.35\}^T$	0.5060	0.5193	0.5445	0.4893	$D_3 > D_2 > D_1 > D_4$
	$\{0.35, 0.15, 0.3, 0.2\}^T$	0.5290	0.4847	0.5981	0.5296	$D_3 > D_4 > D_1 > D_2$
	$\{0.15, 0.2, 0.35, 0.3\}^T$	0.5326	0.5118	0.6158	0.5343	$D_3 > D_4 > D_1 > D_2$
	$\{0.3, 0.35, 0.2, 0.15\}^T$	0.4788	0.4750	0.5202	0.5235	$D_4 > D_3 > D_1 > D_2$

Table 4 Score values obtained by HFSWG operator and the rankings of alternatives

	Weight vector	D_1	D_2	D_3	D_4	Ranking
HFSWG	$\{0.2, 0.3, 0.15, 0.35\}^T$	0.5635	0.6126	0.5538	0.6305	$D_4 > D_2 > D_1 > D_3$
	$\{0.35, 0.15, 0.3, 0.2\}^T$	0.6167	0.5491	0.5579	0.6665	$D_4 > D_1 > D_3 > D_2$
	$\{0.15, 0.2, 0.35, 0.3\}^T$	0.6055	0.5514	0.6263	0.6688	$D_4 > D_3 > D_1 > D_2$
	$\{0.3, 0.35, 0.2, 0.15\}^T$	0.5373	0.5824	0.5076	0.6205	$D_4 > D_2 > D_1 > D_3$

various set of weighting vector. Weighting vector is prepared by the relative experts considering the importance of various symptoms for a particular disease. As per our knowledge, importance of various symptoms might vary with various diseases, so distinct set of weight values are assigned to the weight vector. Result shows the ranking of diseases on various weighting vectors by combining the opinions of all experts using soft aggregation operators.

5 Concluding Remarks

This paper has introduced the concept of HFSS by combining hesitant fuzzy set and soft set theory. HFSS is capable of dealing with several membership values for a particular element and mainly useful in uncertain situations where decision-makers have hesitation to express their opinions. This paper presents hesitant fuzzy soft matrix to represent a hesitant fuzzy soft set. A set of distance measurements methods and score functions based on HFSS are devised with numerical examples. This paper has proposed a couple of aggregation operators in the context of HFSS and used them in MADM problems. Finally, an algorithmic approach is given to explore the application of HFSS in multi-attribute decision-making problems. The validity of the proposed approach has also been illustrated using a practical medical related example. In future, researchers might use HFSS for group decision-making with more imprecise and uncertain information. Researchers may also define the relationship of soft sets with generalized approximation spaces introduced by Skowron and Stepaniuk [26].

References

1. Zadeh, L.A.: Fuzzy sets. Inf. Control **8**, 338–353 (1965)
2. Atanassov, K.: Intuitionistic fuzzy sets. Fuzzy Sets Syst. **20**, 87–96 (1986)
3. Dubois, D., Prade, H.: Fuzzy Sets and Systems: Theory and Applications. Academic Press, New York (1980)
4. Miyamoto, S.: Remarks on basics of fuzzy sets and fuzzy multisets. Fuzzy Sets Syst. **156**, 427–431 (2005)
5. Yager, R.R.: On the theory of bags. Int. J. Gen. Syst. **13**, 23–37 (1986)
6. Miyamoto, S.: Multisets and fuzzy multisets. In: Liu, Z.-Q., Miyamoto, S. (eds.) Soft Computing and Human-Centered Machines, pp. 9–33. Springer, Berlin (2000)
7. Torra, V.: Hesitant fuzzy sets. Int. J. Intell. Syst. **25**, 529–539 (2010)
8. Torra, V., Narukawa, Y.: On hesitant fuzzy sets and decision. In: The 18th IEEE International Conference on Fuzzy Systems, pp. 1378–1382, Jeju Island, Korea (2009)
9. Xu, Z.S., Xia, M.M.: Hesitant fuzzy information aggregation in decision making. Int. J. Approx. Reason. **52**(3), 395–407 (2011)
10. Xia, M., Xu, Z., Chen, N.: Some hesitant fuzzy aggregation operators with their application in group decision making. Group Decis. Negot. pp. 1–21 (2011)
11. Xu, Z.S., Xia, M.M.: Distance and similarity measures for hesitant fuzzy sets. Inf. Sci. **181**(11), 2128–2138 (2011)
12. Xu, Z.S., Xia, M.M.: On distance and correlation measures of hesitant fuzzy information. Int. J. Intell. Syst. **26**(5), 410–425 (2011)
13. Wei, G.W., Zhao, X.F., Wang, H.J., Lin, R.: Hesitant fuzzy choquet integral aggregation operators and their applications to multiple attribute decision making. Inf. Int. Interdiscip. J. **15**(2), 441–448 (2012)
14. Wei, G.W.: Hesitant Fuzzy prioritized operators and their application to multiple attribute group decision making. Knowl.-Based Syst. **31**, 176–182 (2012)
15. Zhu, B., Xu, Z.S., Xia, M.: Hesitant fuzzy geometric Bonferroni means. Inf. Sci. **205**(1), 72–85 (2012)
16. Gu, X., Wang, Y., Yang, B.: A method for hesitant fuzzy multiple attribute decision making and its application to risk investment. J. Converg. Inf. Technol. **6**(6), 282–287 (2011)
17. Molodtsov, D.: Soft set theory-first results. Comput. Math. Appl. **37**, 19–31 (1999)
18. Cagman, N., Enginoglu, S.: Soft matrix theory and its decision making. Comput. Math. Appl. **59**(10), 3308–3314 (2010)
19. Cagman, N., Enginoglu, S.: Soft set theory and uni-int decision making. Eur. J. Oper. Res. **207**(2), 848–855 (2010)
20. Kong, Z., Gao, L., Wang, L.: Comment on a fuzzy soft set theoretic approach to decision making problems. J. Comput. Appl. Math. **223**(2), 540–542 (2009)
21. Roy, A.R., Maji, P.K.: A fuzzy soft set theoretic approach to decision making problems. J. Comput. Appl. Math. **203**(2), 412–418 (2007)
22. Maji, P.K., Roy, A.R., Biswas, R.: An application of soft sets in a decision making problem. Comput. Math. Appl. **44**(8–9), 1077–1083 (2002)

23. Zou, Y., Xiao, Z.: Data analysis approaches of soft sets under incomplete information. Knowl.-Based Syst. **21**(8), 941–945 (2008)
24. Xiao, Z., Gong, K., Zou, Y.: A combined forecasting approach based on fuzzy soft sets. J. Comput. Appl. Math. **228**(1), 326–333 (2009)
25. Kalayathankal, S.J., Singh, G.S.: A fuzzy soft flood alarm model. Math. Comput. Simul. **80**(5), 887–893 (2010)
26. Skowron, A., Stepaniuk, J.: Tolerance approximation spaces. Fundam. Inform. **27**(2–3), 245–253 (1996)

On Fuzzy Ideal Cone Method to Capture Entire Fuzzy Nondominated Set of Fuzzy Multi-criteria Optimization Problems with Fuzzy Parameters

Debdas Ghosh and Debjani Chakraborty

Abstract This paper deals with the computational aspect of the *fuzzy ideal cone method* by Ghosh and Chakraborty, Fuzzy ideal cone: a method to obtain complete fuzzy nondominated set of fuzzy multi-criteria optimization problems with fuzzy parameters, *In*: The Proceedings of IEEE International Conference on Fuzzy Systems 2013, FUZZ IEEE 2013, IEEE Xplore, pp. 1–8 to generate the complete fuzzy nondominated set of a fuzzy multi-criteria optimization problem. In order to formulate the decision feasible region, the concept of inverse points in fuzzy geometry is used. Relation between the fuzzy decision feasible sets evaluated through the inverse points and directly through the extension principle is reported. It is shown that under a certain monotone condition both the decision feasible sets are identical. This result can greatly reduce the computational cost of evaluating the decision feasible region. After evaluating the decision feasible region, criteria feasible region is formulated using the basic fuzzy geometrical ideas. An algorithmic implementation of the fuzzy ideal cone method is presented to find the complete fuzzy nondominated set of the fuzzy criteria feasible region.

Keywords Multiple criteria analysis · Fuzzy nondominated set · Fuzzy geometry · Fuzzy multi-criteria optimization

1 Introduction

In practice, criteria and constraints for decision-making problems in imprecise environment may involve many parameters whose possible values are assigned by a decision-maker (DM) which may be imprecisely or ambiguously known. Usually, these ambiguous parameters are represented by fuzzy numerical values and more appropriately by fuzzy numbers. The resulting fuzzy multi-criteria optimization

D. Ghosh (✉) · D. Chakraborty
Department of Mathematics, Indian Institute of Technology Kharagpur, Kharagpur 721302, West Bengal, India
e-mail: debdas.email@gmail.com

M.K. Chakraborty et al. (eds.), *Facets of Uncertainties and Applications*, Springer Proceedings in Mathematics & Statistics 125,
DOI 10.1007/978-81-322-2301-6_19

problems (FMOPs) whose parameters are fuzzy may be appeared as more realistic than the conventional multi-criteria optimization problems (MOPs).

A general model of a fuzzy multi-criteria optimization problem with fuzzy parameters is described by the following system [3]:

$$\begin{cases} \min & f(x, \widetilde{c}) \\ \text{subject to} & g(x, \widetilde{a}) \precsim \widetilde{b}, \\ & x \in C \subseteq \mathbb{R}^n_{\geqq}, \end{cases} \tag{1}$$

where $f(x, \widetilde{c}) = (f_1(x, \widetilde{c}_1), f_2(x, \widetilde{c}_2), \ldots, f_k(x, \widetilde{c}_k))^t$, $k \geq 2$, $g(x, \widetilde{a}) = (g_1(x, \widetilde{a}_1), g_2(x, \widetilde{a}_2), \ldots, g_m(x, \widetilde{a}_m))^t$, $m \geq 1$, $\widetilde{c}_j = (\widetilde{c}_{j1}, \widetilde{c}_{j2}, \ldots, \widetilde{c}_{jp_j})^t$, $j = 1, 2, \ldots k$, and $\widetilde{a}_i = (\widetilde{a}_{i1}, \widetilde{a}_{i2}, \ldots, \widetilde{a}_{iq_i})^t$, $i = 1, 2, \ldots m$; the parameters $\widetilde{c}_{js}$ and $\widetilde{a}_{it}$ are fuzzy numbers for each $s = 1, 2, \ldots, p_j$ and $t = 1, 2, \ldots, q_i$. The notation $\mathbb{R}^n_{\geqq}$ represents nonnegative hyper-octant of $\mathbb{R}^n$, i.e., $\mathbb{R}^n_{\geqq} = \{x \in \mathbb{R}^n : x = (x_1, x_2, \ldots, x_n)^t, x_i \geq 0$ for all $i = 1, 2, \ldots, n\}$. Let us denote the fuzzy constraint set of (1) by $\widetilde{\mathcal{X}}$, i.e., $\widetilde{\mathcal{X}} = \{x \in C\colon g(x, \widetilde{a}) \precsim \widetilde{b}\}$.

Let us introduce the notation that will be used in the rest of this paper. We use the notation $\widetilde{A}(\alpha)$ to represent α-cut of the fuzzy set $\widetilde{A}$. In particular, the support of $\widetilde{A}$ is denoted by $\widetilde{A}(0)$ and the core or 1-cut of $\widetilde{A}$ is presented by $\widetilde{A}(1)$. The notation $\mu(.|\widetilde{A})$ denotes membership function of the fuzzy set $\widetilde{A}$. $\mathcal{Y}_N$ stands for the nondominated set of $\mathcal{Y} \subset \mathbb{R}^k$ with respect to the usual partial ordering in $\mathbb{R}^k$. An LR-type fuzzy number is denoted by $(a/b/c)_{LR}$ for some reference functions L and R.

In the FMOP (1), we assume that for each x in the fuzzy constraint set $\widetilde{\mathcal{X}}$, $f_j(x, \widetilde{c})$ and $g_i(x, \widetilde{a})$ are continuous fuzzy numbers for all possible i and j and also for each $j \in \{1, 2, \ldots, k\}$, $f_j(x, \widetilde{c}_j(1))$ has minimum value 'zero' on $\widetilde{\mathcal{X}}(1)$.

In fuzzy optimization problem, a proper comparison of fuzzy number valued objective functions, with regard to FMOPs, is not found yet. Many researchers attempted to describe the concept of fuzzy dominance and fuzzy Pareto optimality. We have already reported a detailed overview on the fuzzy dominance concept in our previous paper [3].

From the existing literature, we can notice that several fuzzy dominance relations have been proposed for FMOPs. Methodologies proposed in the above-mentioned various papers may give a particular solution or a part of fuzzy nondominated set. But no research work is focused yet to obtain entire fuzzy nondominated set. Recently, Ghosh and Chakraborty [3] proposed a technique to capture entire nondominated set of FMOPs. However, in [3], we have used the extension principle directly. In this paper, we will use the concept of inverse points of fuzzy geometry [1, 2] to formulate decision and criteria feasible regions of the problem. We also focus on the computational aspect of the proposed technique. Delineation of the paper is as follows.

In the next section, the preliminaries are given which are used throughout the paper. Formulations of fuzzy decision feasible region using inverse points are presented in Sect. 3. An algorithmic implementation of fuzzy ideal cone method to obtain the complete fuzzy nondominated set of a FMOP is proposed in Sect. 4. Lastly, a brief conclusion and a prospect of the future works are given in Sect. 5.

2 Preliminaries

Criteria feasible region $\widetilde{\mathcal{Y}}$ of FMOP (1) can be defined as (see [3])

$$\widetilde{\mathcal{Y}} = \bigcup_{x \in \widetilde{\mathcal{X}}(0)} \hat{f}(x, \widetilde{c}). \tag{2}$$

where $\hat{f}(x_0, \widetilde{c})$ is given by $\mu(y|\hat{f}(x_0, \widetilde{c})) = \min\{\mu(x_0|\widetilde{\mathcal{X}}), \mu(y|f(x_0, \widetilde{c}))\}$, and $\mu(y|f(x_0, \widetilde{c})) = \min\{\mu(y_1|f_1(x_0, \widetilde{c}_1)), \mu(y_2|f_2(x_0, \widetilde{c}_2)), \ldots, \mu(y_k|f_k(x_0, \widetilde{c}_k))\}$.

Definition 1 (*Fuzzy nondominated set* [3]). Let us consider any $x \in \widetilde{\mathcal{X}}$ and $c \in \widetilde{c}$. If the fuzzy region $\widetilde{\mathcal{Y}} \cap (f(x, c) - \mathbb{R}^k_{\geqq})$ is a normal fuzzy set with singleton core $\{f(x, c)\}$, then this intersection region may be said as a nondominated region of $\widetilde{\mathcal{Y}}$. Fuzzy nondominated set, $\widetilde{\mathcal{Y}}_N$ say, of $\widetilde{\mathcal{Y}}$ may be defined by

$$\widetilde{\mathcal{Y}}_N = \bigcup_{\substack{x \in \widetilde{\mathcal{X}}(0) \\ c \in \widetilde{c}(0)}} \Big\{\widetilde{\mathcal{Y}} \cap (f(x, c) - \mathbb{R}^k_{\geq}) : \widetilde{\mathcal{Y}} \cap (f(x, c) - \mathbb{R}^k_{\geqq}) \text{ is a normal fuzzy set with singleton core } \{f(x, c)\}\Big\}.$$

Theorem 1 (See [3]) $\widetilde{\mathcal{Y}}_N = \cup_{y \in \mathcal{Y}(1)_N} (\widetilde{\mathcal{Y}} \cap (y - \mathbb{R}^k_{\geqq}))$.

Definition 2 (*Same and inverse points* [2]). Let x, y be two numbers belonging to the supports of the continuous fuzzy numbers $\widetilde{a}$ and $\widetilde{b}$, respectively. The numbers x and y are said to be same points with respect to $\widetilde{a}$ and $\widetilde{b}$ if:

(i) $\mu(x|\widetilde{a}) = \mu(y|\widetilde{b})$, and
(ii) $x \leq a$ and $y \leq b$, or $x \geq a$ and $y \geq b$, where a, b are midpoints of $\widetilde{a}(1), \widetilde{b}(1)$, respectively.

The numbers x and y are said to be inverse points with respect to $\widetilde{a}$ and $\widetilde{b}$ if $x, -y$ are same points with respect to $\widetilde{a}$ and $-\widetilde{b}$

Using the concept of inverse points, decision feasible region $\widetilde{\mathcal{X}}$ for the FMOP (1) can be formulated as follows.

3 Decision Feasible Set $\widetilde{\mathcal{X}}$

Let us note that constraint of FMOP (1) reads as $\{x \in C : g(x, \widetilde{a}) \precsim \widetilde{b}\}$. We observe that $g(x, \widetilde{a}) \precsim \widetilde{b}$ may be appeared to be equivalent to $g(x, \widetilde{a}) - \widetilde{b} \precsim \widetilde{0}$. Corresponding to each $x \in C$, $g(x, \widetilde{a})$ is a fuzzy number. The substraction $g(x, \widetilde{a}) - \widetilde{b}$ of fuzzy numbers can be done by the concept of inverse points [2]. Thus, let us construct a set $\Omega(\alpha)$ corresponding to each α in [0, 1] as follows:

$$\Omega(\alpha) = \{x \in C : g(x, a) - b \leq 0, \text{ where } g(x, a) \text{ and } b \text{ are inverse}$$

$$\text{points with membership value } \alpha \text{ on } g(x, \widetilde{a}) \text{ and } \widetilde{b}\}.$$

Now let us define membership function of $\widetilde{\mathcal{X}}$ by

$$\mu(x|\widetilde{\mathcal{X}}) = \sup\{\alpha : x \in \Omega(\alpha)\}. \tag{3}$$

To obtain mathematical formulation of $\mu(.|\widetilde{\mathcal{X}})$, one may need its constituent α-cuts, i.e., $\widetilde{\mathcal{X}}(\alpha)$ for each $\alpha \in [0, 1]$. Here natural question may arise whether there is any relation between $\widetilde{\mathcal{X}}(\alpha)$ and $\Omega(\alpha)$. Following theorem investigates the same. Prior to the theorem let us give the following straightforward lemma which will be useful to the theorem.

Lemma 1 *For any* $0 \leq \alpha_1 \leq \alpha_2 \leq 1$, *the set* $\Omega(\alpha_2)$ *is a subset of* $\Omega(\alpha_1)$.

Theorem 2 $\widetilde{\mathcal{X}}(\alpha) = \Omega(\alpha)$ *for all* α *in* [0, 1].

Proof Similar to Theorem 1 of [3] and we skip the proof.

Let $x_1 \in C$. As $g_i(x_i, \widetilde{a}_i)$ is a fuzzy number, there exist two reference functions L_1 and R_1 where L_1 is increasing and left continuous function and R_1 is decreasing and right continuous (see [4], p. 126) such that $g_i(x_1, \widetilde{a}_i) = (g_i^l(x_1)/g_i^m(x_1)/g_i^r(x_1))_{L_1R_1}$.

Similarly, for $x_2 \in C$, $x_2 \neq x_1$, there exist two reference functions L_2 and R_2 such that $g_i(x_2, \widetilde{a}_i) = (g_i^l(x_2)/g_i^m(x_2)/g_i^r(x_2))_{L_2R_2}$.

Apparently, $L_1 = L_2$ and $R_1 = R_2$ since formulation of reference functions depends on reference functions of $\widetilde{a}_i$ and does not depend on the points in C. Let $g_i(x, \widetilde{a}_i) = (g_i^l(x)/g_i^m(x)/g_i^r(x))_{LR}$. Indeed, left spread $g_i^m(x) - g_i^l(x)$ and right spread $g_i^r(x) - g_i^m(x)$ vary point-to-point on C. We also note that $g_i^l(x)$, $g_i^m(x)$ and $g_i^r(x)$ functions, though seems to be dependent on x alone, but they implicitly depend on the constants in $\widetilde{a}_i(0)$. Following formulation gives how to evaluate membership value of any point lies in the support of $g_i(x, \widetilde{a}_i)$.

For each $i = 1, 2, \ldots, m$ and $x \in C$, $g_i(x, \widetilde{a}_i)$ is evaluated in the following way:

$$\mu(y|g_i(x, \widetilde{a}_i)) = \sup_{y=g(x,a)} \pi(c|\widetilde{c}) \text{ where } \pi(c|\widetilde{c}) = \min_{j=1,2,\ldots,q_i} \mu(c|\widetilde{c}_{ij})$$

Theorem 3 *For each $\alpha \in [0, 1]$, let us define a set*

$$\Omega'(\alpha) = \{x \in C : g(x, a) \leq b, a \in \widetilde{a}(\alpha), b \in \widetilde{b}(\alpha)\}.$$

(1) In general $\Omega(\alpha)$ is subset of $\Omega'(\alpha)$.
(2) Let for each $i = 1, 2, \ldots m$, $\widetilde{a}_i(0) = [a_i^L, a_i{}^U]$ and $\widetilde{b}(0) = [b^L, b^U]$. Let L_i and R_i be restrictions of $\mu(.|\widetilde{a}_i)$ on $[a_i^L, \widetilde{a}(1)]$ and $[\widetilde{a}_i(1), a_i^U]$ respectively; L_{m+1} and R_{m+1} be restriction of $\mu(.|\widetilde{b})$ on $[b^L, \widetilde{b}(1)]$ and $[\widetilde{b}(1), b^U]$ respectively. If L_i, R_i are continuous and strictly increasing functions and each component of g is strictly increasing function,[1] then $\Omega(\alpha) = \Omega'(\alpha)$.

Proof We will prove both the results for g to be a single function. When g is a vector function, from the componentwise generalization we can prove the result.

(1) Let $x_0 \in \Omega(\alpha)$. Therefore, there exist $a \in \widetilde{a}(0), b \in \widetilde{b}(0)$ such that $g(x_0, a) \leq b$ where $g(x, a)$ and b are inverse points with membership value α with regard to $g(x_0, \widetilde{a})$ and $\widetilde{b}$. If possible let $\mu(a|\widetilde{a}) < \alpha$. Then according to the evaluation of membership function of $g(x_0, \widetilde{a})$, we get $\mu(g(x_0, a)|g(x_0, \widetilde{a})) < \alpha$. A contradiction arises. This shows that x_0 lies in $\Omega'(\alpha)$. Thus, $\Omega(\alpha) \subseteq \Omega'(\alpha)$.
(2) To prove $\Omega'(\alpha) \subseteq \Omega(\alpha)$, let us take any element $x_0 \in \Omega'(\alpha)$. As $x_0 \in \Omega'(\alpha)$, there exist $a^0 = (a_1^0, a_2^0, \ldots, a_m^0)^t \in (\widetilde{a}_1(\alpha), \widetilde{a}_2(\alpha), \ldots, \widetilde{a}_m(\alpha))^t = \widetilde{a}(\alpha), b^0 \in \widetilde{b}(\alpha)$ such that $g(x^0, a^0\ ) \leq b^0$.
Let
$$\phi(a_1, a_2, \ldots, a_m, b) = g(x^0; a_1, a_2, \ldots, a_m) - b.$$

As g is an strictly increasing function with respect to each a_i and decreasing for b. Suppose $-b = a_{m+1}$, $-\widetilde{b} = \widetilde{a}_{m+1}$ and $-b^0 = -a_{m+1}^0$. Let

$$\begin{aligned} h(a_1, a_2, \ldots, a_m, a_{m+1}) &= \phi(a_1, a_2, \ldots, a_m, -a_{m+1}) \\ &= g(x^0, a_1, a_2, \ldots, a_m) + a_{m+1}. \end{aligned}$$

Then, h is strictly increasing function w.r.t. its each variable. Let $k = h(a_1^0, a_2^0, \ldots, a_{m+1}^0) = g(x^0, a_1^0, a_2^0, \ldots, a_m^0) + a_{m+1}^0 = g(x^0, a^0) - b^0$. Therefore, $k \leq 0$.
Here two cases may arise:

(2.a) $g(x^0, a^0) - b^0 \leq g(x^0, \widetilde{a}(1)) - \widetilde{b}(1)$ or
(2.b) $g(x^0, a^0) - b^0 > g(x^0, \widetilde{a}(1)) - \widetilde{b}(1)$.

(2.a) In this case, $h(a_1^0, a_2^0, \ldots, a_{m+1}^0) \leq h(\widetilde{a}_1(1), \widetilde{a}_2(1), \ldots, \widetilde{a}_{m+1}(1))$. Let us define $\psi = h(L_1^{-1}, L_2^{-1}, \ldots, L_{m+1}^{-1})$. As each L_i^{-1} is continuous and strictly increasing, so is ψ on its domain. Also ψ is one-to-one function.
Let $\beta = \psi^{-1}(k)$ and $a_i^* = L_i^{-1}(\beta), i = 1, 2, \ldots, m + 1$.

[1] a function $g(x_1, x_2, \ldots, x_n)$ is said to be strictly increasing when $x_1 > y_1, x_2 > y_2, \ldots, x_n > y_n$ implies $g(x_1, x_2, \ldots, x_n) > g(y_1, y_2, \ldots, y_n)$.

Therefore, $h(a_1^*, a_2^*, \ldots, a_{m+1}^*) = \psi(\beta) = k = h(a_1^0, a_2^0, \ldots, a_{m+1}^0)$. Moreover $a_i^* \leq \widetilde{a}_i(1)$ and $\beta = \mu(k|h(\widetilde{a}_1, \widetilde{a}_2, \ldots, \widetilde{a}_{m+1})) \geq \alpha$. Therefore $g(x_0, a^*) \leq g(x_0, \widetilde{a}(1))$ and $b^* \geq \widetilde{b}(1)$ where $a^* = (a_1^*, a_2^*, \ldots, a_m^*)$. Also $\mu(g(x_0, a^*)|g(x_0, \widetilde{a})) = \beta = \mu(b^*|\widetilde{b}) \geq \alpha$.
Thus $g(x_0, a^*)$ and b^* are inverse points with membership value β on $g(x_0, \widetilde{a})$ and $\widetilde{b}$, respectively. Again $g(x_0, a^*) - b^* = k \leq 0$. Hence x_0 lies in $\Omega(\beta) \subseteq \Omega(\alpha)$.

(2.b) This case is similar to the case (2.a) with $R_1, R_2, \ldots, R_{m+1}$ in place of $L_1, L_2, \ldots, L_{m+1}$.

Hence under the conditions stated in the theorem, we get $\Omega(\alpha) = \Omega'(\alpha)$.

Corollary 1 *If constraint set $\widetilde{\mathcal{X}}$ of FMOP (1) is fuzzy linear inequality $\widetilde{a}_1 x_1 + \widetilde{a}_2 x_2 + \cdots + \widetilde{a}_n x_n \precsim \widetilde{b}$ where $x = (x_1, x_2, \ldots, x_n)^t \in C \subseteq \mathbb{R}^n_{\geqq}$, then*

$$\widetilde{\mathcal{X}}(\alpha) = \{x \in C : a_1^\alpha x_1 + a_2^\alpha x_2 + \cdots + a_n^\alpha \leq b^\alpha : \textit{where} \, a_1^\alpha, a_2^\alpha, \ldots, a_n^\alpha \, \textit{and} - b^\alpha$$
$$\textit{are same points with respect to} \, \widetilde{a}_1, \widetilde{a}_2, \ldots, \widetilde{a}_n \textit{and} - \widetilde{b}\}.$$

4 On Generation of Entire $\widetilde{\mathcal{Y}}_N$: Fuzzy Ideal Cone Method

To obtain a nondominated point of $\widetilde{\mathcal{Y}}(1)$, according to fuzzy ideal cone method, one must solve the following minimization problem [3] corresponding to a particular $\hat{\beta} \in \mathbb{S}^{k-1}_{\geqq} = \mathbb{S}^{k-1} \cap \mathbb{R}^k_{\geqq}$ (here $\mathbb{S}^{k-1}$ represents the unit ball in $\mathbb{R}^k$):

$$\begin{cases} \min z \\ \text{subject to } z\hat{\beta} \geqq f(x, \widetilde{c}(1)), \\ x \in \widetilde{\mathcal{X}}(1). \end{cases} \tag{4}$$

Due to this method and Theorem 1,

$$\begin{aligned} \widetilde{\mathcal{Y}}_N &= \bigcup_{y \in \widetilde{\mathcal{Y}}(1)_N} (\widetilde{\mathcal{Y}} \cap (y - \mathbb{R}^k_{\geqq})) \\ &= \bigcup_{\hat{\beta} \in \bar{\mathbb{S}}^{k-1}_{\geqq}} (\widetilde{\mathcal{Y}} \cap (z_\beta \hat{\beta} - \mathbb{R}^k_{\geqq})), \end{aligned}$$

where z_β is 'min z' of (4) for $\hat{\beta}$.

Here let us note that $\widetilde{\mathcal{Y}}(0)$ can be computed from the union (refer to Sect. 2)

$$\widetilde{\mathcal{Y}} = \bigcup_{x \in \widetilde{\mathcal{X}}(0)} \hat{f}(x, \widetilde{c}).$$

Corresponding to each $\hat{\beta} \in \bar{\mathbb{S}}^{k-1}_{\geqq}$, the set $\widetilde{\mathcal{Y}} \cap (z_\beta \hat{\beta} - \mathbb{R}^k_{\geqq})$ can be obtained by restricting $\widetilde{\mathcal{Y}}$ on the set $A_\beta = \{y \in \widetilde{\mathcal{Y}}(0) : y \leqq z_\beta \hat{\beta}\}$. Now if we take any $y \in A_\beta$, then $\mu(y | \widetilde{\mathcal{Y}} \cap (z_\beta \hat{\beta} - \mathbb{R}^k_{\geqq})) = \mu(y | \widetilde{\mathcal{Y}})$, since $z_\beta \hat{\beta} - \mathbb{R}^k_{\geqq}$ is a crisp set. For any $\hat{\beta} \in \bar{\mathbb{S}}^{k-1}_{\geqq}$, let us denote $\widetilde{\mathcal{Y}} \cap (z_\beta \hat{\beta} - \mathbb{R}^k_{\geqq})$ as $\widetilde{A}_\beta$. Obviously, membership function of $\widetilde{A}_\beta$ is given by $\mu(y | \widetilde{A}_\beta) = \mu(y | \widetilde{\mathcal{Y}})$ when y in A_β and '0' otherwise. Thus, entire $\widetilde{\mathcal{Y}}_N$ can be obtained by

$$\widetilde{\mathcal{Y}}_N = \bigcup_{\hat{\beta} \in \bar{\mathbb{S}}^{k-1}_{\geqq}} \widetilde{A}_\beta.$$

4.1 Algorithmic Implementation of the Fuzzy Ideal Cone Method

Let us note that any $\hat{\beta} \in \mathbb{S}^{k-1}$ can be expressed by

$$\left(\cos\phi_1, \cos\phi_2 \sin\phi_1, \cos\phi_3 \sin\phi_2 \sin\phi_1, \ldots, \cos\phi_{k-1} \prod_{i=1}^{k-2} \sin\phi_i, \prod_{i=1}^{k-1} \sin\phi_i \right),$$

for $\phi_i \in [0, \frac{\pi}{2}]$, $i = 1, 2, \ldots, (k-1)$. This is well known spherical discretization technique. However, if we discretize each ϕ_i to equal number of subintervals, then set of discretized points will be much congested near the point $(1, 0, \ldots, 0)$. Thus, to get a uniform discretized points on $\mathbb{S}^{k-1}$, let us attempt to divide ϕ_1 by m number of points and ϕ_i by $round(m \prod_{l=1}^{i} \sin\phi_i)$ number of points, for $i = 2, 3, \ldots, k-1$. Here *round* is the rounding function to the nearest integer.

Following Algorithm 1 provides a sequential procedure to obtain complete Pareto set of a tri-criteria problem. In tri-criteria problem, we need to run 3 for loops for each ϕ_i, $i = 1, 2, 3$. For k-criteria problem, we only have to run k for loops for each ϕ_i, $i = 1, 2, \ldots, k$.

In the next, an numerical example has been presented to elaborate the methodology.

Algorithm 1 Algorithm to generate the complete fuzzy nondominated set

Require: Given fuzzy MOP:

$$\begin{cases} \min\ f(x,\widetilde{c}) \\ \text{subject to } g(x,\widetilde{a}) \widetilde{\leq} \widetilde{b}, \\ x \in C \subseteq \mathbb{R}^n_{\geqq}. \end{cases}$$

Final output $\widetilde{\mathcal{Y}}_N$ of the algorithm is the complete fuzzy nondominated set of the problem.

```
Initialize φ1, φ2, and φ3 to 0.
Initialize Ỹ_N ← Ø.
Give m (total number of grid points for φ1)
for φ1 = 0 to π/2 with step length π/(2m) do
  Find m2 = round(m sin φ1)
  for φ2 = 0 to π/2 with step length π/(2m2) do
    Find β̂ = (cos φ1, cos φ2 sin φ1, sin φ2 sin φ1)
    Find z_β which is solution of the following problem for β̂:
```

$$\begin{cases} \min\ z \\ \text{subject to } z\hat{\beta} \geqq f(x,\widetilde{c}(1)), \\ x \in \widetilde{\mathcal{X}}(1). \end{cases}$$

```
    Find Ã_β = Ỹ ∩ (z_β β̂ − R³_≧).
     Set Ỹ_N ← Ỹ_N ∪ Ã_β.
  end for
end for
```

4.2 An Illustrative Example

Example 1 Let us consider the following fuzzy bi-criteria minimization problem:

$$\min \begin{pmatrix} f_1 \\ f_2 \end{pmatrix} = \begin{pmatrix} x_1 + (-\frac{1}{4}/0/\frac{1}{2}) \\ x_2 + (-\frac{1}{4}/0/\frac{1}{2}) \end{pmatrix}$$

$$\text{subject to } \widetilde{4}(x_1-1)^2 + \widetilde{2}(x_2-1)^2 \widetilde{\leq} \widetilde{2},$$

$$-\frac{1}{2} \leq x_1 \leq 1, -\frac{1}{2} \leq x_2 \leq 1,$$

where $\widetilde{2} = (1/2/3)$ and $\widetilde{4} = (2/4/6)$. Same points with respect to $\widetilde{4}$, $\widetilde{2}$ and $-\widetilde{2}$ are

$$2+2\alpha,\ 1+\alpha \text{ and } -(3,-\alpha) \quad \text{or} \quad 6-2\alpha,\ 3-\alpha \text{ and } -(1+\alpha).$$

Therefore according to decision feasible region construction through invere points (Sect. 3), the set $\widetilde{\mathcal{X}}$ is given by

$$\widetilde{\mathcal{X}} = \bigvee_{\alpha\in[0,1]} \Big\{\{(x_1, x_2) \in [-1/2, 1] \times [-1/2, 1] : (2+2\alpha)(x_1-1)^2 + (1+\alpha)(x_2-1)^2 \le 3-\alpha\} \cup \{(x_1, x_2) \in [-1/2, 1] \times [-1/2, 1] : (6-2\alpha)(x_1-1)^2 + (3-\alpha)(x_2-1)^2 \le 1+\alpha\}\Big\}$$

For each $\alpha \in [0, 1]$, α-cut of $\widetilde{\mathcal{X}}$ is determined by

$$\widetilde{\mathcal{X}}(\alpha) = \Omega(\alpha) = \Big\{(x_1, x_2) \in [-1/2, 1] \times [-1/2, 1] : 2(x_1-1)^2 + (x_2-1)^2 \le \frac{3-\alpha}{1+\alpha}\Big\}.$$

For any $(x_1, x_2) \in [-1/2, 1] \times [-1/2, 1]$, closed form of membership function of $\widetilde{\mathcal{X}}$ is as follows:

$$\mu((x_1, x_2)|\widetilde{\mathcal{X}}) = \begin{cases} 1 & \text{if } 2(x_1-1)^2 + (x_2-1)^2 \le 1 \\ \frac{4}{1+2(x_1-1)^2+(x_2-1)^2} - 1 & \text{if } 1 \le 2(x_1-1)^2 + (x_2-1)^2 \le 3 \\ 0 & \text{elsewhere.} \end{cases}$$

The decision feasible region $\widetilde{\mathcal{X}}$ is depicted in Fig. 1. Any point on the part of ellipse $\{(x_1, x_2) \in [-1/2, 1] \times [-1/2, 1] : 2(x_1-1)^2 + (x_2-1)^2 \le \frac{3-\alpha}{1+\alpha}\}$ has membership value α on $\widetilde{\mathcal{X}}$.

Corresponding to each point $x = (x_1, x_2) \in \widetilde{\mathcal{X}}(0)$, $f(x_0, \widetilde{c})$ determines a fuzzy point with support $[x_1 + 1/4, x_1 + 1/2] \times [x_2 - 1/4, x_2 + 1/2]$.

For each $\alpha \in [0, 1]$, α-cut of $f(x_0, \widetilde{c})$ is given by

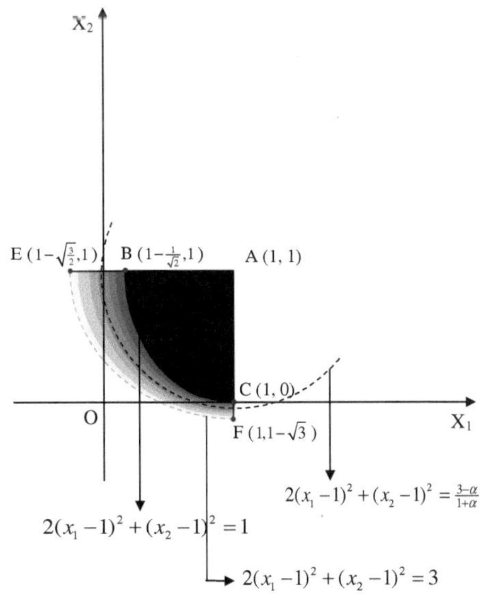

Fig. 1 Fuzzy decision feasible region $\widetilde{\mathcal{X}}$ of the Example 1

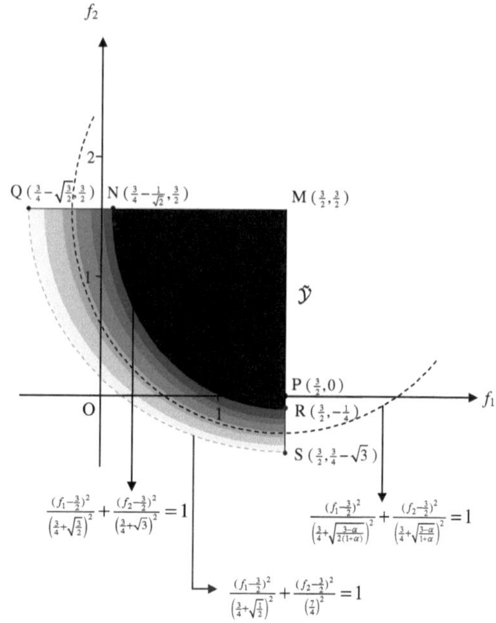

Fig. 2 Criteria feasible region $\widetilde{\mathcal{Y}}$ and nondominated set $\widetilde{\mathcal{Y}}_N$ of the Example 1

$$f(x_0, \widetilde{c})(\alpha) = \left[x_1 + \frac{\alpha - 1}{4}, x_1 + \frac{1-\alpha}{2}\right] \times \left[x_2 + \frac{\alpha - 1}{4}, x_2 + \frac{1-\alpha}{2}\right] = \mathcal{S}(\alpha) \text{ say.}$$

Here corresponding to any (x_1, x_2) in $\widetilde{\mathcal{X}}(0)$ (with $\mu((x_1, x_2)|\widetilde{\mathcal{X}}) = \gamma$ say) the fuzzy set $\hat{f}((x_1, x_2), \widetilde{c})$ is given by its α-cuts for each α in [0, 1] as follows:

$$\hat{f}((x_1, x_2), \widetilde{c})(\alpha) = \begin{cases} \mathcal{S}(\alpha) & \text{if } \alpha \in [0, \gamma] \\ \mathcal{S}(\gamma) & \text{if } \alpha \in [\gamma, 1]. \end{cases}$$

For example if we consider $(x_1, x_2) = (\frac{1}{5}, \frac{1}{2}) \in \widetilde{\mathcal{X}}(0)$, then $\mu((\frac{1}{5}, \frac{1}{2})|\widetilde{\mathcal{X}}) = \frac{400}{253} - 1 = 0.58$ and for each $\alpha \in [0, 1]$:

$$\hat{f}((^1/_5, ^1/_2), \widetilde{c})(\alpha) = \begin{cases} \mathcal{S}(\alpha) & \text{if } \alpha \in [0, 0.58] \\ [0.1, 0.4] \times [0.4, 0.7] & \text{if } \alpha \in [0.58, 1], \end{cases}$$

where $\mathcal{S}(\alpha) = [\frac{5\alpha-1}{20}, \frac{7-5\alpha}{10}] \times [\frac{1+\alpha}{4}, \frac{2-\alpha}{2}]$.

The union $\bigcup_{x \in \widetilde{\mathcal{X}}(0)} \hat{f}((x_1, x_2), \widetilde{c})$ determines the criteria feasible region $\widetilde{\mathcal{Y}}$. Criteria feasible region is shown in the Fig. 2.

We obtain from the proposed method on $\widetilde{\mathcal{Y}}(1)$ that

$$\widetilde{\mathcal{Y}}(1)_N = \left\{(f_1, f_2) \in \left[\frac{3}{4} - \frac{1}{\sqrt{2}}, \frac{3}{2}\right] \times \left[\frac{3}{4} - \sqrt{3}, \frac{3}{2}\right] : \frac{(f_1 - \frac{3}{2})^2}{(\frac{3}{4} + \frac{1}{\sqrt{2}})^2} + \frac{(f_2 - \frac{3}{2})^2}{(\frac{7}{4})^2} = 1\right\}$$

and $\bar{\mathbb{S}}^1 = \{\hat{\beta} = (\cos\theta_\beta, \sin\theta_\beta) : 0 \leq \theta_\beta \leq \frac{\pi}{2}\}$. Entire nondominated set $\widetilde{\mathcal{Y}}_N$ of the considered problem is the fuzzy region, on the support of $\widetilde{\mathcal{Y}}$, lying inside and boundary of the region $QNRSQ$ in the Fig. 2. For each $\alpha \in [0, 1]$, α-cut of $\widetilde{\mathcal{Y}}_N$ is the set

$$\left\{(f_1, f_2) \in \left[\frac{3}{4} - \frac{1}{\sqrt{2}}, \frac{3}{2}\right] \times \left[\frac{3}{4} - \sqrt{3}, \frac{3}{2}\right] : \frac{(f_1 - \frac{3}{2})^2}{(\frac{3}{4} + \sqrt{\frac{3-\alpha}{2(1+\alpha)}})^2} + \frac{(f_2 - \frac{3}{2})^2}{(\frac{3}{4} + \sqrt{\frac{3-\alpha}{1+\alpha}})^2} \leq 1\right\}.$$

5 Discussion and Concluding Remarks

This paper deals with the computational aspect of the fuzzy ideal cone method [3] to generate the complete fuzzy nondominated set of fuzzy multi-criteria optimization problems. Concept of inverse points in fuzzy geometry is being used to formulate decision feasible region. Relation between the fuzzy decision feasible sets evaluated by inverse points and by extension principle has been presented. It is shown that under a certain continuity and a monotone condition on the reference functions of the fuzzy numbers $g_i(x_i, \widetilde{a}_i)$ both the decision feasible sets are equal. This result can greatly reduce computational cost to evaluate decision feasible region, since direct use of extension principle uses only inverse points of $g_i(x_i, \widetilde{a}_i)$ and b_i, while extension principle uses their all possible combinations to obtain fuzzy decisions feasible region. An algorithmic implementation of the fuzzy ideal cone method is presented to find complete fuzzy nondominated set of the fuzzy criteria feasible region. Some more computational aspect of the fuzzy ideal cone method can be obtained in our future research.

Practically, in any decision-making problem, available data sets are used to be inherently imprecise. However, it turns out that solution is always crisp. Usually from the available data set we fit the coefficients of the constraints. Thus, in the considered FMOP (1), coefficients of the constraint set are taken as fuzzy. Again as final decision for any decision-making process is always crisp, so we considered the decision variables as crisp. For more analysis on the proposed method, future research may be focused on the FMOPs with fuzzy variable.

To show effectiveness of the proposed method, sensitivity analysis must be an issue. However, focus on this paper is to explore the computational aspect of the fuzzy ideal cone method and thus we keep the sensitivity analysis as a future research. However, on the prospect of the sensitivity analysis we note that the ideal cone method is solely dependent on the core of the fuzzy decision feasible region (please see Algorithm 1, Step 8) and not on the fuzzy information. Hence,

(i) If the spreads of the coefficients are changed while keeping the cores fixed, a fuzzy nondominated solution generated by the method will be still a fuzzy nondominated solution if it is feasible. For instance, in the Example 1, proposed method will give a reliable result up to any bounded increase of the constants.
(ii) If a fuzzy inequality is added, though it will reduce the feasible region, but still a fuzzy nondominated solution generated by the method will be a fuzzy nondominated solution if it is feasible.
(iii) If we add an extra criteria on the FMOP, though it will change the dimension of the criteria feasible region, the generated fuzzy nondominated set for the problem without the extra criteria will be a subpart of the fuzzy nondominated set for the problem with the extra criteria.

Our detailed research on this sensitivity analysis may be obtained in the future.

Future research may also be done on the real-life applications of the proposed method. Particularly, the procedure to acquire the fuzzy parameters, the scalability issues, etc., can be highlighted.

For a numerical illustration of the proposed method, we have given the Example 1. In the mentioned problem, question may arise that how an increase of the width of the constants will effect the result and how far we can increase the width of the fuzzy constants for reliable results. To answer the question we note that generation of the fuzzy nondominated set depends on the core of the criteria feasible region $\widetilde{\mathcal{Y}}$ (see Theorem 1). Thus if we increase the spreads of the constants, keeping the cores fixed, the result will increase at its imprecise region and not at its core level. Also, a generated nondominated region will be a superset of a nondominated region for the problem without the increase on the coefficients.

We note that proposed method essentially depends on the core set $\widetilde{\mathcal{Y}}(1)$. If $\widetilde{\mathcal{Y}}(1)$ is empty, then problem to obtain entire $\widetilde{\mathcal{Y}}_N$ may be appeared as a challenging task and proposed method cannot work. Future research work may be focused for this extension. Fuzzy dominance under general convex cone may also be focused in the future.

References

1. Chakraborty, D., Ghosh, D.: Analytical fuzzy plane geometry II. Fuzzy Sets Syst. **243**, 84–109 (2014)
2. Ghosh, D., Chakraborty, D.: Analytical fuzzy plane geometry I. Fuzzy Sets Syst. **209**, 66–83 (2013)
3. Ghosh, D., Chakraborty, D.: Fuzzy ideal cone: a method to obtain complete fuzzy non-dominated set of fuzzy multi-criteria optimization problems with fuzzy parameters. In: The Proceedings of IEEE International Conference on Fuzzy Systems 2013, FUZZ IEEE 2013, IEEE Xplore, pp. 1–8
4. Wang, X., Ruan, D., Kerre, E.E.: Mathematics of Fuzziness-Basic Issues, vol. 245. Springer, Berlin (2009)

A Bi-Objective Solid Transportation Model Under Uncertain Environment

Amrit Das and Uttam Kumar Bera

Abstract In this paper, we study a solid transportation problem with uncertain cost and uncertain time, where the supplies, the demands, the conveyance capacities are regarded as uncertain in nature. For the first time we minimize the uncertain transportation time. According to the inverse uncertainty distribution, the model can be transformed into a deterministic form by taking expected value on objective functions and confidence level on the constraint functions. We solve the uncertain solid transportation problem by fuzzy programming technique and using the LINGO 13.0 software. Finally, this paper is illustrated by a numerical example on uncertain solid transportation problem to show the application of the model.

Keywords Uncertain solid transportation problem · Uncertain cost and time · Fuzzy programming technique.

1 Introduction

Transportation models are widely used in system distribution, job assignment, and other problems. In traditional TP, there are usually two kinds of constraints to be considered, namely, source constraint and destination constraint suggested by Balinski [1] in (1961). But in real situation, besides of these two constraints we have to deal with another constrain such as product type constraint or transportation mode constraint. For that reason the traditional TP turns into the solid transportation problem (STP) where we deal with three types of constraints. So as a generalization of the traditional TP, the STP was introduced by Haley [2] in 1962. Recently, the STP obtained much attention and many models and algorithms under both crisp

A. Das · U.K. Bera (✉)
Department of Mathematics, National Institute of Technology Agartala,
Barjala, Jirania, West Tripura 799046, India
e-mail: bera_uttam@yahoo.co.in

A. Das
e-mail: das.amrit12@gmail.com

M.K. Chakraborty et al. (eds.), *Facets of Uncertainties and Applications*,
Springer Proceedings in Mathematics & Statistics 125,
DOI 10.1007/978-81-322-2301-6_20

environment and uncertain environment have been investigated. For examples, Bit et al. [3] presented the fuzzy programming model for a multi-objective STP, Mahapatra et al. [4] investigated a multi-objective stochastic transportation problem involving log-normal, Kaufmann [5] studied two kinds of uncertain STP, that is, the supplies, demands, and conveyance capacities are interval numbers and fuzzy numbers, respectively. Sheng [6] and Pandian et al. [7] provided a new method to find an optimal solution of the STP. More recently, Baidya et al. [8, 9] introduced safety measure in solid transportation problem under different environment.

In reality, due to changes in market supply and demand, weather conditions, road conditions and other uncertainty factors, uncertainty transportation problem is particularly important. Therefore studying uncertainty in transportation problem has both theoretical and practical significances. In order to construct model for STP in uncertain environment, we shall first introduce some knowledge of uncertainty theory. Uncertainty theory was founded by Liu [10] in 2007 and refined by Liu [11–13] in 2009 and 2012 respectively, which is a branch of mathematics based on normality, duality, subadditivity, and product axioms. Now, uncertainty theory has become a mathematical tool to model the indeterminate phenomenon in our real world. It has been developed to a fairly complete mathematical system [14]. So many models had been developed by many researchers in this area. Jimenez et al. [15] investigated uncertain solid transportation problem in 1998. Yuhong Sheng and Kai Yao studied a Transportation Model with Uncertain Costs and Demands in [16, 17]. Yuhong Sheng and Kai Yao presented Fixed Charge Transportation Problem and its Uncertain Programming Model in [18]. Cui and Sheng [19] also presented Uncertain Programming Model for Solid Transportation Problem and so on. In this paper, the STP is modeled based on uncertainty theory. In [20] Minimization of transportation time is considered by Bhatia et al. under crisp environment.

In this paper, we solve a bi-objective solid transportation problem (BOSTP) with uncertain cost and uncertain time, where the supplies, the demands, the conveyance capacities are regarded as fuzzy in nature. One task of this paper is to find a transportation plan such that the transportation cost and time are minimized. For the first time we minimize the uncertain transportation time. According to the inverse uncertainty distribution, the model can be transformed into a deterministic form by taking expected value on objective functions and confidence level on the constraint functions. We solve the uncertain solid transportation problem by fuzzy programming technique and using the LINGO 13.0 software. Finally, this paper is illustrated by a numerical example on uncertain solid transportation problem to show the application of the model.

1.1 Preliminaries

Uncertain Variable:

Definition (Liu [10]) An uncertain variable is a measurable function ξ from an uncertainty space $(\Gamma, \mathcal{L}, \mathcal{M})$ to the set of real numbers, i.e., for any Borel set B of real numbers, the set $\{\xi \in B\} = \{\gamma \in \Gamma | \xi(\gamma) \in B\}$ is an event.

Definition (Liu [10]) The uncertainty distribution Φ of an uncertain variable ξ is defined by $\Phi(x) = \mathcal{M}\{\xi \leq x\}$ for any real number x.

Definition An uncertainty distribution Φ is said to be regular if its inverse function $\Phi^{-1}(\alpha)$ exists and is unique for each $\alpha \in (0, 1)$.

Definition Let ξ be an uncertain variable with regular uncertainty distribution Φ. Then the inverse function $\Phi_{\xi}^{-1}$is called the inverse uncertainty distribution of ξ.

Example: The inverse uncertainty distribution of normal uncertain variable is $\mathcal{N}(e, \sigma)$ is $\Phi^{-1}(\alpha) = e + \frac{\sigma\sqrt{3}}{\pi} \ln \frac{\alpha}{1-\alpha}$.

Definition (Liu [10]) The uncertain variables, $\xi_1, \xi_2, \ldots, \xi_n$ are said to be independent if $\mathcal{M}\{\bigcap_{i=1}^{n}(\xi_i \in B_i)\} = \bigwedge_{i=1}^{n} \mathcal{M}\{(\xi_i \in B_i\}$ for any *Borel* sets B_1,B_2, ..., B_n of real numbers.

Definition (Liu [10]) Let ξ be an uncertain variable. Then the expected value of ξ is defined by E $[\xi] = \int_0^{+\infty} \mathcal{M}\{\xi \geq r\}dr - \int_{-\infty}^{0} \mathcal{M}\{\xi \leq r\}dr$, provided that at least one of the two integrals is finite. Let ξ be uncertain variable with uncertainty distribution Φ.

If the expected value exists, then E $[\xi] = \int_0^1 \Phi^{-1}(\alpha)d\alpha$.

In fact, the expected value operator is linear.

Theorem 1 (Liu [10]) Let $\xi_1, \xi_2, \ldots, \xi_n$ be independent uncertain variables with uncertainty distributions $\Phi_1, \Phi_2, \ldots, \Phi_n$, respectively. If f is a strictly increasing function, then $\xi = f(\xi_1, \xi_2, \ldots, \xi_n)$ is an uncertain variable with inverse uncertainty distribution $\Psi^{-1}(\alpha) = f(\Phi_1^{-1}(\alpha), \Phi_2^{-1}(\alpha), \ldots, \Phi_n^{-1}(\alpha))$.

Theorem 2 (Liu [10]) Let $\xi_1, \xi_2, \ldots, \xi_n$ be independent uncertain variables with uncertainty distributions $\Phi_1, \Phi_2, \ldots, \Phi_n$, respectively. If f is a strictly decreasing function, then $\xi = f(\xi_1, \xi_2, \ldots, \xi_n)$is an uncertain variable with inverse uncertainty distribution $\Psi^{-1}(\alpha) = f(\Phi_1^{-1}(1-\alpha), \Phi_2^{-1}(1-\alpha), \ldots, \Phi_n^{-1}(1-\alpha))$.

2 Uncertain Solid Transportation Model Formulation

Let there are m sources, n destinations and k conveyances of the STP. The amount of products in source i is denoted by a_i, the minimal demand of products in destination j is denoted by b_j, the transportation capacities of conveyance k is denoted by e_k, the unit transportation cost is denoted by ξ_{ijk}, x_{ijk} be the quantity, where $i = 1, 2, \ldots, m$, $j = 1, 2, \ldots, n$, $k = 1, 2, \ldots, k$.

In order to model the above-mentioned uncertain solid transportation problem, the following notations are employed: $y_{ijk} = \begin{cases} 1, if x_{ijk} > 0 \\ 0, otherwise \end{cases}$ where $i = 1, 2, \ldots, m$, $j = 1, 2, \ldots, n$, $k = 1, 2, \ldots, k$, respectively. This implies that, if the transportation activities are assigned from source i to destination j by k conveyance, then the corresponding time will be occurring.

To describe the problems conveniently, we denote the cost objective function and the time objective function of model in the following way,

$$f_1(x,\xi) = \sum_{i=1}^{m}\sum_{j=1}^{n}\sum_{k=1}^{l} \xi_{ijk}\, x_{ijk}$$

$$f_2(x,t) = \sum_{i=1}^{m}\sum_{j=1}^{n}\sum_{k=1}^{l} t_{ijk}\, y(x_{ijk})$$

where x, ξ, t denote the vectors consisting of $x_{ijk}, \xi_{ijk}, t_{ijk}, i = 1, 2, \ldots, m, j = 1, 2, \ldots, n, k = 1, 2, \ldots, l$ respectively. Therefore model Bi-Objective Solid Transportation Problem (BOSTP) can be stated as follows:

$$\min = \sum_{i=1}^{m}\sum_{j=1}^{n}\sum_{k=1}^{l} \xi_{ijk}\, x_{ijk}$$

$$\min = \sum_{i=1}^{m}\sum_{j=1}^{n}\sum_{k=1}^{l} t_{ijk}\, y(x_{ijk}) \tag{1}$$

subject to

$$\sum_{j=1}^{n}\sum_{k=1}^{l} x_{ijk} \le a_i, \quad i = 1, 2, \ldots, m$$

$$\sum_{i=1}^{m}\sum_{j=1}^{n} x_{ijk} \ge b_j, \quad j = 1, 2, \ldots, n$$

$$\sum_{i=1}^{m}\sum_{j=1}^{n} x_{ijk} \le e_k, \quad k = 1, 2, \ldots, l$$

$$x_{ijk} \ge 0, i = 1, 2, \ldots, m, j = 1, 2, \ldots, n, k = 1, 2, \ldots, l.$$

But due to the complexity of the real world, we may always meet uncertain phenomena in constructing mathematical model. For such condition, we generally add the uncertain variables to the model. Hence, in this paper, we assume that the unit cost, transportation time, the capacity of each source and that of each destination are all uncertain variables and denoted by $\widetilde{\xi}_{ijk}, \widetilde{t}_{ijk}, \widetilde{a}_i, \widetilde{b}_j, \widetilde{e}_k$, respectively. Also we assume that all the uncertain variables $\widetilde{a}_i, \widetilde{b}_j, \widetilde{e}_k, \widetilde{\xi}_{ijk}$, and $\widetilde{t}_{ijk}$ are independent. Then the bi-objective STP becomes uncertain bi-objective STP.

The expected-constrained programming model is constructed by [10]. The main idea of this model is to optimize the expected value of the objective function under the chance constraints.

Definition (Liu [21]) Assume that $f(x, \xi)$ is an objective function, and $g_j(x, \xi)$ are constraints functions, $j = 1, 2, \ldots, k$. A solution x is feasible if and only if $\mathcal{M}\{g_j(x, \xi) \leq 0\} \geq \alpha_j$ for $j = 1, 2, \ldots, n$. A solution x^* is an optimal solution to the uncertain programming model if $E[f(x^*, \xi)] \leq E[f(x, \xi)]$ if for any feasible solution x.

By taking the expected value criterion on the objective functions and confidence level on the constraint functions, the above model turns into the following mathematical model:

$$
\begin{aligned}
&\min E\left[\sum_{i=1}^{m}\sum_{j=1}^{n}\sum_{k=1}^{l} \tilde{\xi}_{ijk}\, x_{ijk}\right]\\
&\min E\left[\sum_{i=1}^{m}\sum_{j=1}^{n}\sum_{k=1}^{l} \tilde{t}_{ijk}\, x_{ijk}\right]
\end{aligned}
\tag{2}
$$

subject to

$$
\begin{aligned}
&\mathcal{M}\left\{\sum_{j=1}^{n}\sum_{k=1}^{l} x_{ijk} \leq \tilde{a}_l\right\} \leq \alpha_i,\ i = 1, 2, \ldots, m\\
&\mathcal{M}\left\{\sum_{i=1}^{m}\sum_{k=1}^{l} x_{ijk} \geq \tilde{b}_j\right\} \geq \beta_j,\ j = 1, 2, \ldots, n\\
&\mathcal{M}\left\{\sum_{i=1}^{m}\sum_{j=1}^{n} x_{ijk} \leq \tilde{e}_k\right\} \leq \gamma_k,\ k = 1, 2, \ldots, l
\end{aligned}
$$

$$x_{ijk} \geq 0, \xi_{ijk} \geq 0, t_{ijk} \geq 0, i = 1, 2, \ldots, m,\ j = 1, 2, \ldots, n, k = 1, 2, \ldots, l.$$

where $\alpha_i, \beta_j, \gamma_k$ are specified confidence levels for $i = 1, 2, \ldots, m,\ j = 1, 2, \ldots, n,$ $k = 1, 2, \ldots, l$. The first constraint implies that total amount transported from source should be no more than its supply capacity at the confidence level α_i; the second constraint implies that the total amount transported from source i should satisfy the requirement of destination j at the credibility level β_j; the third constraint states that the total amount transported by conveyance k should be no more than its transportation capacity at the confidence level γ_k.

3 Crisp Equivalences of Models:

Since the proposed model have so many uncertain variables, to solve the models, we have to convert the models into crisp equivalences of models. Here, we shall induce the deterministic form for model taking advantage of some properties of expected value and uncertain measure in uncertainty theory.

Theorem 3 If $\widetilde{a}_i, \widetilde{b}_j, \widetilde{e}_k, \widetilde{\xi}_{ijk}$ and $\widetilde{t}_{ijk}$, are independent uncertain variables with uncertainty distributions $\Phi_{\widetilde{a}_i}$, $\Phi_{\widetilde{b}_j}$, $\Phi_{\widetilde{e}_k}$, $\Phi_{\widetilde{\xi}_{ijk}}$, and $\Phi_{\widetilde{t}_{ijk}}$, respectively, then model (2) is equivalent to the following model

$$\min \sum_{i=1}^{m}\sum_{j=1}^{n}\sum_{k=1}^{l} x_{ijk} \int_0^1 \Phi^{-1}_{\widetilde{\xi}_{ijk}}(\alpha)d\alpha$$

$$\min \sum_{i=1}^{m}\sum_{j=1}^{n}\sum_{k=1}^{l} y_{(x_{ijk})} \int_0^1 \Phi^{-1}_{\widetilde{t}_{ijk}}(\alpha)d\alpha \tag{3}$$

subject to

$$\sum_{j=1}^{n}\sum_{k=1}^{l} x_{ijk} - \Phi^{-1}_{\widetilde{a}_l}(1-\alpha_i) \le 0, \quad i = 1, 2, \ldots, m$$

$$\Phi^{-1}_{\widetilde{b}_j}(\beta_j) - \sum_{i=1}^{m}\sum_{k=1}^{l} x_{ijk} \le 0, \quad j = 1, 2, \ldots, n$$

$$\sum_{i=1}^{m}\sum_{j=1}^{n} x_{ijk} - \Phi^{-1}_{\widetilde{e}_k}(1-\gamma_k) \le 0, \quad k = 1, 2, \ldots, l$$

$$x_{ijk} \ge 0, i = 1, 2, \ldots, m, j = 1, 2, \ldots, n, k = 1, 2, \ldots, l.$$

Proof: Since $\widetilde{\xi}_{ijk}, \widetilde{t}_{ijk}, \widetilde{a}_l, \widetilde{b}_j, \widetilde{e}_k$ are independent uncertain variables with uncertainty distributions $\Phi_{\xi_{ijk}}$, $\Phi_{t_{ijk}}$, $\Phi_{\widetilde{a}_l}$, $\Phi_{\widetilde{b}_j}$, $\Phi_{\widetilde{e}_k}$ respectively. According to the linearity of expected value operator, we have

$$E\left[\sum_{i=1}^{m}\sum_{j=1}^{n}\sum_{k=1}^{l} \widetilde{\xi}_{ijk}x_{ijk}\right] \rightleftharpoons \sum_{i=1}^{m}\sum_{j=1}^{n}\sum_{k=1}^{l} x_{ijk}E[\widetilde{\xi}_{ijk}]$$

where $E[\widetilde{\xi}_{ijk}] = \int_0^1 \Phi^{-1}_{\widetilde{\xi}_{ijk}}(\alpha)d\alpha, i = 1, 2, \ldots, m, j = 1, 2, \ldots, n, k = 1, 2, \ldots, l.$

According to the Theorems 1 and 2, the constraints are converted as follows: the first constraint of the model (2)

$$\mathcal{M}\left\{\sum_{j=1}^{n}\sum_{k=1}^{l} x_{ijk} \le \widetilde{a}_l\right\} \ge \alpha_i, i = 1, 2, \ldots, m,$$

is equivalent to

$$\sum_{j=1}^{n}\sum_{k=1}^{l} x_{ijk} - \Phi^{-1}_{\widetilde{a}_l}(1-\alpha_i) \le 0, \quad i = 1, 2, \ldots, m. \tag{4}$$

the second constraint of the model (2)

$$\mathcal{M}\left\{\sum_{i=1}^{m}\sum_{k=1}^{l} x_{ijk} \geq \widetilde{b}_j\right\} \geq \gamma_j \qquad j = 1, 2, \ldots, n.$$

is equivalent to

$$\Phi_{\widetilde{b}_j}^{-1}(\beta_j) - \sum_{i=1}^{m}\sum_{k=1}^{l} x_{ijk} \leq 0, \quad j = 1, 2, \ldots, n. \tag{5}$$

and the third constraint of the model (2)

$$\mathcal{M}\left\{\sum_{i=1}^{m}\sum_{j=1}^{n} x_{ijk} \leq \widetilde{e}_k\right\} \leq \eta_k, \qquad k = 1, 2, \ldots, l$$

is equivalent to

$$\sum_{i=1}^{m}\sum_{j=1}^{n} x_{ijk} - \Phi_{\widetilde{e}_k}^{-1}(1-\gamma_k) \leq 0, \quad k = 1, 2, \ldots, l. \tag{6}$$

the result follows from immediately. Assume that all uncertain variables are normal uncertain variables,

$$\begin{aligned}\widetilde{\xi}_{ijk} \text{ and } \widetilde{t}_{ijk} &\sim N(e_{ijk}, \sigma_{ijk}), \widetilde{a}_i \sim N(e_i, \sigma_i), \widetilde{b}_j \sim N(e'_j, \sigma'_j), \widetilde{e}_k \\ &\sim N(e''_k, \sigma''_k) i = 1, \ldots, 4, \; j = 1, \ldots, 6, \; k = 1, 2,\end{aligned}$$

4 Techniques to Solve a Crisp Bi-Objective Linear/Nonlinear Problem:

To solve the transformed crisp forms of the model we used the fuzzy programming technique, where we first find the lower bound as L_p and the upper bound as U_p for the pth objective function Z_p, $p = 1, 2, \ldots, P$ here U_p is the highest acceptable level of achievement for objective p, L_p the aspired level of achievement for objective p and $d_p = U_p - L_p$ the degradation allowance for objective p. When the aspiration levels for each of the objective functions have been specified, a fuzzy model is formed and then the fuzzy model is converted into a crisp model. The solution of BOSTP can be obtained by the following steps.

Step-1: Solve the BOSTP and as a single objective STP using each time only one objective and ignore other objective and taking the constraints.

Step-2: From the results of step-1, determine the corresponding value for every objective functions at each solution.

Step-3: Find upper and lower bounds (i.e., U_p and L_p) for pth objective from the two objective values derived in step-2. We construct a payoff matrix, according to every objective w.r.t. each solution. The payoff matrix in the main program gives the set of nondominated solution which should be in the following table:

	$Z_1 Z_2 Z_3$	......	Z_p
$x^{(1)}$	$Z_{11} Z_{12} Z_{13}$	......	Z_{1p}
$x^{(2)}$	$Z_{21} Z_{22} Z_{23}$	......	Z_{2p}
.	.	.	.
.	.	.	.
$x^{(p)}$	$Z_{p1} Z_{p2} Z_{p3}$	......	Z_{pp}

where $x^{(1)}, x^{(2)}, x^{(3)}, \ldots\ldots, x^{(p)}$ is the ideal solution for the objective Z_1, Z_2, Z_3, ..., Z_p respectively.

Let $Z_{ij} = Z_j(x^i)$, $i = 1, 2, \ldots\ldots, p$ and $j = 1, 2 \ldots\ldots, p$ are the minimum value (best) for each objective $Z_r, r = 1, 2, \ldots p$.

Step-4: To find the best (L_r) and worst for each objectives corresponding to the set of solution, i.e., $L_r = Z_{rr}$ and $U_r = max_{r \geq 1} \{Z_{1r}, Z_{2r}, \ldots\ldots, Z_{pr}\}$. For simplicity, $Z_r \leq L_r = 1, 2, 3, \ldots\ldots, p$ and constraints.

Step-5: Then the proposed model converted to the following crisp model:

$$\text{Maximize } \lambda$$
$$\text{subject to, } \quad Z_l + \lambda(U_p - L_p) \leq U_p, p = 1, 2, \ldots, P$$

and the constraints (4)–(6) along with $x_{ijk} \geq 0\, \forall i, j, k$ and $\lambda \geq 0$.

Fuzzy programming technique with exponential membership function (MF):

An exponential membership function is defined by

$$\mu^E(Z_p) = \begin{cases} 1, & \text{if } Z_p \geq L_p \\ \frac{e^{-s\Psi_p(x)-e^{-s}}}{1-e^{-s}}, & \text{if } L_p < Z_p < U_p \\ 0 & \text{if } Z_p \geq U_p \end{cases} \tag{7}$$

where, $\Psi_p(X) = \frac{(Z_p - L_p)}{(U_p - L_p)}$, $p = 1, 2, \ldots, P$, S is a nonzero parameter prescribed by the decision-maker.

Use of exponential MF will give the following equivalent crisp model:

$$\text{Maximize } \lambda$$
$$\text{subject to, } \quad \lambda \leq \frac{e^{-s\Psi_p(X)_{-e^{-s}}}}{1 - e^{-s}}, p = 1, 2, \ldots, P$$

and the constraints (4)–(6) along with $x_{ijk} \geq 0\, \forall i, j, k$ and $\lambda \geq 0$.

Fuzzy programming technique with hyperbolic membership function:

A hyperbolic membership function is defined by

$$\begin{cases} 1, & \text{if } Z_p \leq L_p \\ \frac{1}{2}\frac{e^{\{(U_p+L_p)/2-Z_p(X)\}\alpha_p}-e^{-\{(U_p+L_p)/2-Z_p(X)\}\alpha_p}}{e^{\{(U_p+L_p)/2-Z_p(X)\}\alpha_p}+e^{-\{(U_p+L_p)/2-Z_p(X)\}\alpha_p}}+\frac{1}{2}, & \text{if } L_p < Z_p < U_p \\ 0 & \text{if } Z_p \geq U_p \end{cases}$$

where $\alpha_p = \frac{6}{(U_p-L_p)}$

Use of hyperbolic MF will give the following equivalent crisp model:

$$\begin{aligned} &\text{Maximize } \lambda \\ &\text{subject to,} \quad \lambda \leq \frac{1}{2}\frac{e^{\{(U_p+L_p)/2-Z_p(X)\}\alpha_p}-e^{-\{(U_p+L_p)/2-Z_p(X)\}\alpha_p}}{e^{\{(U_p+L_p)/2-Z_p(X)\}\alpha_p}+e^{-\{(U_p+L_p)/2-Z_p(X)\}\alpha_p}} \\ &\qquad\qquad + \frac{1}{2} p = 1, 2, \ldots, P \end{aligned} \tag{8}$$

and the constraints (4)–(6) along with $x_{ijk} \geq 0 \, \forall i, \ j, \ k$ and $\lambda \geq 0$.

5 Numerical Experiments

Suppose that there are four coal mines to supply the coal for six cities. During the process of transportation, two kinds of conveyances are available to be selected, i.e., train and cargo ship. Now, the task for the decision-maker is to make the transportation plan for the next month in advance such that the transportation cost and the transportation time is minimum. At the beginning of this task, the decision-maker needs to obtain the basic data, such as supply capacity, demand, transportation cost of unit product, transportation time, and so on. In fact, since the transportation plan is made in advance, we generally cannot get these data exactly. For this condition, the usual way is to obtain the uncertain data by means of experience evaluation or expert advice and the corresponding uncertain data are as follows (Tables 1 and 2):

5.1 Input Data

Then the model (3) is equivalent to the following model:

$$\min \sum_{i=1}^{4}\sum_{j=1}^{6}\sum_{k=1}^{2} e_{ijk}\, x_{ijk}$$

$$\min \sum_{i=1}^{4}\sum_{j=1}^{6}\sum_{k=1}^{2} m_{ijk}\, x_{ijk}$$

Table 1 Unit objective parameters by different conveyances

		1	2	3	4	5	6	
1	Costs by train	(16, 2)	(16, 2)	(16, 2)	(15, 1.5)	(16, 2)	(6, 1.5)	(e_{ij1}, σ_{ij1})
2		(6, 1)	(7, 1.5)	(3, 1.5)	(16, 2)	(16, 1.5)	(16, 1.5)	
3		(6, 1.5)	(14, 1.5)	(4, 1.5)	(8, 1.5)	(16, 1.5)	(18, 1.5)	
4		(17, 1.5)	(10, 1.5)	(14, 1.5)	(8, 1.5)	(9, 1.5)	(18, 2)	
1	Times by train	(16, 1)	(16, 1)	(16, 1)	(15, 0.75)	(16, 1)	(6, 0.75)	(e_{ij1}, σ_{ij1})
2		(6, 0.5)	(7, 0.75)	(3, 0.75)	(16, 1)	(16, 0.75)	(16, 0.75)	
3		(6, 0.75)	(14, 0.75)	(4, 0.75)	(8, 0.75)	(16, 0.75)	(18, 0.75)	
4		(17, 0.75)	(10, 0.75)	(14, 0.75)	(8, 0.75)	(9, 0.75)	(18, 1)	
1	Costs by ship	(30, 2)	(30, 2)	(30, 2)	(29, 1.5)	(30, 2)	(20, 1.5)	(e_{ij2}, σ_{ij2})
2		(10, 1)	(21, 1.5)	(17, 1.5)	(30, 2)	(30, 1.5)	(30, 1.5)	
3		(10, 1.5)	(28, 1.5)	(18, 1.5)	(22, 1.5)	(30, 1.5)	(32, 1.5)	
4		(31, 1.5)	(24, 1.5)	(28, 1.5)	(22, 1.5)	(23, 1.5)	(32, 2)	
1	Times by ship	(30, 1)	(30, 1)	(30, 1)	(29, 0.75)	(30, 1)	(20, 0.75)	(e_{ij2}, σ_{ij2})
2		(10, 0.5)	(21, 0.75)	(17, 0.75)	(30, 1)	(30, 0.75)	(30, 0.75)	
3		(10, 0.75)	(28, 0.75)	(18, 0.75)	(22, 0.75)	(30, 0.75)	(32, 0.75)	
4		(31, 0.75)	(24, 0.75)	(28, 0.75)	(22, 0.75)	(23, 0.75)	(32, 1)	

Table 2 Supplies, demands, and conveyance capacities

	1	2	3	4	5	6	
Sources	(25,1.5)	(30,1.5)	(32,2)	(28,2)			(e_i, σ_i)
Demands	(10,1.5)	(14,1)	(22,1)	(18,1)	(16,1)	(12,1)	(e'_j, σ'_i)
Conveyance	(40,1.5)	(60,1)					(e''_k, σ''_k)

subject to:

$$\sum_{j=1}^{6}\sum_{k=1}^{2} x_{ijk} - \left[e_i + \frac{\sigma_i\sqrt{3}}{\pi}\text{In}\frac{1-\beta_i}{\beta_i}\right] \leq 0, i = 1, 2, 3, 4$$

$$\left[e'_j + \frac{\sigma'_j\sqrt{3}}{\pi}\text{In}\frac{\gamma_i}{1-\gamma_j}\right] - \sum_{i=1}^{4}\sum_{k=1}^{2} x_{ijk} \leq 0, j = 1, 2, 3, 4, 5, 6$$

$$\sum_{i=1}^{4}\sum_{j=1}^{6} x_{ijk} - \left[e''_k + \frac{\sigma''_k\sqrt{3}}{\pi}\text{In}\frac{1-\eta_k}{\eta_k}\right] \leq 0, k = 1, 2$$

$$x_{ijk} \geq 0, i = 1, 2, 3, 4, j = 1, 2, 3, 4, 5, 6, k = 1, 2.$$

Therefore with above input data the problem can be reformed as:

$$\begin{aligned}\mathrm{Min} Z_1 &= 16x_{111} + 16x_{121} + 16x_{131} + 15x_{141} + 16x_{151} + 6x_{161} + 30x_{112} \\ &+ 30x_{122} + 30x_{132} + 29x_{142} + 30x_{152} + 20x_{162} + 6x_{211} + 7x_{221} + 3x_{231} + 16x_{241} \\ &+ 16x_{251} + 16x_{261} + 10x_{212} + 21x_{222} + 17x_{232} + 30x_{242} + 30x_{252} + 30x_{262} \\ &+ 6x_{311} + 14x_{321} + 4x_{331} + 8x_{341} + 16x_{351} + 18x_{361} + 10x_{312} + 28x_{322} + 18x_{332} \\ &+ 22x_{342} + 30x_{352} + 32x_{362} + 17x_{411} + 10x_{421} + 14x_{431} + 8x_{441} + 9x_{451} + 18x_{461} \\ &+ 31x_{412} + 24x_{422} + 28x_{432} + 22x_{442} + 23x_{452} + 32x_{462}\end{aligned}$$

$$\begin{aligned}\mathrm{Min} Z_2 &= 18x_{111} + 18x_{121} + 18x_{131} + 17x_{141} + 18x_{151} + 8x_{161} + 38x_{112} \\ &+ 38x_{122} + 38x_{132} + 37x_{142} + 38x_{152} + 28x_{162} + 8x_{211} + 10x_{221} + 6x_{231} + 18x_{241} \\ &+ 18x_{251} + 18x_{261} + 18x_{212} + 29x_{222} + 25x_{232} + 38x_{242} + 38x_{252} + 38x_{262} \\ &+ 8x_{311} + 16x_{321} + 7x_{331} + 11x_{341} + 18x_{351} + 20x_{361} + 18x_{312} + 36x_{322} + 26x_{332} \\ &+ 30x_{342} + 38x_{352} + 40x_{362} + 20x_{411} + 13x_{421} + 16x_{431} + 11x_{441} + 12x_{451} \\ &+ 20x_{461} + 39x_{412} + 32x_{422} + 36x_{432} + 30x_{442} + 31x_{452} + 40x_{462} \text{ subject to}\end{aligned}$$

$$\begin{aligned}&x_{111} + x_{121} + x_{131} + x_{141} + x_{151} + x_{161} + x_{112} + x_{122} + x_{132} + x_{142} + x_{152} \\ &+ x_{162} \le 25 + \frac{1.5 \times \sqrt{3}}{\pi} \mathrm{In} \left(\frac{1 - 0.9}{0.9}\right) \\ &x_{211} + x_{221} + x_{231} + x_{241} + x_{251} + x_{261} + x_{212} + x_{222} + x_{232} + x_{242} + x_{252} \\ &+ x_{262} \le 30 + \frac{1.5 \times \sqrt{3}}{\pi} \mathrm{In} \left(\frac{1 - 0.9}{0.9}\right) \\ &x_{311} + x_{321} + x_{331} + x_{341} + x_{351} + x_{361} + x_{312} + x_{322} + x_{332} + x_{342} + x_{352} \\ &+ x_{362} \le 32 + \frac{1.5 \times \sqrt{3}}{\pi} \mathrm{In} \left(\frac{1 - 0.9}{0.9}\right) \\ &x_{411} + x_{421} + x_{431} + x_{441} + x_{451} + x_{461} + x_{412} + x_{422} + x_{432} + x_{442} + x_{452} \\ &+ x_{462} \le 28 \frac{1.5 \times \sqrt{3}}{\pi} \mathrm{In} \left(\frac{1 - 0.9}{0.9}\right)\end{aligned}$$

$$\begin{aligned}&x_{111} + x_{211} + x_{311} + x_{411} + x_{112} + x_{212} + x_{312} + x_{412} \ge 10 \\ &+ \frac{1.5 \times \sqrt{3}}{\pi} \mathrm{In} \left(\frac{0.9}{1 - 0.9}\right) \\ &x_{121} + x_{221} + x_{321} + x_{421} + x_{122} + x_{222} + x_{322} + x_{422} \ge 14\end{aligned}$$

$$+\frac{1\times\sqrt{3}}{\pi}\text{In}\left(\frac{0.9}{1-0.9}\right)$$

$$x_{131}+x_{231}+x_{331}+x_{431}+x_{132}+x_{232}+x_{332}+x_{432}\geq 22+\frac{1\times\sqrt{3}}{\pi}\text{In}\left(\frac{0.9}{1-0.9}\right)$$

$$x_{141}+x_{241}+x_{341}+x_{441}+x_{142}+x_{242}+x_{342}+x_{442}\geq 18+\frac{1\times\sqrt{3}}{\pi}\text{In}\left(\frac{0.9}{1-0.9}\right)$$

$$x_{151}+x_{251}+x_{351}+x_{451}+x_{152}+x_{252}+x_{352}+x_{452}\geq 16+\frac{1\times\sqrt{3}}{\pi}\text{In}\left(\frac{0.9}{1-0.9}\right)$$

$$x_{161}+x_{261}+x_{361}+x_{461}+x_{162}+x_{262}+x_{362}+x_{462}\geq 12+\frac{1\times\sqrt{3}}{\pi}\text{In}\left(\frac{0.9}{1-0.9}\right)$$

$$x_{111}+x_{121}+x_{131}+x_{141}+x_{151}+x_{161}+x_{211}+x_{221}+x_{231}+x_{241}+x_{251}+x_{261}+x_{311}+x_{321}+x_{331}+x_{341}+x_{351}+x_{361}+x_{411}+x_{421}+x_{431}+x_{441}+x_{451}+x_{461}\leq 40+\frac{1.5\times\sqrt{3}}{\pi}\text{In}\left(\frac{1-0.9}{0.9}\right)$$

$$x_{112}+x_{122}+x_{132}+x_{142}+x_{152}+x_{162}+x_{212}+x_{222}+x_{232}+x_{242}+x_{252}+x_{262}+x_{312}+x_{322}+x_{332}+x_{342}+x_{352}+x_{362}+x_{412}+x_{422}+x_{432}+x_{442}+x_{452}+x_{462}\leq 60+\frac{1\times\sqrt{3}}{\pi}\text{In}\left(\frac{1-0.9}{0.9}\right)$$

For all $i, j, k, x_{ijk}\geq 0$;

Next to solve the problem we use the LINGO 13.0 software and the procedure for that is discussed here.

Solution Methodologies:

Using the fuzzy programming technique first we find out the minimum and maximum values of the first objective function ignoring the second objective function. Similarly we find the minimum and maximum values for the second objective function to form the payoff matrix as follows:

	Z_1	Z_2
min	4106.792	4392.708
max	4184.332	4636.050

Then we get, $L_1 = \min(4106.792, 4184.332) = 4106.792$, $L_2 = \min(4392.708, 4636.050) = 4392.708$, $U_1 = \max(4106.792, 4184.332) = 4184.332$ and $U_2 = \max(4392.708, 4636.050) = 4636.050$.
If we use linear *membership function*, then crisp model can be presented as follows:

$$\begin{aligned} &\text{Max } \lambda \\ &\text{subject to,} \quad Z_2 + \lambda(U_1 - L_1) \leq U_1 \\ &\qquad\qquad Z_1 + \lambda(U_2 - L_2) \leq U_2 \quad 0 \leq \lambda \leq 1. \end{aligned}$$

and the constraints (4) to (6) along with $x_{ijk} \geq 0$ for all i, j, k.

Result with linear, exponential and hyperbolic membership functions

Using the linear MF, exponential MF given by (7) and hyperbolic membership functions given by (8), respectively, and proceedings as before, we get the following optimal results:

MF	Optimal cost (Z_{1^*})	Optimal time (Z_{2^*})	x_{ijk^*}	λ^*
Linear MF	4128.53	4460.92	$x_{122} = 2.41$, $x_{142} = 4.48$, $x_{162} = 16.29$, $x_{222} = 6.56$, $x_{341} = 6.37$, $x_{312} = 23.21$, $x_{441} = 8.37$, $x_{451} = 17.21$, $x_{221} = 6.24$, and all others x_{ijk} are zero	0.72
Exponential MF	4145.56	4440.1	$x_{122} = 4.31$, $x_{142} = 1.26$, $x_{441} = 11.47$, $x_{222} = 4.89$, $x_{341} = 6.48$, $x_{312} = 23.09$, $x_{451} = 14.11$, $x_{221} = 6.01$, $x_{152} = 3.10$, $x_{162} = 13.47$, $x_{231} = 0.12$ and all others x_{ijk} are zero	0.38
Hyperbolic MF	4145.56	4440.1	$x_{122} = 4.31$, $x_{142} = 2.12$, $x_{162} = 13.93$, $x_{222} = 9.38$, $x_{341} = 8.72$, $x_{312} = 20.85$, $x_{441} = 8.37$, $x_{451} = 17.21$, $x_{221} = 1.53$, $x_{231} = 2.36$ and others x_{ijk} are zero	0.50

6 Conclusion

This paper mainly investigated a new uncertain cost and uncertain time solid transportation problem based on uncertainty theory. As a result, a decision model under criteria was presented. The construction of expected-constrained programming model was according to the idea of expected value of the objective under the chance constraints

In this paper, BOSTP under uncertain environment is solved by using fuzzy programming technique with linear, exponential, and hyperbolic membership functions. It has been found that for BOSTP under uncertain environment with multi-objective functions the optimal solutions do not change if we use exponential and hyperbolic membership functions but is different compared to if we use a linear membership function. For the problem we find that the first objective functions, i.e., Z_1 is minimum with respect to the linear membership function and Z_2 is minimum when the membership function is nonlinear.

References

1. Balinski, M.L.: Fixed-cost transportation problems. Naval Res. Logistics Q. **8**, 41–54 (1961)
2. Haley, K.B.: The solid transportation problem. Oper. Res. **11**, 446–448 (1962)
3. Bit, A.K., Biswal, M.P., Alam, S.S.: Fuzzy programming approach to multi-objective solid transportation problem. Fuzzy Sets Syst. **57**, 183–194 (1993)
4. Mahapatra, D., Roy, S., Biswal, M.: Multi-objective stochastic transportation problem involving log-normal. J. Phys. Sci. **14**, 63–76 (2010)
5. Kaufmann, A.: Introduction to The Theory of Fuzzy Subsets, vol. I. Academic Press, New York (1975)
6. Sheng, Y.: A solid transportation model based on uncertainty theory. Information **15**(12), 342–348 (2012)
7. Pandian, P., Anuradha, D.: A new approach for solving solid transportation problems. Appl. Math. Sci. **4**, 3603–3610 (2010)
8. Baidya, A., Bera, U.K., Maiti, M.: Multi-item solid transportation problem with safety measure, OPSEARCH. Springer (2013). doi:10.1007/s12597-013-0129-2
9. Baidya, A., Bera, U.K., Maiti, M.: Multi-item interval valued solid transportation problem with safety measure under fuzzy-stochastic environment. Int. J. Transp. Secur., Springer (2013) doi:10.1007/s12198-013-0109-z
10. Liu, B.: Uncertainty Theory: A Branch of Mathematics for Modeling Human Uncertainty. Springer, Berlin (2010)
11. Liu, B.: Some research problems in uncertainty theory. J. Uncertain. Syst. **3**, 3–10 (2009)
12. Liu, B.: Theory and Practice of Uncertain Programming, 2nd edn. Springer, Berlin (2009)
13. Liu, B.: Why is there a need for uncertainty theory? J. Uncertain Syst. **6**, 3–10 (2012)
14. Liu, Y., Ha, M.: Expected value of function of uncertain variables. J. Uncertain Syst. **4**, 181–186 (2010)
15. Jimenez, F., Verdegay, J.L.: Uncertain solid transportation problems. Fuzzy Sets Syst. **100**, 45–57 (1998)
16. Sheng, Y., Yao, K.: Fixed charge transportation problem in uncertain environment. Ind. Eng. Manag. Syst. **11**, 183–187 (2012)
17. Sheng, Y., Yao, K.: A transportation model with uncertain costs and demands. Inform. Int. Interdiscip. J. **15**, 3179–3186 (2012)

18. Yang, L., Feng, Y.: A bi-criteria solid transportation problem with fixed charge under stochastic environment. Appl. Math. Model. **31**, 2668–2683 (2007)
19. Cui, Q., Sheng, Y.: Uncertain programming model for solid transportation problem. Information in press (2012)
20. Bhatia, H., Swarup, K., Puri, M.: Time minimizing solid transportation problem. Math. Oper. Stat. **7**, 395–403 (1976)
21. Liu, B.: Uncertainty Theory, 2nd edn. Springer, Berlin (2007)

A Food Web Population Model in Deterministic and Stochastic Environment

D. Sadhukhan, B. Mondal and M. Maiti

Abstract This paper deals with the selective harvesting of two species from a food chain model of three species, in which prey and predator obey the Gompertz law of growth. Initially the dynamical behaviour of the system was studied under deterministic environment. In deterministic case, the local stability of the system was also studied; we investigated the condition of global stability and the existence of the bionomic equilibrium was examined. The optimal harvesting policy is studied with the help of Pontryagin's maximum principle. In the second part of the problem, we investigated the stability condition of the system under stochastic environment. Then a comparison is made between deterministic and stochastic cases.

Keywords Prey-predator-superpredator · Gompertz growth law · Optimal harvesting · Wiener process · Stochastic stability

1 Introduction

Harvesting of multispecies food chain system especially in fisheries is an important branch of study in modern day population biology. The pioneering work in this field was first done by Clark [1]. Clark also worked on selective harvesting in a fishery, consisting of two competitive species. Brauer and Soudack [2, 3], Myerscough et al. [4], Dia and Tang [5], Xiao and Ruan [6], etc. also discussed the constant rate of harvesting in population dynamics. Recently, Chaudhuri [7, 8], Kar and Chaudhuri [9, 10] and Purohit and Chaudhuri [11] studied the combined harvesting of two

D. Sadhukhan (✉)
Darjeeling Government College, Darjeeling 734101, West Bengal, India
e-mail: dipankarsadhukhan@yahoo.com

B. Mondal
Raja N.L.K. Womens College, Midnapore, West Bengal, India

M. Maiti
Department Of Applied Mathematics, Vidyasagar University, Midnapore, West Bengal, India

M.K. Chakraborty et al. (eds.), *Facets of Uncertainties and Applications*, Springer Proceedings in Mathematics & Statistics 125,
DOI 10.1007/978-81-322-2301-6_21

competitive species and discussed the dynamics of optimization of the harvesting policy.

In this type of study, Chaudhuri and SahaRay [12] developed a model in which combined harvesting of prey and predator was discussed where some prey hide in refuges. Dubey et al. [13], Kar [14], etc., studied harvesting of some prey–predator systems by dividing the space into reserved and unreserved areas. There are also some papers of harvesting of mature species by Kar and Matsuda [15], Kar and Pahari [16] and Kar [17] and others by incorporating time delay.

With the improvement of the study in population dynamics, researchers have investigated the effect of environmental fluctuation in the harvesting of species; this phenomenon has been studied by Kar and Chaudhuri [18], Beddington and May [19], May et al. [20], Ludwig [21], Reed [22], Hanson [23], etc. There are also some harvesting models under random environment studied by Braumann [24], Turelli [25], etc. Very few researchers have developed the harvesting phenomena in a food chain model of three species [26, 27].

To our knowledge, almost in all food chain models of three species, no researcher has considered the Gompertz law of growth. In our present work, we have considered a food chain model of three species—prey, predator and super-predator considering Gompertz growth law and harvesting is considered only on prey and predator as we are not interested in the harvesting of super predator. In the deterministic case, we found the local stability condition and also the global stability conditions by selecting suitable Lyapunov function. Also, we discuss the optimal harvesting policy using Pontryagin's maximum principle. Again in the second part of the problem, we investigated the stability conditions of the system under stochastic environment.

2 Notations

(i) x_1, x_2 and x_3 are, respectively, the number of prey, predator and super-predator at time t.
(ii) k_1, k_2 and k_3 are respective environmental carrying capacities of prey, predator and super predator.
(iii) r_1, r_2 and r_3 are the intrinsic growth rates of prey, predator and super predator respectively.
(iv) q_1 and q_2 are the respective catchability coefficients of prey and predator.
(v) E is the common catching effort.
(vi) α_{12} and α_{13} are predator response rates toward the prey and super-predator respectively.
(vii) α_{23} super predator are the response rates toward the predator.
(viii) α_{21} and α_{31} are the rates of conversion of prey to predator and super-predator respectively.
(ix) α_{32} is the rate of conversion of predator to super predator.
(x) α_{12}, α_{13} and α_{23} are the predation coefficients.
(xi) α_{31}, α_{32} and α_{21} are conversion parameters.

(xii) C is the constant fishing cost per unit effort.
(xiii) p_1 and p_2 are constant price per unit biomass of first and second species respectively.

3 Model Formulation

As ideal living conditions are normally prevailed in the initial stage, there should be very rapid growth initially. Thereafter, as the population grows, the limitation of resources forces the growth rate to decline and the population gradually approaches the saturation level. Compared to logistic law, the Gompertz law exhibits faster early growth, but a slower approach to the asymptote, with longer period of linear growth about the point of inflexion.

When the fish population size becomes considerably large, it tends to maintain stronger pressure on the newly produced biomass through cannibalism which is more effective on larger egg- aggregates formed by overcrowding of eggs in large fish populations. Also, there will be intraspecific competition amongst individuals in the population for the use of limited resources available in the habitat. These effects, coupled together, should retard the growth of the population to a large extent and as a result, the population size should approach its asymptote rapidly as in the case of the logistic model. These retarding effects are, however, counterbalanced to some extent by group movement (*cf.* Sutinen [28]), which is a special behavioural characteristic of a fish population. As a result, the approach of the population size to asymptote is slowed down. This peculiar feature of faster early growth and slower approach to the asymptote are reflected in the Gompertz law of growth (*cf.* Pradhan [29]).

Considering this advantage of Gompertz law of growth, in this section we develop a general food chain model with selective harvesting, within which prey and predator follow Gompertz law of growth and harvesting is allowed for these prey and predators. But we have not considered the harvesting for super-predator, as in many coastal areas, super-predators such as shark and whale harvesting are banned.

The governing equations describing the system are as follows:

$$\begin{aligned}
\frac{dx_1}{dt} &= r_1x_1\ln\frac{k_1}{x_1} - \alpha_{12}x_1x_2 - \alpha_{13}x_1x_3 - q_1Ex_1\\
\frac{dx_2}{dt} &= r_2x_2\ln\frac{k_2}{x_2} + \alpha_{21}x_1x_2 - \alpha_{23}x_2x_3 - q_2Ex_2\\
\frac{dx_3}{dt} &= r_3x_3 + \alpha_{31}x_1x_3 + \alpha_{32}x_2x_3 - x_3^2
\end{aligned} \tag{1}$$

where $0 \le x_1 \le k_1, 0 \le x_2 \le k_2, 0 \le x_3 \le k_3$., and $\alpha_{12}, \alpha_{13}, \alpha_{21}, \alpha_{23}, \alpha_{31}$ and α_{32} are positive constants. The catch rate functions q_1Ex_1 and q_2Ex_2 are based on CPUE (CATCH-PER-UNIT EFFORT).

4 The Steady States and Stability Analysis of the System

4.1 The Steady States

The steady states of the above system (1) are obtained solving the equations. The possible states, i.e. points may be assumed as: $P_0(0, 0, 0)$, $P_1(0, x_{21}, x_{31})$, $P_2(x_{12}, 0, x_{32})$, $P_3(x_{13}, x_{23}, 0)$, $P_4(0, 0, x_{34})$, $P_5(0, x_{25}, 0)$, $P_6(x_{16}, 0, 0)$, $P_7(x_1^*, x_2^*, x_3^*)$.

The nontrivial steady state $P_7(x_1^*, x_2^*, x_3^*)$, is given by $\frac{dx_1}{dt} = \frac{dx_2}{dt} = \frac{dx_3}{dt} = 0$ and is obtained by solving the following equations simultaneously:

$$\lambda_1 \ln \frac{K_1}{x_1} - \alpha_{12}x_2 - \alpha_{13}x_3 - q_1 E = 0 \tag{2}$$

$$\lambda_2 \ln \frac{K_2}{x_2} + \alpha_{21}x_1 - \alpha_{23}x_3 - q_2 E = 0 \tag{3}$$

$$r_3 + \alpha_{31}x_1 + \alpha_{32}x_2 - x_3 = 0 \tag{4}$$

Now to get the nontrivial steady state $P_7(x_1^*, x_2^*, x_3^*)$ for a given set of parametric values, we have from Eqs. (2) and (3)

$$\frac{r_1 \ln \frac{K_1}{x_1} - \alpha_{12}x_2 - \alpha_{13}x_3}{q_1} = \frac{r_2 \ln \frac{K_2}{x_2} + \alpha_{21}x_1 - \alpha_{23}x_3}{q_2} \tag{5}$$

Therefore the biological steady state $P_7(x_1^*, x_2^*, x_3^*)$ must satisfy the biological equilibrium path (4) and (5).

4.2 Local Stability

The variational matrix $V(x_1, x_2, x_3)$ is given as

$$V(x_1, x_2, x_3) = \begin{bmatrix} V_{11} & V_{12} & V_{13} \\ V_{21} & V_{22} & V_{23} \\ V_{31} & V_{32} & V_{33} \end{bmatrix} \tag{6}$$

where

$$\begin{aligned}
V_{11} &= r_1 \ln \tfrac{k_1}{x_1} - r_1 - \alpha_{12}x_2 - \alpha_{13}x_3 - q_1 E, \\
V_{22} &= r_2 \log \tfrac{k_2}{x_2} - r_2 + \alpha_{21}x_1 - \alpha_{23}x_3 - q_2 E, \\
V_{33} &= r_3 + \alpha_{31}x_1 + \alpha_{32}x_2 - 2x_3, \\
V_{12} &= -\alpha_{12}x_1,\ V_{13} = -\alpha_{13}x_1, \\
V_{23} &= -\alpha_{23}x_2,\ V_{21} = \alpha_{21}x_2, \\
V_{31} &= \alpha_{31}x_3 \text{ and } V_{32} = \alpha_{32}x_3
\end{aligned} \tag{7}$$

For the point P_3: The characteristic equation for $V\,(x_{13}, x_{23}, 0)$ is given as

$$[\mu - (r_3 + \alpha_{31}x_{13} + \alpha_{32}x_{23})]\left[\mu^2 + (r_1 + r_2)\,\mu + (r_1r_2 + \alpha_{12}\alpha_{21}x_{13}x_{23})\right] = 0$$

One of the eigenvalues of the variational matrix $V\,(x_{13}, x_{23}, 0)$ is $r_3 + \alpha_{31}x_{13} + \alpha_{32}x_{23}\ (> 0)$. So, the system is unstable at $P_3\,(x_{13}, x_{23}, 0)$.

For the point P_7: The characteristic equation for $V\left(x_1^*, x_2^*, x_3^*\right)$ is given by
$\mu^3 + a_1'\mu^2 + a_2'\mu + a_3' = 0$
where, $a_1' = r_1 + r_2 + x_3^*$,
$a_2' = \left(r_1r_2 + \alpha_{12}\alpha_{21}x_1^*x_2^*\right) + \left(r_2x_3^* + \alpha_{23}\alpha_{32}x_2^*x_3^*\right) + \left(r_1x_3^* + \alpha_{13}\alpha_{31}x_1^*x_3^*\right)$
and
$a_3' = r_1r_2x_3^* + r_1\alpha_{23}\alpha_{32}x_2^*x_3^* + x_3^*\alpha_{12}\alpha_{21}x_1^*x_2^* + r_2\alpha_{13}\alpha_{31}x_1^*x_3^*$
$-\alpha_{12}\alpha_{23}\alpha_{31}x_1^*x_2^*x_3^* + \alpha_{13}\alpha_{21}\alpha_{32}x_1^*x_2^*x_3^*$
Therefore as $a'_1 > 0$, so by *Routh-Hurwitz* condition, P_7 will be stable if $\begin{vmatrix} a_1' & a_3' \\ 1 & a_2' \end{vmatrix} > 0.$

4.3 Global Stability

In this section we prove the global stability of the governing system by constructing a suitable Lyapunov function.

Theorem 1 *The interior equilibrium point* P_7 *is globally asymptotically stable if*

$$(\alpha_{21} - \alpha_{12})^2 < \frac{r_1r_2}{(x_1 - x_1^*)(x_2 - x_2^*)} \ln\frac{x_1}{x_1^*} \ln\frac{x_2}{x_2^*},$$

$$(\alpha_{31} - \alpha_{13})^2 < \frac{r_1}{(x_1 - x_1^*)} \ln\frac{x_1}{x_1^*}$$

and

$$(\alpha_{32} - \alpha_{23})^2 < \frac{r_2}{(x_2 - x_2^*)} \ln\frac{x_2}{x_2^*}$$

Proof Let us consider a Lyapunov function

$$L(x_1, x_2, x_3) = \left(x_1 - x_1^*\right) - x_1^* \ln \frac{x_1}{x_1^*} + \left(x_2 - x_2^*\right) - x_2^* \ln \frac{x_2}{x_2^*} + \left(x_3 - x_3^*\right) - x_3^* \ln \frac{x_3}{x_3^*} \quad (8)$$

Obviously, $L(x_1, x_2, x_3)$ is positive definite and continuous $\forall x_1, x_2, x_3 > 0$

$$\frac{dL}{dt} = \sum \frac{\partial L}{\partial x_i}\frac{dx_i}{dt} = \frac{\partial L}{\partial x_1}\frac{dx_1}{dt} + \frac{\partial L}{\partial x_2}\frac{dx_2}{dt} + \frac{\partial L}{\partial x_3}\frac{dx_3}{dt} \quad (9)$$

After simplification, we have

$$\frac{dL}{dt} = -\frac{1}{2}b_{11}\left(x_1 - x_1^*\right)^2 + b_{12}\left(x_1 - x_1^*\right)\left(x_2 - x_2^*\right) - \frac{1}{2}b_{22}\left(x_2 - x_2^*\right)^2$$

$$-\frac{1}{2}b_{11}\left(x_1 - x_1^*\right)^2 + b_{13}\left(x_1 - x_1^*\right)\left(x_3 - x_3^*\right) - \frac{1}{2}b_{33}\left(x_3 - x_3^*\right)^2$$

$$-\frac{1}{2}b_{22}\left(x_2 - x_2^*\right)^2 + b_{23}\left(x_2 - x_2^*\right)\left(x_3 - x_3^*\right) - \frac{1}{2}b_{33}\left(x_3 - x_3^*\right)^2$$

where,

$b_{11} = \frac{r_1}{(x_1 - x_1^*)} \ln \frac{x_1}{x_1^*}$, $b_{22} = \frac{r_2}{(x_2 - x_2^*)} \ln \frac{x_2}{x_2^*}$, $b_{33} = 1$,
$b_{12} = \alpha_{21} - \alpha_{12}$, $b_{23} = \alpha_{32} - \alpha_{23}$, $b_{13} = \alpha_{31} - \alpha_{13}$

Therefore the system will be globally asymptotic stable if

$$b_{12}^2 < b_{11}b_{22},\ b_{13}^2 < b_{11}b_{33} \text{ and } b_{23}^2 < b_{22}b_{33} \quad (10)$$

Hence the result.

5 Bioeconomic Equilibrium

The term bionomic(bioeconomic) equilibrium is an amalgamation of the concept of biological equilibrium as well as economic equilibrium. As already said, a biological equilibrium is given by $\dot{x}_1 = 0, \dot{x}_2 = 0, \dot{x}_3 = 0$. The economic equilibrium is said to be achieved when TR (the total revenue obtained by selling the harvested biomass) equals TC (the total cost for the effort devoted to harvesting).
The economic rent (net revenue) at any time is given by

$$\pi(x_1, x_2, x_3, E) = (p_1q_1x_1 + p_2q_2x_2 - C)E \quad (11)$$

Therefore the economic equilibrium will follow the path given by the equation

$$p_1q_1x_1 + p_2q_2x_2 - C = 0 \quad (12)$$

We treat Eq. (12) as an economic equilibrium path. The bionomic equilibrium point (x_{1b}, x_{2b}, x_{3b}) is the solution of Eqs. (4) and (5) together with Eq. (12) for any given E and other values of the parameters.

6 Optimal Harvesting Policy

In this section, the present value J of continuous time-stream of revenues is given by

$$J = \int_0^\infty e^{-\delta t} \pi (x_1, x_2, x_3, E, t)\, dt \tag{13}$$

where $\pi (x_1, x_2, x_3, E, t) = (p_1 q_1 x_1 + p_2 q_2 x_2 - C)E$ and δ denotes the annual discount rate. Now we have to maximize J subject to the system of Eq. (1) by Pontryagin's Maximal Principle [1]. The control variable $E(t)$ is subjected to the constraints $0 \leq E(t) \leq E_{\max}$ so that $V_t = [0, E_{\max}]$ is the control set and $E_{\max}$ is a feasible upper limit for the harvesting effort.

The Hamiltonian for the problem is given as

$$\begin{aligned} H &= e^{-\delta_1 t} (p_1 q_1 x_1 + p_2 q_2 x_2 - c)\, E \\ &+ \mu_1 (t) \left[r_1 x_1 \ln \tfrac{k_1}{x_1} - \alpha_{12} x_1 x_2 - \alpha_{13} x_1 x_3 - q_1 E x_1 \right] \\ &+ \mu_2 (t) \left[r_2 x_2 \ln \tfrac{k_2}{x_2} + \alpha_{21} x_1 x_2 - \alpha_{23} x_2 x_3 - q_2 E x_2 \right] \\ &+ \mu_3 (t) \left[r_3 x_3 + \alpha_{31} x_1 x_3 + \alpha_{32} x_2 x_3 - x_3^2 \right] \end{aligned} \tag{14}$$

The adjoint equations are

$$\frac{d\mu_1}{dt} = -\frac{\partial H}{\partial x_1}, \frac{d\mu_2}{dt} = -\frac{\partial H}{\partial x_2}, \frac{d\mu_3}{dt} = -\frac{\partial H}{\partial x_3} \tag{15}$$

Therefore,

$$\begin{aligned} \tfrac{d\mu_1}{dt} &= r_1 \mu_1 - \mu_2 \alpha_{21} x_2 - \mu_3 \alpha_{31} x_3 - p_1 q_1 E e^{-\delta_1 t} \\ \tfrac{d\mu_2}{dt} &= \mu_1 \alpha_{12} x_1 + r_2 \mu_2 - \mu_3 \alpha_{32} x_3 - p_2 q_2 E e^{-\delta_1 t} \\ \tfrac{d\mu_3}{dt} &= \mu_1 \alpha_{13} x_1 + \mu_2 \alpha_{23} x_2 + x_3 \mu_3 \end{aligned} \tag{16}$$

The solution of the above system of linear differential equations is given as

$$\mu_1 = A_1 e^{m_1 t} + A_2 e^{m_2 t} + A_3 e^{m_3 t} + \frac{M_1}{N} e^{-\delta t} \tag{17}$$

where m_1, m_2 and m_3 are the roots of the cubic equation

$$a_0 m^3 + a_1 m^2 + a_2 m + a_3 = 0 \tag{18}$$

with,

$$a_0 = 1 \tag{19}$$

$$a_1 = -(r_1 + r_2 + x_3) \tag{20}$$

$$a_2 = (r_1 r_2 + \alpha_{12}\alpha_{21} x_1 x_2) + (r_2 x_3 + \alpha_{23}\alpha_{32} x_2 x_3) + (r_1 x_3 + \alpha_{13}\alpha_{31} x_1 x_3) \tag{21}$$

$$\begin{aligned} a_3 = -(&r_1 r_2 x_3 + r_1 \alpha_{23}\alpha_{32} x_2 x_3 + x_3 \alpha_{12}\alpha_{21} x_1 x_2 + r_2 \alpha_{13}\alpha_{31} x_1 x_3 \\ &-\alpha_{12}\alpha_{23}\alpha_{31} x_1 x_2 x_3 + \alpha_{13}\alpha_{21}\alpha_{32} x_1 x_2 x_3) \end{aligned} \tag{22}$$

μ_1 is bounded if $m_i < 0, i = 1, 2, 3$ or $A_i^t = 0$.
The Hurwitz matrix is

$$\begin{pmatrix} a_1 & 1 & 0 \\ a_3 & a_2 & a_1 \\ 0 & 0 & a_3 \end{pmatrix} \quad \text{and} \quad \Delta_1 = a_1, \Delta_2 = a_1 a_2 - a_3, \Delta_3 = a_3 (a_1 a_2 - a_3) \tag{23}$$

Therefore, the roots of the cubic equation are real negative or complex conjugate having negative real parts iff $\Delta_1, \Delta_2, \Delta_3$ are all greater than zero.

But $\Delta_1 < 0$, so it is difficult to check whether $m_i < 0$, therefore we take $A_i = 0$. Then

$$\mu_1(t) = \frac{M_1}{N} e^{-\delta t} \tag{24}$$

By similar process we get

$$\mu_2(t) = \frac{M_2}{N} e^{-\delta t} \tag{25}$$

and

$$\mu_3(t) = \frac{M_3}{N} e^{-\delta t} \tag{26}$$

where,

$$M_1 = -E \begin{bmatrix} p_1 q_1 \left\{ \delta_1^2 + \delta_1 (r_2 + x_3) + r_2 x_3 + \alpha_{23} \alpha_{32} x_2 x_3 \right\} \\ + p_2 q_2 \{ \alpha_{21} x_2 \delta_1 + \alpha_{21} x_2 x_3 - \alpha_{31} \alpha_{23} \} \end{bmatrix} \tag{27}$$

$$M_2 = -E \begin{bmatrix} p_2 q_2 \left\{ \delta_1^2 + \delta_1 (r_1 x_3) + r_1 x_3 + \alpha_{13} \alpha_{31} x_1 x_3 \right\} \\ + p_1 q_1 \{ \alpha_{12} x_1 \delta_1 - \alpha_{12} x_1 x_3 - \alpha_{13} \alpha_{32} \} \end{bmatrix} \tag{28}$$

$$M_3 = -E \left[p_1 q_1 \{ -\alpha_{13} x_1 \delta_1 + \alpha_{13} x_1 r_2 + \alpha_{12} \alpha_{23} \} \right] \tag{29}$$

and

$$N = -\left(a_0 \delta^3 - a_1 \delta^2 + a_2 \delta - a_3 \right) \neq 0 \tag{30}$$

We find that the shadow prices $\mu_i(t) e^{R_L t}, i = 1, 2, 3.$ of the three species remain bounded as $t \to \infty$ and hence satisfy the transversality condition at ∞.

The Hamiltonian must be maximized for $E \in [0, E_{\max}]$. Assuming that the control constraints $0 \leq E \leq E_{\max}$ are not binding (that is, the optimal equilibrium does not occur either at (E=0 or, $E = E_{\max}$), so we consider the singular control.

Therefore,

$$\frac{\partial H}{\partial E} = e^{-\delta t} (p_1 q_1 x_1 + p_2 q_2 x_2 - C) - \mu_1 q_1 x_1 - \mu_2 q_2 x_2 = 0 \tag{31}$$

or,

$$e^{-\delta t} \frac{d\pi}{dE} = \mu_1 q_1 x_1 + \mu_2 q_2 x_2 \tag{32}$$

As we know from (11) that,

$$\frac{d\pi}{dE} = (p_1 q_1 x_1 + p_2 q_2 x_2 - C) \tag{33}$$

This Eq. (33) indicates that the total user cost of harvest per unit effort must be equal to the discounted value of the future profit at the steady-state effort level [1].

Now from (28) and (30), we get

$$e^{-\delta t}\left(p_1 q_1 x_1 + p_2 q_2 x_2 - C\right) = \mu_1 q_1 x_1 + \mu_2 q_2 x_2 \tag{34}$$

Substituting the values of μ_1 and μ_2, Eq. (34) reduces to

$$\left(p_1 - \frac{M_1}{N}\right) q_1 x + \left(p_2 - \frac{M_2}{N}\right) q_2 y = C \tag{35}$$

The above Eq. (35) together with Eq. (1) gives the optimal equilibrium population densities as $x_1 = x_{1\delta}$, $x_2 = x_{2\delta}$ and $x_3 = x_{3\delta}$. Now when $\delta \to \infty$, the above equation leads to the result

$$p_1 q_1 x_\infty + p_2 q_2 y_\infty = C \tag{36}$$

which gives that $\pi(x_{1\infty}, x_{2\infty}, x_{3\infty}, E) = 0$.

Using (35), we get

$$\pi = \left(p_1 q_1 x + p_2 q_2 y - C\right) E = \frac{\left(M_1 q_1 x + M_2 q_2 y\right) E}{N} \tag{37}$$

As each M_1 and M_2 is of $o(\delta)$ and N is of $o\left(\delta^2\right)$, therefore π is of $o\left(\delta^{-1}\right)$. Thus π is a decreasing function of $\delta (\geq 0)$. We therefore conclude that $\delta = 0$ leads to maximization of π.

7 Stochastic Model

In this model, we allow stochastic perturbations of the variables x_1, x_2, x_3 around their values at the positive equilibrium $\left(x_1^*, x_2^*, x_3^*\right)$ in R_+^3, in the case when it is feasible and locally asymptotically stable. The local stability of $\left(x_1^*, x_2^*, x_3^*\right)$ is implied by the condition of existence of $\left(x_1^*, x_2^*, x_3^*\right)$. So, in system (1), we assume that stochastic perturbations of variables around their equilibrium values $\left(x_1^*, x_2^*, x_3^*\right)$ are of white noise type, which is proportional to the distance of x_1, x_2, x_3 from the values x_1^*, x_2^*, x_3^*. So the system (1) reduces to

$$\begin{aligned}
dx_1 &= \left(r_1 x_1 \ln \tfrac{k_1}{x_1} - \alpha_{12} x_1 x_2 - \alpha_{13} x_1 x_3 - q_1 E x_1\right) dt + \sigma_1 \left(x_1 - x_1^*\right) d\xi_t^1 \\
dx_2 &= \left(r_2 x_2 \ln \tfrac{k_2}{x_2} + \alpha_{21} x_1 x_2 - \alpha_{23} x_2 x_3 - q_2 E x_2\right) dt + \sigma_2 \left(x_2 - x_2^*\right) d\xi_t^2 \\
dx_3 &= \left(r_3 x_3 + \alpha_{31} x_1 x_3 + \alpha_{32} x_2 x_3 - x_3^2\right) dt + \sigma_3 \left(x_3 - x_3^*\right) d\xi_t^3
\end{aligned} \tag{38}$$

where $\sigma_i,\ i = 1, 2, 3$ are real constants, $\xi_t^i = \xi_i(t),\ i = 1, 2, 3$ are independent from each other standard Wiener processes [30–32]. Now we investigate the asymptotic stability behaviour of the equilibrium point $\left(x_1^*, x_2^*, x_3^*\right)$ under such kind of stochasticity for the system (38) which is Ito stochastic differential system.

8 Stochastic Stability of the Positive Equilibrium

Now as the system (1) has the equilibrium point $\left(x_1^*, x_2^*, x_3^*\right)$ in R_+^3, then the stochastic differential system (38) can be centred at its positive equilibrium $\left(x_1^*, x_2^*, x_3^*\right)$ using change of variables as

$$u_1 = x_1 - x_1^*,\ u_2 = x_2 - x_2^*,\ u_3 = x_3 - x_3^* \tag{39}$$

Therefore, the linearized stochastic differential equations around $\left(x_1^*, x_2^*, x_3^*\right)$ are of the form

$$du(t) = f(u(t))\,dt + g(u(t))\,d\xi(t) \tag{40}$$

where $u(t) = col\,(u_1(t), u_2(t), u_3(t))$ and

$$f(u(t)) = \begin{bmatrix} -r_1 & -\alpha_{12}x_1^* & -\alpha_{13}x_1^* \\ \alpha_{21}x_2^* & -r_2 & -\alpha_{23}x_2^* \\ \alpha_{31}x_3^* & \alpha_{32}x_3^* & -x_3^* \end{bmatrix} u(t), \tag{41}$$

$$g(u) = \begin{bmatrix} \sigma_1 u_1 & 0 & 0 \\ 0 & \sigma_2 u_2 & 0 \\ 0 & 0 & \sigma_3 u_3 \end{bmatrix} \tag{42}$$

In Eq. (40), the positive equilibrium $\left(x_1^*, x_2^*, x_3^*\right)$ corresponds to the trivial solution $u(t) = 0$.

Let U be the set $U = (t \geq t_0) \times R^n, t_0 \epsilon R^+$. Hence $V \epsilon C_2^0(U)$ is a twice continuously differentiable function with respect to u and a continuous function with respect to t (cf. Afanas'ev [33]).

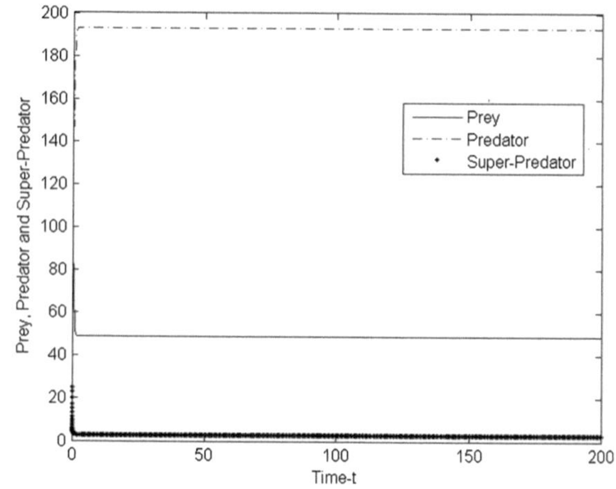

Fig. 1 Variation of the populations against time, beginning with $x_1 = 50$, $x_2 = 50$ *and* $x_3 = 25$

Now, with reference to (40),

$$LV(t,u) = \frac{\partial V(t,u)}{\partial t} + f^T(u)\frac{\partial V(t,u)}{\partial u} + \frac{1}{2}Tr\left[g^T(u)\frac{\partial^2 V(t,u)}{\partial u^2}g(u)\right] \tag{43}$$

where $\frac{\partial V}{\partial u} = Col\left(\frac{\partial V}{\partial u_1}, \frac{\partial V}{\partial u_2}, \frac{\partial V}{\partial u_3}\right)$, $\frac{\partial^2 V(t,u)}{\partial u^2} = \left(\frac{\partial^2 V}{\partial u_i \partial u_j}\right) i, j = 1, 2, 3$ and T denotes transposition (Fig. 1).

Theorem 2 *Suppose there exists a function* $V(t,u) \epsilon C_2^0(U)$ *satisfying the inequalities*

$$\begin{aligned} &K_1|u|^p \le V(t,u) \le K_2|u|^p, \\ &LV(t,u) \le -K_3|u|^p,\ K_i > 0,\ p > 0 \end{aligned} \tag{44}$$

Then the trivial solution of (40) is exponentially p-stable for $t \geq 0$.

If in (44), $p = 2$, then the trivial solutions of (40) is globally asymptotically stable in probability (cf. Afanas'ev [33]).

Theorem 3 *Let* $r_1 > \frac{1}{2}\sigma_1^2, r_2 > \frac{1}{2}\sigma_2^2, x_3^* > \frac{1}{2}\sigma_3^2$. *Then the zero solution of (40) is asymptotically mean square stable.*

Proof Let us consider the Lyapunov function

$$L(u) = \frac{1}{2}\left[w_1 u_1^2 + w_2 u_2^2 + w_3 u_3^2\right] \tag{45}$$

where w_i are real positive constants. The first inequalities of (44) is obvious for $p = 2$.

Again

$$\begin{aligned}\mathbf{L}L(u) = {} & w_1\left(-r_1u_1 - \alpha_{12}x_1^*u_2 - \alpha_{13}x_1^*u_3\right)u_1 \\ & + w_2\left(\alpha_{21}x_2^*u_1 - r_2u_2 - \alpha_{23}x_2^*u_3\right)u_2 \\ & + w_3\left(\alpha_{31}x_3^*u_1 + \alpha_{32}x_3^*u_2 - x_3^*u_3\right)u_3 \\ & + \tfrac{1}{2}Tr\left[g^T(u)\,\tfrac{\partial^2 L(t,u)}{\partial u^2}g(u)\right]\end{aligned} \tag{46}$$

where $\frac{\partial^2 L}{\partial u^2} = \begin{bmatrix} w_1 & 0 & 0 \\ 0 & w_2 & 0 \\ 0 & 0 & w_3 \end{bmatrix}$.

Therefore $g^T(u)\frac{\partial^2 L(t,u)}{\partial u^2}g(u) = \begin{bmatrix} w_1\sigma_1^2u_1^2 & 0 & 0 \\ 0 & w_2\sigma_2^2u_2^2 & 0 \\ 0 & 0 & w_3\sigma_3^2u_3^2 \end{bmatrix}$

with

$$\frac{1}{2}Tr\left[g^T(u)\,\frac{\partial^2 L(t,u)}{\partial u^2}g(u)\right] = \frac{1}{2}\left[w_1\sigma_1^2u_1^2 + w_2\sigma_2^2u_2^2 + w_3\sigma_3^2u_3^2\right] \tag{47}$$

Fig. 2 Phase-space trajectory

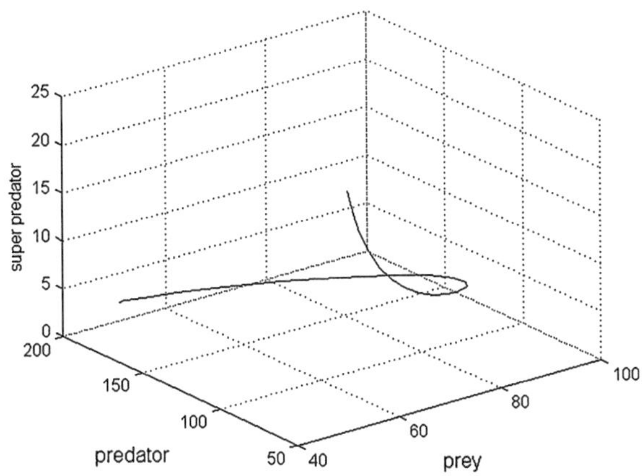

Now choosing $w_1\alpha_{12}x_1^* = w_2\alpha_{21}x_2^*$, $w_2\alpha_{23}x_2^* = w_3\alpha_{32}x_3^*$ and $w_1\alpha_{13}x_1^* = w_3\alpha_{31}x_3^*$ in (44) and using (47) in (46) we have

$$\mathbf{L}L\,(u) = -\left(r_1 - \frac{1}{2}\sigma_1^2\right) w_1 u_1^2 - \left(r_2 - \frac{1}{2}\sigma_2^2\right) w_2 u_2^2 - \left(x_3^* - \frac{1}{2}\sigma_3^2\right) w_3 u_3^2 \quad (48)$$

This completes the proof of the theorem (Fig. 2).

9 Numerical Experiments

Let $r_1 = 6.09$, $r_2 = 4.07$, $r_3 = 1.6$, $k_1 = 300$, $k_2 = 200$, $\alpha_{12} = 0.05$, $\alpha_{13} = 0.06$, $\alpha_{21} = 0.005$, $\alpha_{23} = 0.05$, $\alpha_{31} = 0.006$, $\alpha_{32} = 0.05$, $q_1 = 0.05$, $q_2 = 0.01$, and $E = 25$.

With these following set of data, the stability diagram and the Phase-space trajectory are as follows:

10 Conclusion

In this work, a prey-predator-super predator aquatic model is formulated considering selective harvesting of prey and predators under deterministic and stochastic environments. In deterministic case, we investigate local stability, Global stability, bionomic equilibrium, and optimal harvesting for the system and under stochastic conditions, the stability criteria for the system using Lyapunnov function are studied. It is to be noted that for the present system, when growth rates increase, the asymptotic mean square stability property is achieved.

In real-life projects this type of interaction model is most suitable to discuss the marine food web system considering sardine, menhaden, etc., as prey with tuna, Leerfish, Bluefish as predator and shark as super-predator.

This model is scalable to more complex dynamical systems by introducing Gompertz law of growth of prey-predator species, selective harvesting of prey-predator only, but not super predator and optimal harvesting policy in deterministic environment. The uncertainty behaviours of ecosystem model by introducing Winner process in stochastic environment is discussed.

In deterministic environment, we have considered Holling type-I response function between prey-predator, prey-super predator and predator-super predator and also harvesting term which is the product of common effort, catchability co-efficient and available biomass in differential equation to discuss complex dynamical system. In stochastic environment, complex dynamical system is discussed by Ito stochastic differential equation and stochastic perturbation of the variables.

In this model, numerical simulation is discussed using the data which is similar in the published paper in highly rated journal in this area. At present, we are not in a position to discuss the model with life data due to nonavailability. If it is available in the near future, then it will be discussed in our model formulation.

However, to increase the complexity of this type of food web system as per nature's demand one can introduce time delay, difference equation, delay differential equation, fuzziness, randomness, etc. in different parameters of our model.

Acknowledgments This work is financially supported by the major research project of Dr. Biswajit Mondal vide memo no. 42-14/2013(SR) dated 12.03.2014 and also by C.P.E. fund of U.G.C. and D.S.T.-F.I.S.T. fund of Raja N.L.Khan women's College, Midnapore, West Bengal.

References

1. Clark, C.W.: The Optimal Management of Renewable Resources. Mathematical Bioeconomic. Wiley, New York (1976)
2. Braure, F., Soudack, A.C.: Stability regions and transition phenomena for harvested predator prey systems. J. Math. Biol. **7**, 319–337 (1979)
3. Braure, F., Soudack, A.C.: Stability regions in predator prey systems with constant—rate prey harvesting. J. Math. Biol. **8**, 55–71 (1979)
4. Myerscough, M.R., Gray, B.F., Hogarth, W.L., Norbury, J.: An analysis of an ordinary differential equation model for a two species predator-prey system with harvesting and stocking. J. Math. Biol. **30**, 389–411 (1992)
5. Dia, G., Tang, M.: Coexistence region and global dynamics of a harvested predator prey systems. SIAM J. Appl. Math. **58**, 193–210 (1998)
6. Xiao, D., Ruan, S.: Bogdanov-Takens bifurcations in predator-prey systems with constant rate harvesting. Fields Inst. Commun. **21**, 493–506 (1999)
7. Chaudhuri, K.S.: A bioeconomic model of harvesting a multispecies fishery. Ecol. Model. **32**, 267–279 (1986)
8. Chaudhuri, K.S.: Dynamic optimization of combined harvesting of a two species fishery. Ecol. Model. **41**, 21–25 (1987)
9. Kar, T.K., Chaudhuri, K.S.: On non selective harvesting of a multispecies fishery. Int. J. Math. Educ. Sci. Technol. **33**(4), 543–556 (2002)
10. Kar, T.K., Chaudhuri, K.S.: Regulation of a prey-predator fishery by taxation. Int. J. Math. Educ. Sci. Technol. **34**(3), 543–556 (2003)
11. Purohit, D., Chaudhuri, K.S.: A bioeconomic model of non selective harvesting of two competing fish species. Int. J. Math. Educ. Sci. Technol. **33**(4), 543–556 (2002)
12. Chaudhuri, K.S., SahaRay, S.: On the combined harvesting of a prey-predator system. J. Biol. Syst. **4**, 376–389 (1996)
13. Dubey, B., Chandra, P., Sinha, P.: A model of fishery resource with reserve area. Nonlinear Anal. Real World Appl. **4**, 625–637 (2003)
14. Kar, T.K., Mishra, S.: Influence of prey reserve in a prey-predator fishery. Nonlinear Anal. **65**, 1725–1735 (2006)
15. Kar, T.K., Matsuda, H.: Controllability of a harvested prey-predator system with time delay. J. Biol. Syst. **14**(2), 243–254 (2006)
16. Kar, T.K., Pahari, U.K.: Modelling and analysis of a prey-predator system with age-structure and harvesting. Nonlinear Anal. **8**, 601–609 (2007)
17. Kar, T.K.: Selective harvesting in a prey-predator fishery with time delay. Math. Comput. Model. **38**, 449–458 (2003)

18. Kar, T.K., Chaudhuri, K.S.: On selective harvesting of two competing fish species in the presence of environmental fluctuation. Nat. Resour. Model. **17**(3), 251–273 (2004)
19. Beddington, J.R., May, R.M.: Harvesting populations in a randomly fluctuation environment. Science **197**, 463–465 (1977)
20. May, R.M., Beddington, J.R., Horwood, J.W., Shepherd, J.G.: Exploiting natural populations in an uncertain world. Math. Biosci. **42**, 219–252 (1978)
21. Ludwig, D.A.: Harvesting strategies for randomly fluctuating populations. J. du Conseil International pour l'Exploration de la Mer **39**, 166–174 (1980)
22. Reed, W.J.: The steady state of a stochastic harvesting model. Math. Biosci. **41**, 273–307 (1978)
23. Hanson, F.B.: Analysis of a singular fluctuation delay equation that arises from the stochastic modelling of population growth. J. Math. Appl. **96**, 405–419 (1982)
24. Braumann, C.A.: Variable effort fishing models in random environments. Math. Biosci. **156**, 1–19 (1999)
25. Turelli, M.: A reexamination of stability in random varying environments with comments on stochastic theory of limiting similarity. Theor. Popul. Biol. **13**, 244–267 (1978)
26. Kar, T.K., Chaudhuri, K.S.: Harvesting in a two prey one predator fishery: a bioeconomic model. Aust. N. Z. Ind. Appl. Math. J. **45**, 443–456 (2004)
27. Sadhukhan, D., Sahoo, L.N., Mondal, B., Maiti, M.: Food chain model with optimal harvesting in fuzzy environment. J. Appl. Math. Comput. **1**(2), 1–2 (2009)
28. Sutinen, J.G., Andersen, P.: The economics of fisheries law enforcement. Land Econ. **61**, 387–397 (1985)
29. Pradhan, T., Chaudhuri, K.S.: Bioeconomic modelling of a single species fishery with Gompertz law of growth. J. Biol. Syst. **6**(4), 393–409 (1998)
30. Gikhman, I.I., Skorokhod, A.V.: The Theory of Stochastic Processes, vol. I. Springer, Berlin (1974)
31. Gikhman, I.I., Skorokhod, A.V.: The Theory of Stochastic Processes, vol. II. Springer, Berlin (1975)
32. Gikhman, I.I., Skorokhod, A.V.: The Theory of Stochastic Processes, vol. III. Springer, Berlin (1979)
33. Afanas'ev, V.N., Kolmanowski, V.B., Nosov, V.R.: Mathematical Theory of Global Systems Design. Kluwer Academic, Dordrecht (1996)

Computational Method for High-Order Weighted Fuzzy Time Series Forecasting Based on Multiple Partitions

Sukhdev Singh Gangwar and Sanjay Kumar

Abstract In this paper, we present modified version of computational algorithm given by Gangwar and Kumar (Expert Syst Appl 39:12158–12164, 2012 [5]) for higher order weighted fuzzy time series with multiple partitioning to enhance the accuracy in forecasting. The developed method provides a better approach to enhance the accuracy in forecasted values. The proposed method was implemented on the historical student enrollments data of University of Alabama. The suitability of the developed method has been examined in comparison with other models in terms of mean square and average forecasting errors to show its superiority.

Keywords Fuzzy time series · Enrollment · Weighted · Fuzzy logical relations · Linguistic variable

1 Introduction

The knowledge of forecasting based on available time series data is one of the core components in planning and decision-making. One of the major limitations of conventional mathematical and statistical models is not to address the forecasting problem in which historical data are imprecise and vague. Concept of fuzzy set theory introduced by Zadeh [21, 22] was applied by Song and Chissom [15–17] to forecast the historical enrollments of the University of Alabama. Chen [1] proposed high-order fuzzy time series modes for forecasting the enrollments. Own and Yu [10] presented a heuristic higher order model by introducing a heuristic function to incorporate the heuristic knowledge to improve TAIFEX forecast.

S.S. Gangwar · S. Kumar (✉)
Department of Mathematics, Statistics & Computer Science,
G. B. Pant University of Agriculture & Technology, Pantnagar 263145,
Uttarakhand, India
e-mail: skruhela@hotmail.com

S.S. Gangwar
e-mail: gangwarsukhdev@gmail.com

M.K. Chakraborty et al. (eds.), *Facets of Uncertainties and Applications*,
Springer Proceedings in Mathematics & Statistics 125,
DOI 10.1007/978-81-322-2301-6_22

Yu [20] introduced weighted fuzzy time series forecasting model to tackle issues of recurrence and weighting in fuzzy time series forecasting. Cheng et al. [3] proposed a trend-based fuzzy time series model to improve the forecast accuracy in information and communication technologies products. Lee et al. [8] proposed the weighted adoption and the difference between actual data toward midpoint interval based on fuzzy time series. Ismail and Efendi [6] proposed the development of weighted fuzzy time series based on a collection of variation of the chronological number in the fuzzy logical group. Suhartono and Lee [18] proposed a new hybrid model based on the Winter's model and weighted fuzzy time series to improve the forecast accuracy in trend and seasonal data.

Singh [11–14] presented fixed difference parameters-based computational algorithm for forecasting with fuzzy time series. Gangwar and Kumar [5] proposed a computational method of forecasting based on multiple partitioning and higher order fuzzy time series. Joshi and Kumar [7] also presented another computational method based on difference parameters and considering the order of fuzzy logical relation as variable.

In this paper, we propose an enhanced version of computational algorithm proposed by Gangwar and Kumar [5] for weighted higher order fuzzy time series forecasting with multiple partitions. An innovative scheme to assign weight of fuzzy logical relations (FLRs) is also proposed in the present study. The objective of this study is to improve forecasting accuracy by using higher order weighted fuzzy time series, multiple partitioning of universe of discourse, and difference parameters as relations for forecasting. The proposed algorithm minimizes the time of generating relational equations by using complex min–max composition operations and various defuzzification process. The proposed method has been implemented on benchmarking problem of forecasting the historical student enrollments data of University of Alabama and compared with the various other recent methods proposed by Singh [14], Gangwar and Kumar [5] and Joshi and Kumar [7].

2 Fuzzy Set and Fuzzy Time Series

Definitions of fuzzy set and fuzzy time series given by [15–17, 21] are described as follows:

Definition 1 A fuzzy set A_i defined on the Universe of discourse, $U = \{u_1, u_2, u_3, \ldots, u_n, \}$, is represented as follows:

$$A_i = \mu_{A_i}(u_1)/u_1 + \mu_{A_i}(u_2)/u_2 + \mu_{A_i}(u_3)/u_3 + \cdots + \mu_{A_i}(u_n)/u_n \quad (1)$$

where μ_{A_i} is the membership function of fuzzy set A_i, $\mu_{A_i} : U \rightarrow [0, 1]$

Definition 2 Suppose $Y(t)$, $(t = \ldots, 0, 1, 2, \ldots)$ be the Universe of discourse and $Y(t) \subseteq R$. Assume that fuzzy sets, $f_i(t)$, $(i = 1, 2, \ldots)$ are defined in the Universe of discourse $Y(t)$. $F(t)$, collection of $f_i(t)$ is known as fuzzy time series on $Y(t)$.

Definition 3 If $F(t)$ is caused only by $F(t-1)$, i.e., $F(t-1) \rightarrow F(t)$, then this can be expressed by the following fuzzy relational equation:

$$F(t) = F(t-1) \; o \; R(t, t-1) \tag{2}$$

where the symbol "o" is Max–Min composition operator. The relation $R(t, t-1)$ is a fuzzy relation between $F(t)$ and $F(t-1)$ and is called the first order model of $F(t)$.

Definition 4 If $F(t)$ is caused by more fuzzy sets, $F(t-n), F(t-n+1), \ldots, F(t-1)$, the fuzzy relationship is represented by $A_{i_1}, A_{i_2}, A_{i_3} \ldots, A_{i_n} \rightarrow A_j$

$$\text{here, } F(t-n) = A_{i_1}, F(t-n+1) = A_{i_2}, \ldots, F(t-1) = A_{i_n} \tag{3}$$

This relationship is called nth order fuzzy time series model.

Definition 5 Let $F(t)$ be a fuzzy time series and $R(t, t-1)$ be a first order model of $F(t)$. If $R(t, t-1) = R(t-1, t-2)$ for any time t, the $F(t)$ is named a time-invariant fuzzy time series. But if $R(t, t-1)$ is time dependent, that is, $R(t, t-1)$ may be different from $R(t-1, t-2)$ for any time$t$ then $F(t)$ is called time-variant fuzzy time series.

3 Weighted Fuzzy Time Series

Yu [20] proposed the weighted fuzzy time series model to resolve recurrence of fuzzy relationships and weighting problems in fuzzy time series forecasting. In most of the weighted models, the recurrences of each fuzzy logical relation (FLR) are taken into account. From this viewpoint, the FLRs of various recurrences are assigned different weights. In present study weights assigned to recent FLR are higher than that of older FLR keeping the fact that the recent FLR has more importance than the previous ones. The novelty of the proposed weighted fuzzy time series forecasting method is described as follows:

The probability of appearance of the most recent FLR ($t = n$) in the near future is higher than others, hence the highest weight w_{n-1} is assigned for the most recent FLR ($t = n$). On the other hand, the probability of appearance of the most aged FLR ($t = n$) in the near future is lower than in the case of the others, hence the lowest weight of 1 is assigned for the most aged FLR ($t = n$). The proposed method of assigning the weight to FLRs is explained as follows:

Suppose $A_1 \xrightarrow{w_1=1} A_2$, $A_2 \xrightarrow{w_2=2} A_3$, $A_3 \xrightarrow{w_3=3} A_4, \ldots, A_n \xrightarrow{w_n=n} A_{n+1}$ are the FLRs used for forecast A_{n+2} with corresponding weights. The sum of the weight of each FLR should be standardized, hence computation for weight can be determined as follows:

$$W(t) = [w'_1, w'_2, \ldots, w'_n] = \left[\frac{w_1}{\sum_{h=1}^{n} w_h}, \frac{w_2}{\sum_{h=1}^{n} w_h}, \ldots, \frac{w_n}{\sum_{h=1}^{n} w_h} \right] \tag{4}$$

where n is the number of FLR used to forecast for A_{n+2}.
$W(t) = [w'_1, w'_2, \ldots, w'_n]$ should satisfy the necessary condition

$$\sum_{h=1}^{n} w'_h = 1 \tag{5}$$

4 Proposed Method and Computational Algorithm

Proposed method uses the ratio formula given by Gangwar and Kumar [5] for determining the number of partitions. For each partition, the weighted fuzzy relations and difference parameters are defined by following rules:

(i) In order to forecast the enrollment for third year (1973), we implement $A_1 \stackrel{w_1=1}{\longrightarrow} A_2$. In this case $w_1 \times |E_2 - E_1|$ is used.

(ii) In order to forecast the enrollment for fourth year (1974), we implement $A_1 \stackrel{w_1=1}{\longrightarrow} A_2$ & $A_2 \stackrel{w_2=2}{\longrightarrow} A_3$. In this case $\frac{w_1}{w_1+w_2} \times |E_2 - E_1|$ & $\frac{w_2}{w_1+w_2} \times |E_3 - E_2|$ are used.

(iii) In order to forecast the enrollment for fifth year (1975), we implement $A_1 \stackrel{w_1=1}{\longrightarrow} A_2$, $A_2 \stackrel{w_2=2}{\longrightarrow} A_3$ & $A_3 \stackrel{w_3=3}{\longrightarrow} A_4$. In this case $\frac{w_1}{w_1+w_2+w_3} \times |E_2 - E_1|$, $\frac{w_2}{w_1+w_2+w_3} \times |E_3 - E_2|$ & $\frac{w_3}{w_1+w_2+w_3} \times |E_4 - E_3|$ are used.

Similarly, In order to forecast the enrollment for $(n+2)$th year, we implement $A_1 \stackrel{w_1=1}{\longrightarrow} A_2, A_2 \stackrel{w_2=2}{\longrightarrow} A_3, A_3 \stackrel{w_3=3}{\longrightarrow} A_4, \ldots, A_n \stackrel{w_n=n}{\longrightarrow} A_{n+1}$. In this case $\frac{w_1}{w_1+w_2+w_3+\cdots+w_n} \times |E_2 - E_1|$, $\frac{w_2}{w_1+w_2+w_3+\cdots+w_n} \times |E_3 - E_2|$, $\frac{w_3}{w_1+w_2+w_3+\cdots+w_n} \times |E_4 - E_3|, \ldots,$ $\frac{w_n}{w_1+w_2+w_3+\cdots+w_n} \times |E_{n+1} - E_n|$ are used.

Stepwise computational procedure for forecasting historical time series data of year $n+2$, given as follows:

Step 1: Define the Universe of discourse, U based on the range of available time series data, by the rule $U = [E_{min} - D_1, E_{max} + D_2]$ where D_1 and D_2 are two proper positive numbers are selected randomly to accommodate the complete time series data.

Step 2: The Universe of discourse U is partitioned into equal length of intervals: $u_1, u_2, \ldots, u_m$.
Step 3: Construct the fuzzy sets A_i in accordance with the number of intervals in step 2 and apply the triangular membership function to each fuzzy set constructed.
Step 4: Fuzzify the data by choosing maximum membership and establish the fuzzy logical relationships by the rule proposed by [5].
Step 5: Repartition the whole fuzzy time series using the ratio formula proposed by [5].
Step 6: Rules for forecasting
Some notations used are defined as follows:

$[*A_j]$ is corresponding interval u_j for which membership in A_j is supreme (i.e., 1). $L[*A_j]$ and $U[*A_j]$ is the lower and upper bound of interval u_j. $M[*A_i]$ and $M[*A_j]$ is the mid value of the interval u_i and u_j, respectively, having supremum value in A_i and A_j.
For a fuzzy logical relation $A_i \rightarrow A_j$

A_i and $A_j$is the fuzzified enrollment of year n and $(n+1)$, respectively. E_i, E_{i-1}, E_{i-2}, E_{i-c}, and $E_{i-(c+1)}$ are the actual enrollment of the years n, $(n-1)$, $(n-2)$, $(n-c)$, and $(n-(c+1))$, respectively and F_j is the crisp forecasted enrollment of the year $(n+1)$.

The proposed method utilizes the historical data of year 1 to n for framing rules to implement on fuzzy logical relation, $A_i \rightarrow A_j$, whereA_i, the current state, is the fuzzified enrollments of year n and A_j, the next state, is fuzzified enrollments of year $n+1$. The proposed method for forecasting is mentioned as computational algorithms for assigning the weight and generating the relations between the time series data of year 1 to n for forecasting the enrollment of year $n+1$ in each partition. The developed computational algorithm uses the weighted differences in enrollment of past n years and have been considered a fuzzy parameter in framing the fuzzy rules to impose on current year fuzzified enrollment to get forecast of next year enrollments.

Computational algorithm: The proposed computational algorithm used starting from $k=1$ (first partition) to $k=K$ (last partition). Within $k=1$ we used data starting from $i=2$ to N (end of time series data for each partition) and obtained fuzzy logical relation $A_i \rightarrow A_j$ for year i to $i+1$.

Compute parameters D_i for corresponding year i by using the following expression:

$$D_i = \left| \frac{(i-1)}{\sum_{w=1}^{i-1} w} |E_i - E_{i-1}| - \left[\sum_{c=1}^{i-1} \left\{ \frac{(i-(c+1))}{\sum_{w=1}^{i-1} w} |E_{i-c} - E_{i-(c+1)}| \right\} \right] \right| \quad (6)$$

$P=0$ and $Q=0$
For $a=2$ to i
Calculate F_{ia} and FF_{ia} by using the following expression:

$$F_{ia} = M[{}^*A_i]+(a^*D_i)/(a-1)$$
$$FF_{ia} = M[{}^*A_i]-(a^*D_i)/(a-1)$$

If F_{ia} is greater than and equal to lower bound of interval u_j having supremum value in A_j and less than and equal to upper bound of interval u_j having supremum value in A_j then

$$\left.\begin{aligned} P &= P + (a-1)/(i-1)^* F_{ia} \\ Q &= Q + (a-1)/(i-1) \end{aligned}\right\} \tag{7}$$

If FF_{ia} is greater than and equal to lower bound of interval u_j having supremum value in A_j and less than and equal to upper bound of interval u_j having supremum value in A_j then

$$\left.\begin{aligned} P &= P + (a-1)/(i-1)^* FF_{ia} \\ Q &= Q + (a-1)/(i-1) \end{aligned}\right\} \tag{8}$$

Finally, using P, Q, F_{ia}, FF_{ia} and $M[{}^*A_j]$, we calculate F_j given by following expression:

$$F_j = (P + M({}^*A_j))/(Q+1) \tag{9}$$

The above process is repeated to end term of time series data for each partition.

Step 7: Mean square error (MSE) and Average forecasting error (AFE) are common tools which are used in fuzzy time series forecasting to verify the performance. The MSE and AFE are defined as follows:

$$\text{MSE} = \frac{\sum_{i=1}^{n} (actual\ value_i - forecasted\ value_i)^2}{n} \tag{10}$$

$$\text{Forecasting error (in \%)} = \frac{|forecasted\ \text{-}\ actual\ value|}{actual\ value} \times 100 \tag{11}$$

$$\text{AFE (in \%)} = \frac{sum\ of\ forecasting\ error}{numbers\ of\ errors} \tag{12}$$

5 Implementation of Proposed Method

The stepwise implementation of proposed method on the time series data of student enrollments at University of Alabama is given as follows:

Step 1: Universe of discourse $U = [13{,}000, 20{,}000]$ is defined for the available time series data.

Step 2: The Universe of discourse is partitioned into following seven intervals:

u_1=[13,000, 14,000] u_2 = [14,000, 15,000] u_3 = [15,000, 16,000]
u_4 = [16,000, 17,000] u_5 = [17,000, 18,000] u_6 = [18,000, 19,000]
u_7 = [19,000, 20,000]

Step 3: Seven fuzzy sets $A_1, A_2, \ldots, A_7$ are define on the universe of discourse U and the membership grades to these fuzzy sets are defined as follows:

$$\begin{aligned}
A_1 &= 1/u_1 + 0.5/u_2 + 0/u_3 + 0/u_4 + 0/u_5 + 0/u_6 + 0/u_7,\\
A_2 &= 0.5/u_1 + 1/u_2 + 0.5/u_3 + 0/u_4 + 0/u_5 + 0/u_6 + 0/u_7,\\
A_3 &= 0/u_1 + 0.5/u_2 + 1/u_3 + 0.5/u_4 + 0/u_5 + 0/u_6 + 0/u_7,\\
A_4 &= 0/u_1 + 0/u_2 + 0.5/u_3 + 1/u_4 + 0.5/u_5 + 0/u_6 + 0/u_7,\\
A_5 &= 0/u_1 + 0/u_2 + 0/u_3 + 0.5/u_4 + 1/u_5 + 0.5/u_6 + 0/u_7,\\
A_6 &= 0/u_1 + 0/u_2 + 0/u_3 + 0/u_4 + 0.5/u_5 + 1/u_6 + 0.5/u_7,\\
A_7 &= 0/u_1 + 0/u_2 + 0/u_3 + 0/u_4 + 0/u_5 + 0.5/u_6 + 1/u_7.
\end{aligned}$$

Step 4: The fuzzified historical time series data of enrollments are obtained and fuzzy logical relations are established (Table 1).
Step 5: Using the ratio formula given by Gangwar and Kumar [5], the time series data again partitioned in three parts. First partition contains enrollments from 1971 to 1978, second partition contains enrollments from 1979 to 1985, and third partition contains enrollments from 1986 to 1992.
Step 6: The computations for the enrollments of University of Alabama have been carried out by using the proposed model (computational algorithm) given in Sect. 4. The results obtained are placed in the Table 2 along with results of other models.
Step 7: Mean square error (MSE) and Average Forecasting Error is calculated and are placed in Table 3 to compare the accuracy in forecasted values of proposed model with other models.

Table 1 Actual and fuzzified enrollments of University of Alabama

Year	Actual	Fuzzified	Year	Actual	Fuzzified
1971	13,055	A1	1982	15,433	A3
1972	13,563	A1	1983	15,497	A3
1973	13,867	A1	1984	15,145	A3
1974	14,696	A2	1985	15,163	A3
1975	15,460	A3	1986	15,984	A3
1976	15,311	A3	1987	16,859	A4
1977	15,603	A3	1988	18,150	A6
1978	15,861	A3	1989	18,970	A6
1979	16,807	A4	1990	19,328	A7
1980	16,919	A4	1991	19,337	A7
1981	16,388	A4	1992	18,876	A6

Table 2 Enrollments forecast by proposed method and other various methods

Partition	Year	Actual	Singh [14]	Joshi and Kumar [7]	Gangwar and Kumar [5]	Proposed
1	1971	13,055	–	–	–	–
	1972	13,563	–	–	–	–
	1973	13,867	–	–	13,500	13,500
	1974	14,696	14,331	14,544	14,500	14,500
	1975	15,460	15,489	15,504	15,500	15,500
	1976	15,311	15,463	15,456	15,500	15,500
	1977	15,603	15,412	15,599	15,500	15,500
	1978	15,861	15,559	15,723	15,500	15,500
2	1979	16,807	16,500	16,482	–	–
	1980	16,919	16,616	16,603	–	–
	1981	16,388	16,516	16,340	16,500	16,500
	1982	15,433	15,538	15,356	15,500	15,622
	1983	15,497	15,440	15,408	15,500	15,500
	1984	15,145	15,497	15,425	15,500	15,500
	1985	15,163	15,280	15,395	15,500	15,500
3	1986	15,984	15,351	15,471	–	–
	1987	16,859	16,395	16,573	–	–
	1988	18,150	18,500	18,683	18,500	18,375
	1989	18,970	18,376	18,646	18,500	18,500
	1990	19,328	19,366	19,373	19,337	19,500
	1991	19,337	19,407	–	19,500	19,500
	1992	18,876	18,604	–	18,704	18,763

Table 3 Comparison of MSE and AFE of proposed method with other methods

Method	Singh [14]	Joshi and Kumar [7]	Gangwar and Kumar [5]	Proposed
MSE	95,306	67943.47	62976.63	61229.44
AFE	1.5319	1.264196	1.269981	1.309517

6 Conclusion

In this paper, we have proposed a computational method for high-order fuzzy time series with a new method of assigning weight to fuzzy logical relations used in forecast. The proposed method is an enhanced version of the computational algorithm given by Gangwar and Kumar [5]. The proposed method has been tested for forecasting efficiency on the historical time series data of enrollments of University of Alabama and has a comparative study with some of existing methods. Even though

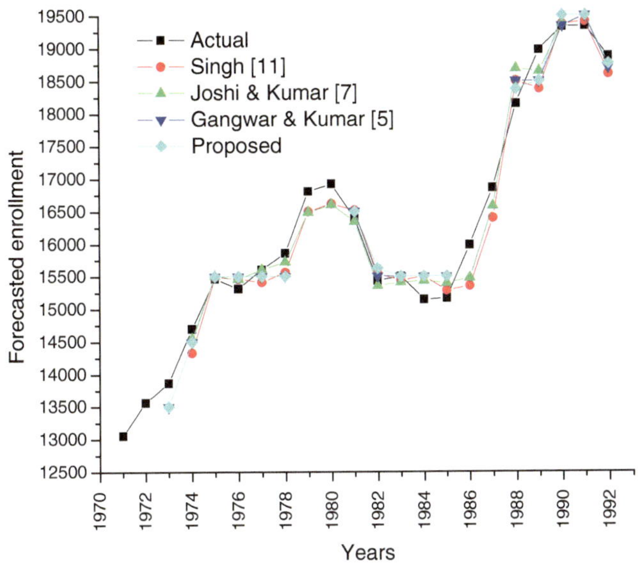

Fig. 1 Actual enrollments versus forecasted enrollments

the AFE in forecasting enrollments using proposed method is slightly higher than the method given by Joshi and Kumar [7] and Gangwar and Kumar [5], but still it is lower than the method given by Singh [14]. The proposed method outperforms with Singh [14], Joshi and Kumar [7], and Gangwar and Kumar [5] in terms of MSE. As the work of Gangwar and Kumar [5] outperformed the work of [2, 4, 9, 11, 13, 19] indirectly, we can conclude that proposed model outperforms these models in forecasting the enrollments at University of Alabama (Fig. 1).

References

1. Chen, S.M.: Forecasting enrollments based on high-order fuzzy time series. Cybern. Syst. Int. J. **33**, 1–16 (2002)
2. Chen, S.M., Tanuwijaya, K.: Fuzzy forecasting based on high-order fuzzy logical relationships and automatic clustering techniques. Expert Syst. Appl. **38**, 15425–15437 (2011)
3. Cheng, C.H., Chen, Y.S., Wu, Y.L.: Forecasting innovation diffusion of products using trend weighted fuzzy time series model. Expert Syst. Appl. **36**, 1826–1832 (2009). doi:10.1016/j.eswa.2007.12.041
4. Cheng, C.H., Wang, J.W., Cheng, G.W.: Multi-attribute fuzzy time series method based on fuzzy clustering. Expert Syst. Appl. **34**, 1235–1242 (2008)
5. Gangwar, S.S., Kumar, S.: Partitions based computational method for high-order fuzzy time series forecasting. Expert Syst. Appl. **39**, 12158–12164 (2012). doi:10.1016/j.eswa.2012.04.039
6. Ismail, Z., Efendi, R.: Enrollment forecasting based on modified weighted fuzzy time series. J. Artif. Intell. **4**(1), 110–118 (2011)

7. Joshi, B.P., Kumar, S.: A computational method for fuzzy time series forecasting based on difference parameters. Int. J. Model. Simul. Sci. Comput. **4**(1), 1250023–1250035 (2013). doi:10.1142/S1793962312500237
8. Lee, M.H., Efendi, R., Ismail, Z.: Modified weighted for enrollment forecasting based on fuzzy time series. Matematika **25**(1), 67–78 (2009)
9. Liu, H.T.: An improved fuzzy time series forecasting method using trapezoidal fuzzy numbers. Fuzzy Optim. Decis. Mak. **6**, 63–80 (2007)
10. Own, C.M., Yu, P.T.: Forecasting fuzzy time series on a heuristic high-order model. Cybern. Syst. Int. J. **36**, 705–717 (2005)
11. Singh, S.R.: A robust method of forecasting based on fuzzy time series. Appl. Math. Comput. **188**, 472–484 (2007a)
12. Singh, S.R.: A simple method of forecasting based on fuzzy time series. Appl. Math. Comput. **186**, 330–339 (2007b)
13. Singh, S.R.: A simple time variant method for fuzzy time series forecasting. Cybern. Syst. Int. J. **38**, 305–321 (2007c). doi:10.1080/01969720601187354
14. Singh, S.R.: A computational method of forecasting based on fuzzy time series. Math. Comput. Simul. **79**, 539–554 (2008)
15. Song, Q., Chissom, B.S.: Fuzzy time series and its models. Fuzzy Sets Syst. **54**, 269–277 (1993a)
16. Song, Q., Chissom, B.S.: Forecasting enrollments with fuzzy time series—part I. Fuzzy Sets Syst. **54**, 1–9 (1993b). doi:10.1016/0165-0114(93)90355-L
17. Song, Q., Chissom, B.S.: Forecasting enrollments with fuzzy time series—part II. Fuzzy Sets Syst. **64**, 1–8 (1994)
18. Suhartono, S., Lee, M.H.: A hybrid approach based on winter's model and weighted fuzzy time series for forecasting trend and seasonal data. J. Math. Stat. **7**(3), 177–183 (2011)
19. Wong, W.K., Bai, E., Chu, A.W.C.: Adaptive time variant models for fuzzy time series forecasting. IEEE Trans. Syst. Man Cybern. Part B Cybern. **40**(6), 1531–1542 (2010)
20. Yu, H.K.: Weighted fuzzy time series models for TAIEX forecasting. Phys. A. **349**, 609–624 (2005). doi:10.1016/j.physa.2004.11.006
21. Zadeh, L.A.: Fuzzy set. Inf. Control **8**, 338–353 (1965)
22. Zadeh, L.A.: The concept of a linguistic variable and its application to approximate reasoning. Part I. Inf. Sci. **8**, 199–249 (1975)

Portfolio Selection with Possibilistic Kurtosis

Sheikh Ahmed Hossain and Rupak Bhattacharyya

Abstract This paper proposes a new approach for modeling multiple objective portfolio selection problem by applying weighted possibilistic moments of trapezoidal fuzzy numbers. The proposed model allows the decision-maker to select the suitable portfolio taking into account the impreciseness to the market scenarios. Here, the objectives are to (i) maximize the expected portfolio return, (ii) minimize the portfolio variance, (iii) maximize the portfolio skewness, and (iv) minimize the portfolio kurtosis for the risky investor. The proposed model has been solved by Zimmermann's fuzzy goal programming technique. The model is illustrated by a numerical example using data extracted from the Bombay Stock Exchange.

Keywords Fuzzy portfolio selection · Possibilistic measures · Mean · Variance · Skewness · Kurtosis · Zimmerman's fuzzy goal programming

1 Introduction

A number of authors(e.g., [1–5]) have proposed to select stock portfolios on the basis of the first three moments of return distributions, rather than the first two (mean and variance) proposed by Markowitz [6] in 1952. The third moment of return distribution is called skewness. Researchers interested in skewness believe investors should prefer positive skewness. All else constant, they should prefer portfolios with a larger probability of very large payoffs. This is not only logical, but also consistent with some empirical evidence that investors exhibit this preference. If the three moments are important to the investor, then the portfolio problem is represented

S.A. Hossain (✉)
Department of Mathematics, Brahmananda Keshab Chandra College, 111/2 B.T. Road, Bonhooghly, Kolkata 700108, West Bengal, India
e-mail: skahossainmath@gmail.com

R. Bhattacharyya
Department of Mathematics, Bijoy Krishna Girls' College, 5/3 Mahatma Gandhi Road, Howrah 711101, West Bengal, India
e-mail: mathsrup@gmail.com

M.K. Chakraborty et al. (eds.), *Facets of Uncertainties and Applications*, Springer Proceedings in Mathematics & Statistics 125,
DOI 10.1007/978-81-322-2301-6_23

in three-dimensional space with mean on one axis, variance on the second, and skewness on the third. The efficient set would be the outer shell of the feasible set with maximum mean, minimum variance, and maximum skewness. However, it is evident that the measures of a return distribution, mean, variance, and skewness cannot form a complete design about the distribution. In addition to these measures, we should consider one more measure which Prof. Karl Pearson calls the *Convexity of a curve* or *Kurtosis*. Kurtosis enables us to have an idea about the flatness or peakness of the curve. It is measured by the coefficient β_2 or its derivation γ_2 given by $\beta_2 = \left(\frac{\mu_4}{\mu_2^2}\right)$, $\gamma_2 = \beta_2 - 3$ [μ_i being the ith order moment]. Curve which is neither flat nor peaked is called the normal curve or mesokurtic curve and for such a curve $\beta_2 = 3$, i.e., $\gamma_2 = 0$. Curve which is flatter than the normal curve is known as platykurtic and for such a curve, $\beta_2 < 3$, i.e.,$\gamma_2 < 0$. Curve which is more peaked than the normal curve is known as leptokurtic and for such a curve $\beta_2 > 3$, i.e., $\gamma_2 > 0$. The high kurtosis (fat tails) in return distribution suggests that periods of stability are interspersed by rapid change.

Two distributions may have the same average, dispersion, and skewness, yet in one there may be high concentration of values near the mode, showing a sharper peak in frequency curve than the other. The classical capital market theory, like the bulk of economics, is based on the equilibrium system articulated so well by Alfred Marshall, the father of modern economics in the 1890s. This view is based on the idea that economics is like Newtonian physics, with well-defined cause–effect relationships.

Empirical evidence suggests that the classical capital market theory falls short in the following ways:

- The distribution of stock returns exhibit a high degree of kurtosis. This means that the tails of the distribution are fatter and the mean of the distribution is higher than what is predicted by a normal distribution. In other words, it means that periods of relatively modest changes are interspersed with periods of booms and busts.
- Financial returns are predictable to some extent.
- Risk and return are not related in a linear manner.
- Investors are prone to make systematic errors in their judgment and trade excessively.

The mean–variance decision criterion by Markowitz [6] is inadequate for allocating wealth when we deal with the funds to be invested in the stock market. Not only are the return distributions asymmetric and leptokurtic, they also display significant coskewness and cokurtosis with the return of other asset classes due to the option-like features of alternative investments. Different approaches have been developed in the financial literature to incorporate the individual preferences for higher order moments into the optimal security allocation problem. Davies et al. [7] and Berenyi ([8, 9]) use the goal programming approach to determine the set of the mean–variance–skewness–kurtosis efficient funds of hedge funds.

Different from [10–12], after recalling the definition of mean, variance, semi-variance, and skewness, this paper considers the kurtosis for portfolio selection with possibilitic fuzzy risk factors. Several empirical studies show that portfolio returns have fat tails. Generally, investors would prefer a portfolio return with smaller kurtosis which indicates the leptokurtosis (fat tails or thin tails) when the mean value, the variance, and the asymmetry are the same. The paper is organized as follows: In Sect. 2, we recall the weighted possibilistic measures of means, variance, skewness of a trapezoidal fuzzy variable. Then we introduce the possibilistic measure of kurtosis for a trapezoidal fuzzy number. In Sect. 3, we have proposed a tetra-objective optimization model for portfolio selection problems. In Sect. 4, we discuss Zimmerman's goal programming method for multiple objective optimization. In Sect. 5, a case study has been done to illustrate our model. In Sect. 6, some concluding remarks are specified.

2 Weighted Possibilistic Measures of Mean, Variance, and Skewness of Trapezoidal Fuzzy Numbers

In this section some basic ideas of fuzzy sets and possibilistic measures of fuzzy sets are discussed. We also introduce possibilistic measure of fourth-order moment followed by possibilistic measure of kurtosis for trapezoidal fuzzy numbers.

Definition 1 A fuzzy set $\tilde{A}$ in $U \subset \mathbb{R}$, where $\mathbb{R}$ is the set of all real numbers, is an ordered paired set $\tilde{A} = \{(x, \mu_{\tilde{A}}(x)) : x \in \mathbb{R}\}$, where $\mu_{\tilde{A}}(x)$ is the membership function of x and $0 \leq \mu_A(x) \leq 1$.

Definition 2 An α-cut of a fuzzy set $\tilde{A}$ is a crisp set $\tilde{A}_\alpha$ that contains all the elements in U and that has membership values in $\tilde{A}$ greater than or equal to α, i.e., $\tilde{A}_\alpha = \{x \in U : \mu_{\tilde{A}}(x) \geq \alpha\}$.

Definition 3 A fuzzy number $\tilde{A} = (a, b, c, d)$ is called a trapezoidal fuzzy number (Tr.F.N.) with core $[b, c]$ if its membership function has the following form:

$$\mu_{\tilde{A}}(x) = \begin{cases} \frac{x-a}{b-a} & for\ x \in [a, b] \\ 1 & for\ x \in [b, c] \\ \frac{d-x}{d-c} & for\ x \in [c, d] \\ 0 & otherwise. \end{cases}$$

Its α-level sets are $\tilde{A}_\alpha = [\underline{a}(\alpha), \bar{a}(\alpha)] = [a + (b - a)\alpha, d - (d - c)\alpha]$.

Definition 4 Let $\tilde{A}_\alpha = [\underline{a}(\alpha), \bar{a}(\alpha)]$ be a α-cut of a fuzzy number $\tilde{A}_\alpha$ and $f(\alpha)$ be a weighted function. Also, let D_L and D_U be two real numbers such that $D_L \leq D_U$. Then nth weighted double possibilistic moments of fuzzy number $\tilde{A}$ about points D_L and D_U are defined as:

$$M_n^{(D_L, D_U)}(\tilde{A}) = \frac{1}{2} \int_0^1 f(\alpha) \left[(\underline{a}(\alpha) - D_L)^n + (\bar{a}(\alpha) - D_U)^n\right] d\alpha, n = 1, 2, 3, \ldots$$

If $D_L = D_U = m(\tilde{A})$, where $m(\tilde{A})$ is the possibilistic mean of the fuzzy number $\tilde{A}$ and $m(\tilde{A})$ is given by $m(\tilde{A}) = \frac{1}{2}\int_0^1 f(\alpha)(\underline{a}(\alpha) + \bar{a}(\alpha))d\alpha$. If $f(\alpha) = 2\alpha$, then $M_n(\tilde{A}) = \int_0^1 \alpha[(\underline{a}(\alpha) - m(\tilde{A}))^n + (\bar{a}(\alpha) - m(\tilde{A}))^n]d\alpha, n = 1, 2, 3, \ldots$

The second, third, and fourth probalistic moments are, respectively, given as

$M_2(\tilde{A}) = \int_0^1 \alpha[(\underline{a}(\alpha) - m(\tilde{A}))^2 + (\bar{a}(\alpha) - m(\tilde{A}))^2]d\alpha$

$M_3(\tilde{A}) = \int_0^1 \alpha[(\underline{a}(\alpha) - m(\tilde{A}))^3 + (\bar{a}(\alpha) - m(\tilde{A}))^3]d\alpha$

$M_4(\tilde{A}) = \int_0^1 \alpha[(\underline{a}(\alpha) - m(\tilde{A}))^4 + (\bar{a}(\alpha) - m(\tilde{A}))^4]d\alpha$

Definition 5 The weighted possibilistic skewness (WPS) of the fuzzy number $\tilde{A}$ is defined by $\gamma_1 = \frac{M_3(\tilde{A})}{\left(\sqrt{M_2(\tilde{A})}\right)^3}$.

Definition 6 The weighted possibilistic kurtosis (WPK) of the fuzzy number $\tilde{A}$ is defined by $\gamma_2 = \frac{M_4(\tilde{A})}{\left(\sqrt{M_2(\tilde{A})}\right)^4}$.

Theorem 1 *Let $\tilde{A} = (a, b, c, d)$ be a trapezoidal fuzzy number. Then the weighted possibilistic mean, variance, and skewness of $\tilde{A}$ are, respectively, given by*

$WPM = \frac{1}{6}[a + 2(b + c) + d]$

$WPV = \frac{1}{36}[2(a^2 + d^2) + 5(b^2 + c^2) + 2(ab + cd - da) - 4(ac + bd) - 8bc]$

$WPS = \frac{1}{5}[19(a^3 + d^3) + 26(b^3 + c^3) - 15ad(a + d) - 30bc(b + c) + 60bc.$
$(a + d) + 30ad(b + c) - 12(a^2b + cd^2) - 30(a^2c + bd^2) - 33(ab^2 + c^2d)$
$-15(a^2c+b^2d)]/[2(a^2+d^2)+5(b^2+c^2)+2(ab+cd-da)-4(ac+bd)-8bc]^{\frac{3}{2}}$

Proof For proof refer to Battacharyya et al. [4].

Theorem 2 *Let $\tilde{A} = (a, b, c, d)$ be a trapezoidal fuzzy number. Then the weighted possibilistic kurtosis of $\tilde{A}$ is given as*

$WPK = 1296[\frac{1}{72}(b - a)^2(d - c)^2 + \frac{3}{8}b^2c^2 - \frac{1}{6}bc[(b - c)^2 + (b - a)(d - c)$
$+(d - c)^2] - \frac{1}{4}bc[b^2 + c^2 - (b - c)(b - a + d - c)] - \frac{1}{18}(b - a)(b - c)$
$(d - c)(b - a + d - c) + \frac{5}{432}[(b - a)^4 + (d - c)^4] + \frac{1}{16}(b^4 + c^4) + \frac{1}{12}$
$(b^2 + c^2)[(b - a)^2 + (b - a)(d - c) + (d - c)^2] - \frac{1}{12}(b^3 - c^3)(b - a + d - c)$
$+\frac{2}{135}(b - a)(d - c)[(b - c)^2 + ((d - c)^2]/[2(a^2 + d^2) + 5(b^2 + c^2) + 2$
$(ab + cd - da) - 4(ac + bd) - 8bc]^2.$

Proof We have $\tilde{A} = (a, b, c, d)$ to be a trapezoidal fuzzy number. Its α-level sets are $\tilde{A}_\alpha = [\underline{a}(\alpha), \bar{a}(\alpha)] = [a + (b - a)\alpha, d - (d - c)\alpha]$. Then the weighted probabilistic fourth-order moment is given as

$M_4(\tilde{A}) = \int_0^1 \alpha[(\underline{a}(\alpha) - m(\tilde{A}))^4 + (\bar{a}(\alpha) - m(\tilde{A}))^4]d\alpha$

$= \int_0^1 \alpha[(a + (b - a)\alpha - \frac{1}{6}\{a + 2(b + c) + d\})^4 + (d - (d - c)\alpha - \frac{1}{6}\{a + 2$
$(b + c) + d\})^4]d\alpha$

$= \frac{1}{72}(b - a)^2(d - c)^2 + \frac{3}{8}b^2c^2 - \frac{1}{6}bc[(b - c)^2 + (b - a)(d - c) + (d - c)^2] -$
$\frac{1}{4}bc[b^2 + c^2 - (b - c)(b - a + d - c)] - \frac{1}{18}(b - a)(b - c)(d - c)(b - a + d - c) +$
$\frac{5}{432}[(b - a)^4 + (d - c)^4] + \frac{1}{16}(b^4 + c^4) + \frac{1}{12}(b^2 + c^2)[(b - a)^2 + (b - a)(d - c) +$

$(d-c)^2] - \frac{1}{12}(b^3-c^3)(b-a+d-c) + \frac{2}{135}(b-a)(d-c)[(b-c)^2+((d-c)^2]$

By definition WPK $= \gamma_2 = \frac{M_4(\tilde{A})}{(\sqrt{M_2(\tilde{A})})^4}$.

Hence the result follows.

3 Weighted Possibilistic Mean–Variance–Skewness–Kurtosis Models for Portfolio Selection

Empirically, it is evident that return from stocks is not fixed, rather range bound. Analyzing the ranges of different stocks we can find that returns can be considered as fuzzy numbers. Let $\tilde{r}_i$ be a fuzzy number representing the return of ith security. Let x_i be the portion of the total capital invested in security i, $i = 1, 2, \ldots, n$. Then $\frac{p_i+d_i-p_i'}{p_i}$, where p_i, p_i', d_i are, respectively, closing price at previous year, closing price at next year, and dividend paid for ith security calculates a particular return.

Theorem 3 *Let $\tilde{r}_i = (a_i, b_i, c_i, d_i)$ be independent trapezoidal fuzzy numbers and $\tilde{\boldsymbol{r}} = (\tilde{r}_1, \tilde{r}_2, \ldots, \tilde{r}_n)$ n component row vector and $\tilde{\boldsymbol{x}} = (x_1, x_2, \ldots, x_n)'$ n component column vector. The weighted possibilistic mean, variance, skewness, and kurtosis of fuzzy number are, respectively, given as*

$E = E(\tilde{\boldsymbol{r}}\boldsymbol{x}) = \frac{1}{6}\left[\sum_{i=1}^n a_i x_i + 2\left(\sum_{i=1}^n b_i x_i + \sum_{i=1}^n c_i x_i\right) + \sum_{i=1}^n d_i x_i\right]$

$V = V(\tilde{\boldsymbol{r}}\boldsymbol{x}) = \frac{1}{36}[2((\sum_{i=1}^n a_i x_i)^2 + (\sum_{i=1}^n d_i x_i)^2) + 5((\sum_{i=1}^n b_i x_i)^2 + (\sum_{i=1}^n c_i x_i)^2) + 2((\sum_{i=1}^n a_i x_i)(\sum_{i=1}^n b_i x_i) + (\sum_{i=1}^n c_i x_i)(\sum_{i=1}^n d_i x_i) - (\sum_{i=1}^n d_i x_i)(\sum_{i=1}^n a_i x_i)) - 4((\sum_{i=1}^n a_i x_i)(\sum_{i=1}^n c_i x_i) + (\sum_{i=1}^n b_i x_i)(\sum_{i=1}^n d_i x_i)) - 8(\sum_{i=1}^n b_i x_i)(\sum_{i=1}^n c_i x_i)]$

$S = S(\tilde{\boldsymbol{r}}\boldsymbol{x}) = \frac{1}{5}[19((\sum_{i=1}^n a_i x_i)^3 + (\sum_{i=1}^n d_i x_i)^3) + 26((\sum_{i=1}^n b_i x_i)^3 + (\sum_{i=1}^n c_i x_i)^3) - 15(\sum_{i=1}^n a_i x_i)(\sum_{i=1}^n d_i x_i)((\sum_{i=1}^n a_i x_i) + (\sum_{i=1}^n d_i x_i)) - 30(\sum_{i=1}^n b_i x_i)(\sum_{i=1}^n c_i x_i)((\sum_{i=1}^n b_i x_i) + (\sum_{i=1}^n c_i x_i)) + 60(\sum_{i=1}^n b_i x_i)(\sum_{i=1}^n c_i x_i)((\sum_{i=1}^n a_i x_i) + (\sum_{i=1}^n d_i x_i)) + 30(\sum_{i=1}^n a_i x_i)(\sum_{i=1}^n d_i x_i)((\sum_{i=1}^n b_i x_i) + (\sum_{i=1}^n c_i x_i)) - 12((\sum_{i=1}^n a_i x_i)^2(\sum_{i=1}^n b_i x_i) + (\sum_{i=1}^n c_i x_i)(\sum_{i=1}^n d_i x_i)^2) - 30((\sum_{i=1}^n a_i x_i)^2(\sum_{i=1}^n c_i x_i) + (\sum_{i=1}^n b_i x_i)(\sum_{i=1}^n d_i x_i)^2) - 33((\sum_{i=1}^n a_i x_i)(\sum_{i=1}^n b_i x_i)^2 + (\sum_{i=1}^n c_i x_i)^2(\sum_{i=1}^n d_i x_i)) - 15((\sum_{i=1}^n a_i x_i)^2(\sum_{i=1}^n c_i x_i) + (\sum_{i=1}^n b_i x_i)^2(\sum_{i=1}^n d_i x_i))] / [2((\sum_{i=1}^n a_i x_i)^2 + (\sum_{i=1}^n d_i x_i)^2) + 5((\sum_{i=1}^n b_i x_i)^2 + (\sum_{i=1}^n c_i x_i)^2) + 2((\sum_{i=1}^n a_i x_i)(\sum_{i=1}^n b_i x_i) + (\sum_{i=1}^n c_i x_i)(\sum_{i=1}^n d_i x_i) - (\sum_{i=1}^n d_i x_i)(\sum_{i=1}^n a_i x_i)) - 4((\sum_{i=1}^n a_i x_i)(\sum_{i=1}^n c_i x_i) + (\sum_{i=1}^n b_i x_i)(\sum_{i=1}^n d_i x_i)) - 8(\sum_{i=1}^n b_i x_i)(\sum_{i=1}^n c_i x_i)]^{\frac{3}{2}}$

$K = K(\tilde{\boldsymbol{r}}\boldsymbol{x}) = 1296[\frac{1}{72}((\sum_{i=1}^n b_i x_i) - (\sum_{i=1}^n a_i x_i))^2((\sum_{i=1}^n d_i x_i) - (\sum_{i=1}^n c_i x_i))^2 + \frac{3}{8}(\sum_{i=1}^n b_i x_i)^2(\sum_{i=1}^n c_i x_i)^2 - \frac{1}{6}(\sum_{i=1}^n b_i x_i)(\sum_{i=1}^n c_i x_i)[((\sum_{i=1}^n b_i x_i) - (\sum_{i=1}^n c_i x_i))^2 + ((\sum_{i=1}^n b_i x_i) - (\sum_{i=1}^n a_i x_i))((\sum_{i=1}^n d_i x_i) - (\sum_{i=1}^n c_i x_i)) + ((\sum_{i=1}^n d_i x_i) - (\sum_{i=1}^n c_i x_i))^2] - \frac{1}{4}(\sum_{i=1}^n b_i x_i)(\sum_{i=1}^n c_i x_i)[(\sum_{i=1}^n b_i x_i)^2 + (\sum_{i=1}^n c_i x_i)^2 - ((\sum_{i=1}^n b_i x_i) - (\sum_{i=1}^n c_i x_i))((\sum_{i=1}^n b_i x_i) - (\sum_{i=1}^n a_i x_i) + (\sum_{i=1}^n d_i x_i) - (\sum_{i=1}^n c_i x_i))] - \frac{1}{18}((\sum_{i=1}^n b_i x_i) - (\sum_{i=1}^n a_i x_i))((\sum_{i=1}^n b_i x_i) - (\sum_{i=1}^n c_i x_i))((\sum_{i=1}^n d_i x_i)$

$- (\sum_{i=1}^{n} c_i x_i))((\sum_{i=1}^{n} b_i x_i) - (\sum_{i=1}^{n} a_i x_i) + (\sum_{i=1}^{n} d_i x_i) - (\sum_{i=1}^{n} c_i x_i)) + \frac{5}{432}$ $[((\sum_{i=1}^{n} b_i x_i) - (\sum_{i=1}^{n} a_i x_i))^4 + ((\sum_{i=1}^{n} d_i x_i) - (\sum_{i=1}^{n} c_i x_i))^4] + \frac{1}{16}((\sum_{i=1}^{n} b_i x_i)^4 +$ $(\sum_{i=1}^{n} c_i x_i)^4) + \frac{1}{12}((\sum_{i=1}^{n} b_i x_i)^2 + (\sum_{i=1}^{n} c_i x_i)^2)[((\sum_{i=1}^{n} b_i x_i) - (\sum_{i=1}^{n} a_i x_i))^2 +$ $((\sum_{i=1}^{n} b_i x_i) - (\sum_{i=1}^{n} a_i x_i))((\sum_{i=1}^{n} d_i x_i) - (\sum_{i=1}^{n} c_i x_i)) + ((\sum_{i=1}^{n} d_i x_i) - (\sum_{i=1}^{n} c_i$ $x_i))^2] - \frac{1}{12}((\sum_{i=1}^{n} b_i x_i)^3 - (\sum_{i=1}^{n} c_i x_i)^3)((\sum_{i=1}^{n} b_i x_i) - (\sum_{i=1}^{n} a_i x_i) + (\sum_{i=1}^{n} d_i x_i)$ $- (\sum_{i=1}^{n} c_i x_i)) + \frac{2}{135}((\sum_{i=1}^{n} b_i x_i) - (\sum_{i=1}^{n} a_i x_i))((\sum_{i=1}^{n} d_i x_i) - (\sum_{i=1}^{n} c_i x_i))$ $[((\sum_{i=1}^{n} b_i x_i) - (\sum_{i=1}^{n} c_i x_i))^2 + (((\sum_{i=1}^{n} d_i x_i) - (\sum_{i=1}^{n} c_i x_i))^2]]/[2((\sum_{i=1}^{n} a_i x_i)^2 +$ $(\sum_{i=1}^{n} d_i x_i)^2) + 5((\sum_{i=1}^{n} b_i x_i)^2 + (\sum_{i=1}^{n} c_i x_i)^2) + 2((\sum_{i=1}^{n} a_i x_i)(\sum_{i=1}^{n} b_i x_i) +$ $(\sum_{i=1}^{n} c_i x_i)(\sum_{i=1}^{n} d_i x_i) - (\sum_{i=1}^{n} d_i x_i)(\sum_{i=1}^{n} a_i x_i)) - 4((\sum_{i=1}^{n} a_i x_i)(\sum_{i=1}^{n} c_i x_i) +$ $(\sum_{i=1}^{n} b_i x_i)(\sum_{i=1}^{n} d_i x_i)) - 8(\sum_{i=1}^{n} b_i x_i)(\sum_{i=1}^{n} c_i x_i)]^2.$

Proof We have

$$\tilde{\boldsymbol{r}}\boldsymbol{x} = \tilde{r}_1 x_1 + \tilde{r}_2 x_2 + \cdots + \tilde{r}_n x_n$$

$$= (\sum_{i=1}^{n} a_i x_i, \sum_{i=1}^{n} b_i x_i, \sum_{i=1}^{n} c_i x_i, \sum_{i=1}^{n} d_i x_i)$$

$$= (a, b, c, d) \text{ (say)},$$

where $a = \sum_{i=1}^{n} a_i x_i,\ b = \sum_{i=1}^{n} b_i x_i,\ c = \sum_{i=1}^{n} c_i x_i,\ d = \sum_{i=1}^{n} d_i x_i$

Hence, proof of the theorem immediately follows from Theorems 1 and 2.

3.1 Proposed Multi-Objective Optimization Model

In this section we have proposed a multi-objective optimization model consisting of four objectives, viz., maximization of return(E), minimization of risk (variance)(V), maximization of skewnwss (S), and minimization of kurtosis subject to the constraint that the sum of all portions of shares is equal to one.

$$\left.\begin{array}{l} Maximize\ \tilde{E}(\tilde{\boldsymbol{r}}\boldsymbol{x}) = \tilde{E}(\tilde{r}_1 x_1 + \tilde{r}_2 x_2 + \cdots + \tilde{r}_n x_n) \\ Minimize\ \tilde{V}(\tilde{\boldsymbol{r}}\boldsymbol{x}) = \tilde{V}(\tilde{r}_1 x_1 + \tilde{r}_2 x_2 + \cdots + \tilde{r}_n x_n) \\ Maximize\ \tilde{S}(\tilde{\boldsymbol{r}}\boldsymbol{x}) = \tilde{S}(\tilde{r}_1 x_1 + \tilde{r}_2 x_2 + \cdots + \tilde{r}_n x_n) \\ Minimize\ \tilde{K}(\tilde{\boldsymbol{r}}\boldsymbol{x}) = \tilde{K}(\tilde{r}_1 x_1 + \tilde{r}_2 x_2 + \cdots + \tilde{r}_n x_n) \\ x_1 + x_2 + \cdots + x_n = 1 \\ x_i \geq 0, i = 1, 2, \ldots, n. \end{array}\right\} \quad (1)$$

4 Solution Methodology: Zimmerman's Fuzzy Goal Programming

A general multi-objective nonlinear programming problem is of the following form:

$$\begin{aligned}&max/min[f_1(x), f_2(x), \ldots, f_K(x)]\\&subject\ to\\&x \in X = \{x : g_s(x)(\leq, =, \geq)0, s = 1, 2, \ldots, m\}\end{aligned} \quad (2)$$

where $f_a(x)$ are objective functions for maximization, $a \in A$ and $f_b(x)$ are objective functions for minimization $b \in B$, A, B being two exhaustive subsets of the index set 1, 2, ...K and x being the decision variable. It is noted that all functions $f_k(x)$ and $g_i(x)$ ($k = 1, 2, \ldots$ K and $i = 1, 2, \ldots$ m) can be linear or nonlinear. In the past two decades, many fuzzy programming techniques have been developed for solving multi-objective optimization problems. In this area, Zimmermann [13] first shows that fuzzy programming technique can be used satisfactorily to solve the multi-objective programming problem using maxmin operator of Bellman and Zadeh [14].

The steps of the fuzzy programming technique are as follows:

Step 1. Each objective function $f_k(x)$ of the MOP problem is optimized separately subject to the constraints of the problem. Let these optimum values be $f_k^*(x)(k = 1, 2, \ldots K)$.

Step 2. For each optimal solution of the K single-objective programming problem solved in step 1, find the value of the remaining objective functions and construct a payoff matrix of order $K \times K$ as given in Table 1.

Step 3. Evaluate
$f_k^L = Min\{f_k(x^1), f_k(x^2), \ldots, f_k(x^K)\}$ and $f_k^U = Max\{f_k(x^1), f_k(x^2), \ldots, f_k(x^K)\}$ for all $k = 1, 2, \ldots, K$.

Step 4. Form the membership functions $\mu_{f_i}(x),\ i \in A$ and $\mu_{f_j}(x),\ j \in B$, respectively, for the maximization objective functions $\mu_{f_i}(x),\ i \in A$ and minimization objective function $\mu_{f_j}(x),\ j \in B$, where $AUB = \{1, 2, \ldots, K\}$ in the linear form as follows.

Table 1 Payoff matrix

Solution	$f_1(x)$	$f_2(x)$	...	$f_K(x)$
(x^1)	$f_1^*(x^1)$	$f_2(x^1)$	...	$f_K(x^1)$
(x^2)	$f_1(x^2)$	$f_2^*(x^2)$	...	$f_K(x^2)$
...	...	...	...	...
(x^K)	$f_1(x^K)$	$f_2(x^K)$	...	$f_K^*(x^K)$

$$\mu_{f_i}(x) = \begin{cases} 1 & \text{if } f_i(x) > f_i^U \\ \frac{f_i(x) - f_i^L}{f_i^U - f_i^L} & \text{if } f_i^L \le f_i(x) \le f_i^U, \; for\ all\ i \in A \\ 0 & \text{if } f_i(x) < f_i^L \end{cases}$$

$$\mu_{f_j}(x) = \begin{cases} 1 & \text{if } f_i(x) > f_i^U \\ \frac{f_j^U - f_j(x)}{f_j^U - f_j^L} & \text{if } f_j^L \le f_j(x) \le f_j^U, \; for\ all\ j \in B \\ 0 & \text{if } f_j(x) < f_j^L \end{cases}$$

Step 5. Using the above membership functions formulate and solve the crisp nonlinear programming model following the methods due to Zimmermann [13].

4.1 Zimmermann's Model

If w_1, w_2, w_3 and w_4 are the intuitive crisp weights for the portfolio mean(E), variance(V), skewness(S) , and kurtosis(K), respectively, then for different models the problem (1) can be formulated as follows:

$$\begin{cases} Maximize\ \ \alpha \\ such\ that \\ w_1(\frac{E-E^U}{E^U-E^L}) = \alpha, \; w_2(\frac{V^U-V}{V^U-V^L}) = \alpha, \; w_3(\frac{S-S^U}{S^U-S^L}) = \alpha, \; w_4(\frac{K^U-K}{K^U-K^L}) = \alpha \\ x \in X \\ 0 \le \alpha \le 1, \;\; w_1 + w_2 + w_3 + w_4 = 1 \end{cases} \tag{3}$$

Zimmerman's fuzzy goal programming is a pre-emptive fuzzy goal programming method where the priorities of the goals are considered to be the same (e.g., α).

5 Case Study: Bombay Stock Exchange (BSE)

In this section we apply the proposed portfolio selection models on the data set extracted from the Bombay Stock Exchange (BSE). BSE is the oldest stock exchange in Asia with a rich heritage of over 133 years of existence. What is now popularly known as BSE was established as The Native Share & Stock Brokers' Association in 1875. It is the first stock exchange in India which obtained permanent recognition (in 1956) from the Government of India under the Securities Contracts (Regulation) Act (SCRA) 1956. With demutualization, the stock exchange has two of the world's prominent exchanges, Deutsche Borse and Singapore Exchange, as its strategic partners. Today, BSE is the world's number one exchange in terms of the number of listed companies and the world's fifth in handling of transactions through its electronic trading system. The companies listed on BSE command a total market capitalization of USD trillion 1.06 as of July 2009. The BSE Index, SENSEX, is India's first and most popular stock market benchmark index. Sensex is tracked worldwide. It constitutes 30 stocks representing 12 major sectors. It is constructed on a free-float methodology, and is sensitive to market movements and market realities. Apart from the SENSEX, BSE offers 23 indices, including 13 sectoral indices. We have taken monthly share price data for 60 months (March 2003–February 2008) of just five companies which are included in Bombay Stock Exchange (BSE) index. Though any finite number of stocks can be considered, we have taken only five stocks to reduce the complexity of representation.

5.1 Example

Let us consider the following multi-objective portfolio selection problem.

$$\left.\begin{array}{l} Maximize\ \tilde{E}(\tilde{\boldsymbol{r}}\boldsymbol{x}) = \tilde{E}(\tilde{r}_1x_1 + \tilde{r}_2x_2 + \tilde{r}_3x_3 + \tilde{r}_4x_4 + \tilde{r}_5x_5) \\ Minimize\ \tilde{V}(\tilde{\boldsymbol{r}}\boldsymbol{x}) = \tilde{V}(\tilde{r}_1x_1 + \tilde{r}_2x_2 + \tilde{r}_3x_3 + \tilde{r}_4x_4 + \tilde{r}_5x_5) \\ Maximize\ \tilde{S}(\tilde{\boldsymbol{r}}\boldsymbol{x}) = \tilde{S}(\tilde{r}_1x_1 + \tilde{r}_2x_2 + \tilde{r}_3x_3 + \tilde{r}_4x_4 + \tilde{r}_5x_5) \\ Minimize\ \tilde{K}(\tilde{\boldsymbol{r}}\boldsymbol{x}) = \tilde{K}(\tilde{r}_1x_1 + \tilde{r}_2x_2 + \tilde{r}_3x_3 + \tilde{r}_4x_4 + \tilde{r}_5x_5) \\ x_1 + x_2 + x_3 + x_4 + x_5 = 1 \\ x_i \geq 0,\ i = 1, 2, 3, 4, 5. \end{array}\right\} \quad (4)$$

The Zimmermann's model for multi-objective decision-making(as described in Sect. 4 is used to solve the example. Using Theorem 3 and using the data given in Table 2, we can find the payoff matrix (Table 3).

Table 2 Fuzzy returns of stocks listed at Bombay Stock Exchange (BSE)

Company	Variables	Fuzzy return in trapezoidal form
Reliance energy	REL (x_1)	(−0.008, 0.0223, 0.0501, 0.0673)
Larsen & Tubro	LT (x_2)	(−0.0031, 0.0287, 0.0611, 0.0866)
BHEL	BHEL (x_3)	(−0.0020, 0.0282, 0.0581, 0.0832)
Tata steel	TISCO (x_4)	(0.0086, 0.0296, 0.0410, 0.0525)
State Bank of India	SBI (x_5)	(−0.100, 0.0217, 0.0576, 0.0789)

Table 3 Payoff matrix on the basis of data in Table 2

Solution	Return(E)	Variance(V)	Skewness	Kurtosis(K)
(0.00, 0.40, 0.37, 0.23, 0.00)	0.04094	0.000512	0.00000592	1.1595
(0.37, 0.00, 0.23, 0.40, 0.00)	0.03577	0.000354	0.00000000	1.1666
(0.00, 0.37, 0.40, 0.23, 0.00)	0.04090	0.000511	0.00000592	1.1430
(0.00, 0.20, 0.40, 0.40, 0.00)	0.03920	0.000407	0.00040700	1.1400

Applying Zimmermann's method we obtain the following solution:

$$\mathrm{E} = 0.039377, \mathrm{V} = 0.0004147, \mathrm{S} = 0.000003345, \mathrm{K} = 1.142835$$
$$x_1 = 0.00, x_2 = 0.33, x_3 = 0.27, x_4 = 0.40, x_5 = 0.00.$$

6 Conclusion

In this paper, we have used the fuzzy possibilistic measure of kurtosis to model a new possibilistic mean–variance–skewness–kurtosis stock portfolio selection model. Zimmerman's fuzzy goal programming method is applied to convert the tetra-objective programming problem into a single-objective programming problem. Data of 60 months from BSE of five stocks are used for testing the effectiveness of the proposed model. In the future we will apply the model on a larger data set. We will also compare the proposed model with other established models of portfolio selection problem. For simulation, genetic algorithm or ant colony optimization algorithm can be used.

References

1. Samuelson, P.: The fundamental approximation theorem of portfolio analysis in terms of means, variances an higher moments. Rev. Econ. Stud. **25**, 65–86 (1958)
2. Bhattacharyya, R., Kar, M.B., Kar, S., Dutta Majumder, D.: Mean- entropy—skewness fuzzy portfolio selection by credibility theory approach. Chaudhury, S. et al. (Eds.) In: PReMI LNCS 5909, pp. 603–608 (2009)
3. Bhattacharyya, R., Kar, S., Dutta Majumder, D.: Fuzzy mean-variance-skewness portfolio selection models by interval analysis. Comput. Math. Appl. **61**, 126–137 (2011)
4. Bhattacharyya, R., Kar, S.: Possibilistic mean-variance-skewness portfolio selection models. Int. J. Oper. Res. **8**, 44–56 (2011)
5. Bhattacharyya, R., Kar, S.: Multi-objective fuzzy optimization for portfolio selection: an embedding theorem approach. Turk. J. Fuzzy Syst. **2**, 14–35 (2011)

6. Markowitz, H.: Portfolio selection. J. Financ. **7**, 77–91 (1952)
7. Davies, R., Kat, H., Lu S.: Fund of hedge funds portfolio selection: a multiple objective approach. Working Paper, ISMA Center (2004)
8. Berenyi, Z.: Performance of leveraged assets funds., Working Paper University of Mnich (2001)
9. Berenyi, Z.: Measuring hedges funds, risk with moment based variance-equivalent measures. In: Adcock- Jurczenko-Maillet (ed.) Multi Moment Asset Pricing Models and Related Topic, Springer, New York (2005)
10. Huang, X.: Mean-semivariance models for fuzzy portfolio selection. J. Comput. Appl. Math. **217**(1), 1–8 (2008)
11. Li, X., Qin, Z., Kar, S.: Mean-variance-skewness model for portfolio selection with fuzzy returns. Eur. J. Oper. Res. **202**(1), 239–247 (2010)
12. Louis, A.F., Jules, S.K., Christian, D.T.: Kurtosis and Semi-kurtosis for Portfolios Selection with Fuzzy Returns. In: Int. Statistical Inst. Proceedings of the 58th World Statistical Congress, Dublin (Session CPS027) (2011)
13. Zimmermann, H.J.: Fuzzy Linear Programming with several objective functions. Fuzzy Sets Syst. **1**, 46–55 (1978)
14. Bellman, R.E., Zadeh, L.A.: Decision-making in a fuzzy environment. Manag. Sci. **17**(4), 141–164 (1970)

Conflicting Bifuzzy Preference Relations Based Method for Multi Criteria Decision Making Problems

Deepa Joshi and Sanjay Kumar

Abstract In this paper, we present conflicting bifuzzy preference relations-based method for multi-criteria decision-making (MCDM) problem. In proposed method linear programming model is used to obtain optimal weights of criteria by utilizing criteria weights and score function. The best alternative is selected in accordance with the value of weighting function. In order to examine the impact of criteria weights on final ranking sensitivity analysis is performed. A numerical example is also given to clarify the developed approach and to demonstrate its effectiveness.

Keywords MCDM · Conflicting bifuzzy preference relations · Linear programming · Score function · Sensitivity analysis

1 Introduction

MCDM problem is the process of finding the best alternative from all the feasible alternatives after qualitative or quantitative assessment of a finite set of interdependent or independent criteria. Desirable alternative can be chosen by providing preference information in terms of exact numerical value but due to lack of knowledge about background and conflicting nature of alternatives, it is difficult for decision-maker to use the exact values to express their preference information about alternatives or criteria. To deal with such cases fuzzy preference relations [3] are used. These relation is based on fuzzy set proposed by Zadeh [9]. To provide the degree that an alternative not priority to another, Szmidt and Kacprzyk [6] generalized the fuzzy preference relation to intuitionistic fuzzy preference relations and Xu [8] also introduced the concept of intuitionistic fuzzy preference relation based on intuitionistic

D. Joshi (✉) · S. Kumar
Department of Mathematics, Statistics & Computer Science, G. B. Pant University of Agriculture & Technology, Pantnagar 263145, Uttarakhand, India
e-mail: deepajoshi.6nov@gmail.com

S. Kumar
e-mail: skruhela@hotmail.com

M.K. Chakraborty et al. (eds.), *Facets of Uncertainties and Applications*, Springer Proceedings in Mathematics & Statistics 125,
DOI 10.1007/978-81-322-2301-6_24

fuzzy set proposed by Atanassov [1]. Intuitionistic fuzzy sets are characterized by two functions, the membership function (μ_A) and the nonmembership function (ν_A) with condition $0 \leq \mu_A(x) + \nu_A(x) \leq 1$. Sometimes, this inequality cannot handle the condition when sum of membership and nonmembership is greater than one, i.e., if performance of a candidate is 'good' is 0.8 in reality, it does not mean that 'poor' performance is always 0.2 but it can be more than 0.2. Based on these arguments Zamali et al. [10] proposed a new concept conflicting bifuzzy set (CBFS), in this set the membership and nonmembership are not compliments to each other, their sum can also be greater than one but cannot be more than two. So we can say that conflicting bifuzzy sets (CBFS) are an extension of intuitionistic fuzzy set and preference relations based on these sets are proposed by Naim et al. [5]. By conflicting bifuzzy preference relations both positive and negative aspects will be considered simultaneously in judgment process, while previous only takes into account the positive aspect without considering negative aspect.

In this paper, we propose a method for solving multi-criteria decision making (MCDM) problem based on conflicting bifuzzy preference relations. We have also presented a linear programming model to obtain optimal weights with help of score function and for final ranking weighting function is used. Sensitivity analysis is also done to see the impact of change of weights on final ranking. The proposed method has been implemented on decision-making problem to determine the best company for investment of money based on some independent criteria. Final ranking is compared for different set of criteria weights for testing the validity of final ranking results.

2 Preliminaries

In this section, basic definitions of fuzzy set by Zadeh [9], intuitionistic fuzzy set as generalization of fuzzy set by Atanassov [1], conflicting bifuzzy set by Zamali et al. [10], and their properties are presented. We also described different type of preference relations as follows:

Definition 1 A fuzzy set A in the Universe of discourse X is characterized by membership function $\mu_A : X \to [0, 1]$. A fuzzy set A is represented by following ordered pair:

$$A = \{(x, \mu_A(x)) : \forall x \in X\} \tag{1}$$

where μ_A is the grade of membership of element x in the set A.

Definition 2 An intuitionistic fuzzy set I on a universe X is defined as an object of the following form:

$$I = \{\langle x, \mu_I(x), \nu_I(x)\rangle : \forall x \in X\} \tag{2}$$

where the functions $\mu_I(x) : X \to [0, 1]$ and $\nu_I(x) : X \to [0, 1]$ represent the degree of membership and degree of nonmembership of an element $x \in I \subset X$

respectively. $\pi_I(x) = 1 - \mu_I(x) - \upsilon_I(x)$ is called degree of hesitation with the condition $0 \leq \mu_I(x) + \upsilon_I(x) \leq 1$.

Definition 3 Let a set *X* be fixed set. A conflicting bifuzzy set *A* of *X* is the object having following form:

$$A = \{\langle x, \mu_A(x), \nu_A(x)\rangle : \forall x \in X\} \tag{3}$$

where the functions $\mu_A(x)$ and $\nu_A(x)$ satisfy the inequality $0 \leq \mu_A(x) \leq 1$ and $0 \leq \nu_A(x) \leq 1$, $\mu_A(x)$ and $\nu_A(x)$ represent positive degree and negative degree of x in *A* respectively. Both $\mu_A(x)$ and $\nu_A(x)$ are some numerical values in the unit interval [0,1]. As $\mu_A(x)$ and $\nu_A(x)$ can take any value in [0, 1] therefore *A* should satisfy following inequality $0 < \mu_I(x) + \upsilon_I(x) < 2$.

Conflicting bifuzzy sets can only be considered in certain cases when it is out of intuitionistic condition.

Definition 4 A fuzzy preference relation *R* on the set *X* is represented by a complementary matrix $= (r_{ij})_{n\times n} \subset X \times X$ for all $i, j = 1, 2, \ldots, n$.

Definition 5 An intuitionistic preference relation *B* on a set *X* is represented by a matrix $B = (b_{ij})_{n\times n} \subset X \times X$ with $b_{ij} = \big\langle (x_i, x_j), \mu(x_i, x_j), \nu(x_i, x_j)\big\rangle$ for all $i, j = 1, 2, \ldots, n$. For convenience, let $b_{ij} = (\mu_{ij}, \nu_{ij})$, for all $i, j = 1, 2, \ldots, n$ where b_{ij} is an intuitionistic fuzzy value, composed by the certain degree μ_{ij} to which x_i is preferred to x_j and certain degree ν_{ij} to which x_i is non-preferred to x_i, and $\pi_{ij}(x) = 1 - \mu_{ij}(x) - \nu_{ij}(x)$ is interpreted as the uncertainty degree to which x_i is preferred to x_j and $\pi_{ij}(x) = 1 - \mu_{ij}(x) - \nu_{ij}(x)$ is interpreted as the uncertainty degree to which x_i is preferred to x_j.

Definition 6 Let $A = \{a_1, a_2, a_3, \ldots, a_n\}$ be a finite set of alternatives and $B = \{b_1, b_2, b_3, \ldots, b_n\}$ the set of decision makers. *X* is a matrix of conflicting bifuzzy preference relation (Naim et al. [5]) represented by $X = (x_{ij})_{n\times n} \subset A \times A$ for all $x_{ij} = \big\langle (a_i, a_j), \mu(a_i, a_j), \nu(a_i, a_j)\big\rangle$ for all $i, j = 1, 2, \ldots, n$ where x_{ij} is a conflicting bifuzzy value, composed by the certainty degree μ_{ij} to which a_i is positively preferred to a_j and certainty degree ν_{ij} to which a_i is negatively preferred to a_j, and $0 < \mu_A(a) + \nu_A(a) < 2$.

In general, a conflicting bifuzzy preference relation *P* is a bifuzzy subset of $A \times A$ which characterized by the following membership function (Naim et al. [5]):

$$\mu_{ij}(A_i, A_j) = \begin{cases} 1, & \text{if } A_i \text{ is positive definitely preferred to } A_j \\ c \in (0.5, 1), & \text{if } A_i \text{ is povitive slightly preferred to } A_j \\ 0.5, & \text{if there is no preference (indifference)} \\ d \in (0.5, 1), & \text{if } A_j \text{ is positive slightly preferred to } A_i \\ 0, & \text{if } A_j \text{ is positive definitely preferred to } A_i \end{cases}$$

And

$$\nu_{ij}(A_i, A_j) = \begin{cases} 1, & \text{if } A_i \text{ is negative definitely preferred to } A_j \\ c \in (0.5, 1), & \text{if } A_i \text{ is negative slightly preferred to } A_j \\ 0.5, & \text{if there is no preference (indifference)} \\ d \in (0.5, 1), & \text{if } A_j \text{ is negative slightly preferred to } A_i \\ 0, & \text{if } A_j \text{ is negative definitely preferred to } A_i \end{cases}$$

2.1 Score Function

Wang et al. [7] introduced Score function in fuzzy multi-criteria decision-making in order to do selection and ranking. The greater the value of score functions the higher the degree of appropriateness that alternative satisfies some criteria. Hong and Choi [2] modified earlier score function formula to take into account the unknown part while Wang et al. [7] consider only true and false function. Modified score function by Hong and Choi [2] is given as follows:

Let $x_{ij} = (\mu_{ij}, \nu_{ij})$ be a conflicting bifuzzy preference value. For $\mu_{ij}, \nu_{ij} \in [0, 1]$, $\mu_{ij} + \nu_{ij} < 2$. The score function of x_{ij} can be evaluated by the score function S defined as,

$$S\left(x_{ij}\right) = \mu_{ij} - \nu_{ij} - \frac{1 - \mu_{ij} - \nu_{ij}}{2}$$

$$= \frac{3\mu_{ij} - \nu_{ij} - 1}{2} \tag{4}$$

where $S(x_{ij}) \in [0, 1]$.

The reason for taking modified formula of score function is given as follows:

If we have two conflicting bifuzzy preference values (0.7, 0.5) and (0.6, 0.4), according to the score function given by Hong and Choi [2] score function of both values is 0.2, so we cannot decide which value is greater. By using formula (4) of score function we can compare the incomparable conflicting bifuzzy preference values.

3 Proposed Method

Proposed method uses following linear programming model to calculate optimal weights of criteria.

$$\sum\nolimits_{i=1}^{m} \{S(x_{ij})^* w_j + S(x_{ik})^* w_k \ldots + S(x_{ip})^* w_p\}$$

$$\text{s.t.}\ w_j^l \le w_j \le w_j^u,$$
$$w_k^l \le w_k \le w_k^u,$$
$$\dots\dots\dots\dots\dots\dots\dots\dots$$
$$w_p^l \le w_p \le w_p^u,$$

$$w_j + w_k + \cdots + w_p = 1 \tag{5}$$

where * represents multiplication and $w_j, w_k, \ldots, w_p$ are respectively weights of the criteria $c_j, c_{k\ldots,} c_p$ for $i,\ j = 1, 2 \ldots n,\ k = 2, 3 \ldots n,\ p = 3, 4 \ldots n.$

We can solve Eq. (5) by Simplex method.

Then, the degree of suitability to which alternative A_i satisfies the decision-maker's requirement can be measured by the weighting function W:

$$W(A_i) = \sum_{i=1}^{m} \{S(x_{ij})^* w_j + S(x_{ik})^* w_k + S(x_{ip})^* w_p\} \tag{6}$$

where $W(A_i) \in [0, 1]$.

Step 1: Construct conflicting bifuzzy relations matrix.

$$X^{(k)} = (x_{ij}^{(k)})_{n \times n}$$

where $i, j, k = 1, 2, \ldots n$.

Step 2: We propose following aggregation operator for tuplewise aggregation.

$$X = \frac{1 \cdot x_{11}^{(1)} + 2 \cdot x_{11}^{(2)} + 3 \cdot x_{11}^{(3)} + \ldots n \cdot x_{11}^{(n)}}{\frac{n(n+1)}{2}} \tag{7}$$

where n is the number of decision-makers.

Step 3: Compute score function for each entry of decision matrix.

Step 4: Determine optimal weights of each criteria using following linear programming:

$$\sum_{i=1}^{m} \{S(x_{ij})^* w_j + S(x_{ik})^* w_k + S(x_{ip})^* w_p\}$$

$$\text{s.t.}\ w_j^l \le w_j \le w_j^u,$$
$$w_k^l \le w_k \le w_k^u,$$
$$\dots\dots\dots\dots\dots\dots\dots\dots$$
$$w_p^l \le w_p \le w_p^u,$$

$$w_j + w_k + \cdots + w_p = 1$$

Step 5: Calculate weight function of each alternative using (6).
Step 6: Finally, rank the alternatives according to descending order of weighting function.

4 Implementation of Proposed Method

Suppose an investment company wants to invest a sum of money in best option. Four alternatives in which to invest money are as follows:

(1) Car company, (2) Food company, (3) Computer company, (4) TV company

These four alternatives are assessed for their performance on the basis of following four criteria:

C_1 = risk analysis
C_2 = growth analysis
C_3 = social-political impact analysis
C_4 = environmental impact analysis

D_1, D_2, D_3 are three decision-makers.

Step 1: Conflicting bifuzzy preference relation decision matrices according to three decision-makers are as follows (Table 1):
Step 2: Aggregating tuplewise each entry of decision matrices by using (7).
Taking no. of decision makers $(n) = 3$.

$$X = \frac{1 \cdot x_{11}^{(1)} + 2 \cdot x_{11}^{(2)} + 3 \cdot x_{11}^{(3)}}{6}$$

Table 1 Conflicting bifuzzy preference relation decision matrices

		C_1	C_2	C_3	C_4
$X^{(1)}$	A_1	(0.5, 0.5)	(0.7, 0.4)	(0.6, 0.5)	(0.2, 0.8)
	A_2	(0.9, 0.2)	(0.8, 0.3)	(0.9, 0.3)	(0.6, 0.5)
	A_3	(0.7, 0.3)	(0.9, 0.2)	(0.8, 0.2)	(0.7, 0.3)
	A_4	(0.8, 0.3)	(0.6, 0.5)	(0.5, 0.5)	(0.9, 0.2)
$X^{(2)}$	A_1	(0.7, 0.3)	(0.8, 0.2)	(0.7, 0.3)	(0.5, 0.5)
	A_2	(0.8, 0.2)	(0.5, 0.5)	(0.9, 0.2)	(0.7, 0.3)
	A_3	(0.9, 0.1)	(0.6, 0.5)	(0.6, 0.5)	(0.8, 0.2)
	A_4	(0.7, 0.4)	(0.9, 0.3)	(0.8, 0.2)	(0.6, 0.2)
$X^{(3)}$	A_1	(0.6, 0.4)	(0.9, 0.3)	(0.8, 0.2)	(0.9, 0.1)
	A_2	(0.7, 0.3)	(0.8, 0.2)	(0.6, 0.5)	(0.5, 0.5)
	A_3	(0.9, 0.2)	(0.6, 0.5)	(0.7, 0.3)	(0.9, 0.2)
	A_4	(0.7, 0.1)	(0.5, 0.5)	(0.6, 0.4)	(0.8, 0.3)

Where $X^{(1)}$, $X^{(2)}$, $X^{(3)}$ are decision matrices according to decision-makers D_1, D_2 and D_3, respectively

Table 2 Decision matrix by aggregating tuplewise

		C_1	C_2	C_3	C_4
X	A_1	(0.617, 0.383)	(0.83, 0.283)	(0.73, 0.283)	(0.25, 0.75)
	A_2	(0.77, 0.25)	(0.7, 0.317)	(0.75, 0.37)	(0.583, 0.43)
	A_3	(0.87, 0.183)	(0.65, 0.45)	(0.683, 0.35)	(0.83, 0.2)
	A_4	(0.717, 0.23)	(0.65, 0.43)	(0.65, 0.35)	(0.75, 0.25)

So decision-maker's weights are (1/6, 2/6, 3/6).
Decision matrix by using proposed aggregation operator is as follows given in Table 2:
Step 3: Score function of each entry is given as follows:

$$S(x_{11}) = 0.234,\ S(x_{12}) = 0.4605,\ S(x_{13}) = 0.4535,\ S(x_{14}) = 0.50$$
$$S(x_{21}) = 0.53,\ S(x_{22}) = 0.3915,\ S(x_{23}) = 0.44,\ S(x_{24}) = 0.1595$$
$$S(x_{31}) = 0.7135,\ S(x_{32}) = 0.25,\ S(x_{33}) = 0.3495,\ S(x_{34}) = 0.645$$
$$S(x_{41}) = 0.4605, S(x_{42}) = 0.26,\ S(x_{43}) = 0.30,\ S(x_{44}) = 0.50$$

Step 4: Optimal weights of criteria are computed by following linear programming:

$$\begin{aligned} &1.992^* w_1 + 1.505^* w_2 + 1.543^* w_3 + 0.8045^* w_4 \\ &\text{s.t. } 0.10 \le w_1 \le 0.20 \\ &\quad 0.15 \le w_2 \le 0.25 \\ &\quad 0.20 \le w_3 \le 0.30 \\ &\quad 0.25 \le w_4 \le 0.35 \\ &\quad w_1 + w_2 + w_3 + w_4 = 1 \end{aligned}$$

Using simplex method to solve the above linear programming, its optimal solution can be obtained as:
$w_1 = 0.20,\ w_2 = 0.25,\ w_3 = 0.30,\ w_4 = 0.25$
Step 5: By applying equation (6) we get
$W(A_1) = 0.208,\ W(A_2 = 0.338,\ W(A_3) = 0.464,\ W(A_4) = 0.092$
Step 6: Finally rank the alternatives according to descending order of weighting function as follows:
$A_3 > A_2 > A_1 > A_4$.

5 Sensitivity Analysis

Sensitivity analysis determines how stable ranking is against the sudden changes in the weights of criteria or inputs. In the present study sensitivity analysis is performed by examining the impact of sudden change in criteria weights on the final ranking. We change set of criteria weights four times in this analysis and solve the above linear programming model by Simplex method, we get sets of final rankings as follows:

Weights of criteria	Rank
$W_1 = 0.20$ $W_2 = 0.25$ $W_3 = 0.30$ $W_4 = 0.25$	$A_3 > A_2 > A_1 > A_4$
$W_1 = 0.40$ $W_2 = 0.30$ $W_3 = 0.25$ $W_4 = 0.05$	$A_3 > A_2 > A_1 > A_4$
$W_1 = 0.30$ $W_2 = 0.40$ $W_3 = 0.10$ $W_4 = 0.20$	$A_3 > A_2 > A_4 > A_1$
$W_1 = 0.10$ $W_2 = 0.25$ $W_3 = 0.40$ $W_4 = 0.25$	$A_3 > A_4 > A_2 > A_1$

In above analysis alternative A_3 had the highest ranking.

6 Conclusions

In this paper, we review the limitations of intuitionistic fuzzy sets and discuss the importance of conflicting bifuzzy sets and conflicting bifuzzy preference relations. We propose an aggregation operator to aggregate decision matrices, it is suitable for all types of sets, i.e., fuzzy sets, intuitionistic fuzzy sets, bifuzzy sets. To obtain optimal criteria weights linear programming model and score function is used and to select the best alternative weighting function is used. Finally, we apply proposed approach in decision-making problem for taking investment decisions. To verify the validity of ranking results, sensitivity analysis is done by changing criteria weights four times. Thus the proposed approach is proven accurate, suitable and recommended to be used by decision-makers in MCDM problems.

References

1. Atanasov, K.: Intuitionistic fuzzy sets. Fuzzy Sets Syst. **110**, 87–96 (1986)
2. Hong, D.J., Choi, C.H.: Multicriteria fuzzy decision making problems based on vague set theory. Fuzzy Sets Syst. **114**, 103–113 (2000)
3. Herrera-Viedma, E., Chiclana, F., Herrera, F., Alonso, S.: Group decision making model with incomplete fuzzy preference relations based on additive consistency. IEEE Trans. Syst. Man Cybern.: Part B **37**(1), 176–179 (2007)
4. Hwang, L.C., Yoon, K.: Multiple Attribute Decision Making: Methods and Applications. Springer, New York (1981)
5. Naim, N.S.M., Abdullah, M.L., Wahab, A.F., Che Taib, C.M.I.: A Multi-Criteria decision making method based on intuitionistic fuzzy sets. In: Proceeding of the 5th Asian Conference, Malaysia, pp. 709–716 (2009)

6. Szmidt, E., Kacprzyk, J.: Group Decision Making under intuitionistic fuzzy preference relations. In: Proceeding of 7th IPMU Conference, Paris, pp. 172–178 (1998)
7. Wang, J., Zhang, J., Liu, S.Y.: A new score function for fuzzy MCDM based on vague set theory. Int. J. Comput. Cogn. **4**, 44–48 (2006)
8. Xu, Z.S.: Intuitionistic preference relations and their applications in group decision making. Inf. Sci. **177**, 2363–2379 (2007)
9. Zadeh, L.A.: Fuzzy sets. Inf. Control **8**, 338–353 (1965)
10. Zamali, T., Lazim Abdullah, M., Abu Ossman, M.T.: An introduction to conflicting bifuzzyset theory. Int. J. Math. Stat. **3**(8), 86–95. (ISSN 0973–8347)

A Linear Goal Programming Method for Solving Chance-Constrained Multiobjective Problems with Interval Data Uncertainty

Mousumi Kumar and Bijay Baran Pal

Abstract This paper presents a goal programming (GP) method for modeling and solving multiobjective decision-making problems having interval parameter sets and a set of chance constraints in uncertain environments. In the proposed approach, planned interval goals defined for the objective goals are converted into standard linear goals in GP by using interval arithmetic technique and introducing under- and over-deviational variables to each of them. The chance constraints are also converted into deterministic equivalents and *Taylor series approximation* technique is used to transform the defined quadratic constraints into linear form to solve the problem effectively by employing linear GP method. Then, from the optimistic point of view of decision-maker, the framework of interval-valued GP is addressed to design goal achievement function for minimizing possible deviations concerned with achievement of goals within their target intervals specified in the decision situation. The approach is illustrated by a numerical example.

Keywords Chance-constrained programming · Fractional programming · Goal programming · Interval programming · Taylor series approximation

1 Introduction

Multiobjective decision-making (MODM) is an area of multiple criteria decision making [1] that is concerned with mathematical optimization problems involving more than one objective function to be optimized simultaneously. Actually, most of the human-centered decision-making problems are with multiplicity of objectives. It is worthy to mention here that the rapid rise in human civilization owing to

M. Kumar
Department of Mathematics, Alipurduar College, Alipurduar, 736122, West Bengal, India
e-mail: mousumi886@gmail.com

B.B. Pal (✉)
Department of Mathematics, University of Kalyani, Kalyani, 741235, West Bengal, India
e-mail: bbpal18@hotmail.com

M.K. Chakraborty et al. (eds.), *Facets of Uncertainties and Applications*,
Springer Proceedings in Mathematics & Statistics 125,
DOI 10.1007/978-81-322-2301-6_25

technological mechanization and cultural revolution have led to enormous increase in the volume of various types of objectives in the premises of MODM in the recent years. Here, in a practical decision situation, it is found that objectives are incommensurable in nature and often conflict each other concerning optimization of them in a decision environment.

When decision-making is emphasized, a human decision-maker (DM) plays an important role and the objective of solving a MODM problem is of finding the most preferred solution according to his/her subjective preferences [2, 3]. There are different underling philosophies regarding multiobjective optimization methods [4] in the literature of mathematical programming.

In the area of MODM, the GP approach [5], which is based on the satisficing (coined by the nobel laureate H.A. Simon) philosophy [6], has appeared as a goal-oriented method and robust tool for multiobjective decision analysis in crisp decision environment. Here, in case of using the conventional GP, it may be mentioned that aspiration levels of goals are assigned by DM and that depend on needs and desires in the decision-making horizon. Although, GP is a widely used method for solving problems with multiplicity of objectives, decision errors are directly involved therein most of the times to solve practical decision problems owing to vague in nature of human knowledge about exact values of physical and technical parameters. In such a situation, if crisp values are set as target levels, solution achievement may not comply with the real situation which is actually dictated by the nature of problem as well as the environment of making decision.

To cope with the above situation, the stochastic programming (SP) approach [7], initially introduced by Charnes and Cooper [8] as chance-constrained programming (CCP), that deals with probabilistically defined uncertain data [9] to modeling MODM problems have been studied deeply [10] in the past. In SP method, it is assumed that the probability distributions of random parameters of the problem are known. But, it is not always easy for DM to specify accurate probability distribution [11] to establish an approximate analytical model for measuring performances against simultaneous optimization of objectives of MODM problems owing to unpredictable nature of objectives in uncertain environment.

Again, to deal with the imprecise characteristics of model parameters, fuzzy goal programming (FGP) [12] as an extension of fuzzy programming (FP) [13, 14] in the framework of conventional GP, which is based on the notion of fuzzy sets initially proposed by Zadeh [15], has been implemented to different real-world MODM problems [16–18]. However, to employ such an approach, the membership functions of fuzzily described objectives need be known. But, in some practical decision situations, it becomes difficult to specify an appropriate membership function of fuzzy goals owing to vagueness and nonlinear nature of objectives in decision environment [19].

To overcome the above difficulties, interval programming [20] has appeared as a promising tool to solve MODM problems in inexact environment. In interval programming method, the uncertain model parameters are represented as interval numbers, where an interval defines a set of numbers on a real line enclosed by certain lower and upper bounds in the notion of interval arithmetic [21]. Actually, the defined intervals of parameter sets are regarded as regions within which the parameters possibly take their respective values in inexact environment. Here, the major advantage of the approach is that interval information can be directly communicated into the optimization process and resulting solution [22] without any distributional information that is always required in FP and SP methods. Eventually, the proposed method involves a relatively low computational requirement and leads to more reliable solutions to optimization problems.

Historically, the seeds of interval programming can be traced in the third century BC, when Archimedes (the father of Mathematics) calculated the value of π (pi) with certain lower- and upper bounds ($223/71 < \pi < 22/7$) on it. The interval arithmetic rule was first formalized in a book prepared by Moore [23]. The study on interval analysis and prominent role of interval arithmetic to global optimization has been well documented by Hansen [24].

Now, the GP formulation of a MODM problem with interval parameter sets, called interval-valued GP (IVGP), has been introduced by Inuiguchi and Kume [25]. The methodological aspects of IVGP approach studied in the past have been surveyed by Olivera and Antunes [26].

The IVGP approach to mobile robot path planning [27] and portfolio selection [28] has been studied in the past. The potential use of IVGP to different real-life MODM problems has also been demonstrated by Inuiguchi and Mizoshita [29], Pal et al. [30–32] in the recent past. A genetic algorithm (GA) [33] based IVGP method to patrol manpower allocation problem has also been proposed Pal et al. [34] in the past. Here, it may be mentioned that the success of a GA scheme highly depends on proper selections of reproduction operators and initialization of population [35], and implementation of it creates solution error and frequently trapped into local optima owing to incommensurability and conflicting in nature of objectives as well as their hidden nonlinearities as arises in conventional GP formulation [36] of real-life problems. Furthermore, in some real-life problems, objective functions with fractional criteria are found frequently involved in the decision situations. For instance, cost-benefit analysis in agricultural planning [30], faculty, and other staff allocation problems [32] for minimizing certain ratios of students' enrolments and staff structure within academic units of educational institutions, and others. Here, it may be mentioned that decision trouble frequently arises with fractional criteria [37] in course of solving problems. The IVGP approach with fractional objectives has been studied by Pal et al. [38] and implemented to real-world problems [32, 39] in the recent past. However, deep study on methodologies of IVGP that concerned with solving practical problems is at an early stage.

Now, in practical decision situations, it is to be realized to the fact that both the interval and probabilistic data are frequently involved in MODM problems and both the aspects of interval programming and SP would have to be taken into account for modeling and solving real-life problems. For example, in an agricultural planning problem [40], optimization of production of various seasonal crops involves interval data, whereas seasonal rainfall as a main productive resource is inherently probabilistic in nature in the decision environment.

In this paper, a MODM problem with multiplicity of linear as well as fractional objectives having interval coefficients under a system of chance constraints with normally distributed parameter sets is taken into consideration. In the proposed approach, unlike the consideration of crisp target values of objective goals in conventional GP, target intervals for objectives within which they possibly take their values in an uncertain situation are taken into account. In the sequel of model formulation, chance constraints are converted into their equivalent crisp system constraints by using the notion of means and variances of probabilistically defined random parameters. The nonlinear crisp constraints are then transformed into linear forms by using *Taylor series approximation* technique [41] to avoid any computational complexity with nonlinear system in the solution search process. Again, to formulate the standard GP model, the objectives with planned interval goals of the defined target intervals are transformed into conventional form of linear goals by employing interval arithmetic rule [23] and introducing under- and over-deviational variables to each of them. In the goal achievement function (regret function), both the *minsum* GP [4] and *minmax* GP [42] modeling aspects in GP are taken into account to minimize regrets toward achieving goals values within the specified target intervals in the decision environment. To illustrate the potential use of the approach a numerical example is solved and the model solution is compared with an approach [25] studied previously.

Now, the IVGP formulation of a chance-constrained MODM problem is presented in the Sect. 2.

2 Chance-Constrained IVGP Formultion

The general structure of a chance-constrained interval programming problem with linear and factional can be presented as [43]:

Find $\boldsymbol{X}(x_1, x_2, \ldots, x_n)$ so as to:

$$\textit{Maximize } T_k(\boldsymbol{X}) : \sum_{j=1}^{n} \left[a_{kj}^L, a_{kj}^U\right] x_j + \left[\alpha_k^L,\ \alpha_k^U\right], \quad k \in K_1 \tag{1}$$

$$\textit{Maximize } T_k(\boldsymbol{X}) : \frac{\sum_{j=1}^{n} \left[c_{kj}^L, c_{kj}^U\right] x_j + \left[\mu_k^L, \mu_k^U\right]}{\sum_{j=1}^{n} \left[b_{kj}^L, b_{kj}^U\right] x_j + \left[\rho_k^L, \rho_k^U\right]}, \quad k \in K_2 \tag{2}$$

$$\text{subject to } \boldsymbol{X} \in S\{\boldsymbol{X} \in R^n | Pr[F(\boldsymbol{X}) \begin{pmatrix} \geq \\ \leq \end{pmatrix} h] \geq p, \boldsymbol{X} \geq 0, h \in R^m\}, \tag{3}$$

where $T_k(\boldsymbol{X})$ represents the *k-th* objective, $\boldsymbol{X}$ designates the vector of n decision variables, where $[a_{kj}^L, a_{kj}^U],\ k \in K_1, j = 1, 2, \ldots, n$ and $[c_{kj}^L, c_{kj}^U], [b_{kj}^L, b_{kj}^U],\ k \in K_2, j = 1, 2, \ldots, n$ are the vectors of coefficient intervals associated with linear and fractional objective functions, respectively. $[\alpha_k^L, \alpha_k^U], k \in K_1$, and $[\mu_k^L, \mu_k^U]$ and $[\rho_k^L, \rho_k^U], k \in K_2$, are the constant intervals associated with kth interval goal of the respective linear and fractional objective functions. The superscripts L and U stand for lower- and upper-bounds, respectively, of the define intervals. It is assumed that the feasible region $S\ (\neq \phi)$ is bounded, and $K_1 \cup K_2 = \{1,\ 2, \ldots, K\}$ with $K_1 \cap K_2 = \phi$.

In case of fractional goal expression in (2), it is customary to assume that $\sum_{j=1}^{n}[b_{kj}^L, b_{kj}^U]x_j + [\rho_k^L, \rho_k^U] > 0$ to avoid any undefined situation. '*Pr*' stands for probabilistically defined constraints, *F*(*X*) is a (linear or nonlinear) function representing structural constraints set, h is a resource vector, and $p\ (0 < p < 1)$ is the vector of satisficing probability levels defined for randomness of parameters associated with the constraints.

Now, the notion of interval number and some basic rules of interval arithmetic [23] that are concerned with present IVGP model formulation are discussed in the following section.

2.1 Basic Concept of Interval Number and Arithmetic Operations

(i) *Interval number*

An interval represents a set that consists of all real numbers between a given pair of numbers. It can also be thought of as a segment of a real number line, and an endpoint marks the end of the line segment. In the notion of interval number, both endpoints, either endpoint or neither endpoint can be included in an interval. The term interval means closed interval that is considered here in IVGP formulation of the problem.

An interval can be defined by an ordered pair as [23]:

$$A = \left[a^L, a^U\right] = \left\{a :\ a^L \leq a \leq a^U;\ a \in \Re\right\}, \tag{4}$$

where a^L and a^U are lower- and upper-bounds, respectively, of the interval A on a real line $\Re$, and where L and U stand for lower and upper, respectively.

The notion of a closed interval is depicted in Fig. 1.

Fig. 1 An illustration of closed interval

Here, for a particular case, if $a^L = a^U = a$, then $A = [a, a]$ represents a real number a and is called degenerate interval $[a, a]$. In this sense, $0 = [0,0]$ can be taken into account in the sequel of interval number representation.

Now, the notions of Absolute-*value, Width and Midpoint* of an interval are defined as follows:

- The *absolute-value* of A, denoted $|A|$, is defined the maximum of the absolute values of its endpoints as:

$$|A| = \max\left\{|a^L|,\ |a^U|\right\},\ \text{where } |a| \le |A|, \text{ for every } a \in A. \tag{5}$$

- The *width* of an interval A is defined and denoted by

$$w[A] = \left(a^U - a^L\right) \tag{6}$$

- The *midpoint* $m[A]$ of A is given by

$$m[A] = \frac{1}{2}\left(a^L + a^U\right) \tag{7}$$

The graphic illustrations of absolute value, width, and midpoint are shown in Fig. 2.

(ii) *Basic interval arithmetic operations*

It may be noted that 'interval arithmetic' defines a set of operations on intervals, whereas classical arithmetic defines operations on individual numbers.

There are different types of possible relations between the two defined intervals $A = \left[a^L, a^U\right]$ and $B = \left[b^L, b^U\right]$.

The diagrammatic illustrations of nonoverlapping and overlapping relations between intervals A and B are depicted in Fig. 3a, b.

Now, the general binary operation (*) between A and B can be defined as:

$$\begin{aligned} A^*B &= \{c/c = a^*b,\ a \in A,\ b \in B\} \\ &= \{c/c = a^*b,\ a^L \le a \le a^U,\ b^L \le b \le b^U\} \end{aligned} \tag{8}$$

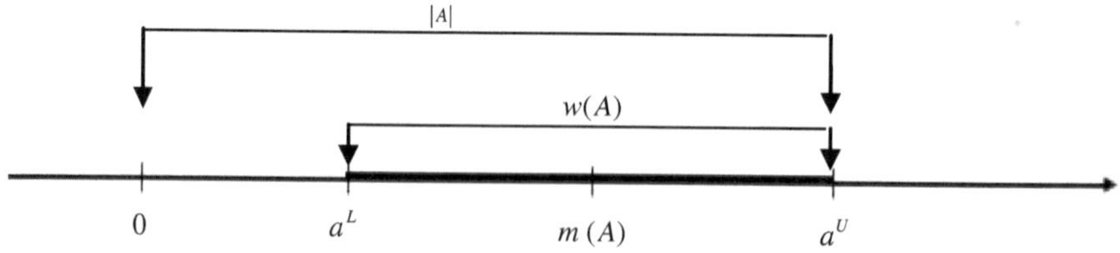

Fig. 2 Graphic illustrations of absolute value, width and midpoint of an interval

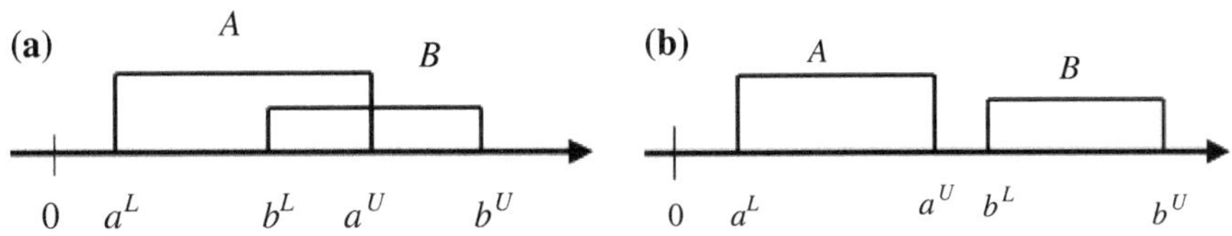

Fig. 3 **a** Relation between two overlapping intervals. **b** Relation between two nonoverlapping intervals

In the notion of interval arithmetic, since $A*B$ determines the region of the value $a*b$, where $a \in A$ and $b \in B$, the operator (*) is called the extended operator.

There are various types of extended operations in the interval arithmetic literature [21]. The possible extended subtraction operation between the non-negative intervals *A* and *B*, which is the most common and useful to modeling the proposed problem, can be defined as

$$A(-)B = A(+)\,(-B) = \left[a^L - b^U,\ a^U - b^L\right]. \tag{9}$$

Then, the resultant intervals for possible extended subtraction operations on the nonoverlapping and overlapping relations defined above are diagrammatically presented in Fig. 4a, b.

Actually, different resultant intervals for possible subtractions can be obtained for different types of relations that exist between *A* and *B*. However, extensive study on interval arithmetic made in the past has also been well documented [44] and widely circulated in the literature.

In multiobjective decision analysis, the other useful interval arithmetic operations are defined as follows:

- The scalar multiplication of A by λ is defined as

$$\lambda A = \begin{cases} \left[\lambda a^L, \lambda a^U\right], & \lambda \geq 0 \\ \left[\lambda a^U, \lambda a^L\right], & \lambda < 0 \end{cases} \tag{10}$$

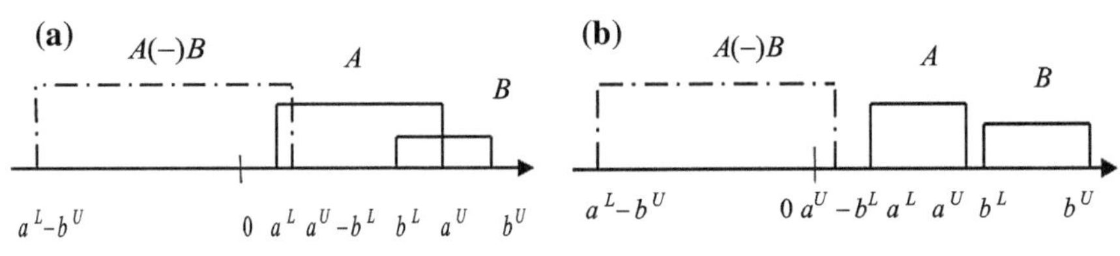

Fig. 4 **a** Possible subtraction operation between two overlapping intervals. **b** Possible subtraction operation between two nonoverlapping intervals

- The variable multiplication of A by x is defined as

$$\left[a^L, a^U\right]x = \left[a^L x,\ a^U x\right]. \tag{11}$$

- If $\{A_j = \left[a_j^L, a_j^U\right],\ j = 1, 2, \ldots, n\}$ be a collection of intervals and $\{x_j(\geq 0); x_j \in \Re,\ j = 1, 2, \ldots, n\}$ be a set of variables, then the possible extended sum of n intervals is given by

$$\left(\sum_{j=1}^{n}\right) A_j x_j = \left[\sum_{j=1}^{n} a_j^L x_j, \sum_{j=1}^{n} a_j^U x_j\right] \tag{12}$$

- The multiplication between A and B can be defined as

$$A.B = \left[a^L, a^U\right]\cdot\left[b^L, b^U\right] = \left[a^L.b^L,\ a^U b^U\right], \text{if}\ \ a^L,\ b^L \geq 0. \tag{13}$$

- The division of A by B can be obtained as

$$A/B = \left[a^L, a^U\right]/\left[b^L, b^U\right] = \left[a^L, a^U\right]\cdot\left[1/b^U,\ 1/b^L\right], \text{if}\ \ 0 \notin \left[b^L, b^U\right] \tag{14}$$

The basic relations and operations of interval numbers along with their constructive diagrammatic representations for practical uses have been discussed in [45] in the past. To formulate the model of the problem within the framework of conventional GP, the objectives with interval coefficients are to be converted into plan-valued objectives by using interval arithmetic rules [23], which define the bounded regions within which the values of objectives (called planned values) possibly can take.

Now, conversions of interval-valued objectives in (1) and (2) into plan-valued objectives are discussed in the following Sect. 2.2.

2.2 *Conversion of Interval Objectives into Plan-Valued Objectives*

Following the operational rules in (11)–(14), the plan-valued objectives corresponding to the defined objectives in (1) and (2) can be obtained as [32]:

$$T_k(\boldsymbol{X}) \colon \left[\sum_{j=1}^{n} a_{kj}^L x_j + \alpha_k^L, \sum_{j=1}^{n} a_{kj}^U x_j + \alpha_k^U\right], \quad k \in K_1 \tag{15}$$

$$T_k(\boldsymbol{X}): \left[\frac{\sum_{j=1}^{n} c_{kj}^{L} x_j + \mu_k^L}{\sum_{j=1}^{n} b_{kj}^{U} x_j + \rho_k^U}, \frac{\sum_{j=1}^{n} c_{kj}^{U} x_j + \mu_k^U}{\sum_{j=1}^{n} b_{kj}^{L} x_j + \rho_k^L}, \right], \quad k \in K_2 \tag{16}$$

Now, in the sequel of model formulation, without loss of generality and for simplicity, it is assumed that the random parameters associated with the system constraints in (3) are normally distributed, which frequently arises in modeling real-life problems.

2.3 Deterministic Equivalents of Chance Constraints

The chance constraints set in (3) with $\geq$ type can be explicitly presented as:

$$Pr\left[\sum_{j=1}^{n} \hat{g}_{ij} x_j \geq \hat{h}_i\right] \geq p_i, \; i = 1, 2, \ldots, m_1; \quad m_1 < m \tag{17}$$

Let $E(\hat{h}_i)$, $E(\hat{g}_{ij})$ and $var(\hat{h}_i)$, $var(\hat{g}_{ij})$ be the means and variances of the random variables $\hat{h}_i$ and $\hat{g}_{ij}$ having the characteristics of normal distribution.

Then, following the notion of distribution function for random variable, the crisp equivalent of the constraints in (3) can be defined.

Let $F(\hat{y})$ be the distribution function of the random variable $\hat{Y}$(say).

Then, since $F(\hat{y})$ is a monotonically nondecreasing function, the value of the corresponding variable $\hat{y}$ can be found as:

$$F^{-1}(\varepsilon) = \{Max\, \hat{y}\, |Pr(\hat{Y} \leq \hat{y}) \leq \varepsilon\}, 0 < \varepsilon < 1, \tag{18}$$

where ε indicates the satisficing level of probability.

Now, since $\hat{g}_{ij}$ and $\hat{h}_i$ are random variables, the conversion of their deterministic equivalent can be described as follows.

$$\text{Let, } \hat{y}_i = \left(\sum_{j=1}^{n} \hat{g}_{ij} x_j - \hat{h}_i\right) \tag{19}$$

Since, $\hat{y}_i$ is linear combination of the normally distributed random variables, it will also follow normal distribution.

Now, the constraints set in (3) with $\geq$ type restrictions can be expressed as:

$$Pr[\hat{y}_i \geq 0] \geq p_i, \; i = 1, 2, \ldots, m_1; \; m_1 \leq m. \tag{20}$$

The expression in (20) can be generalized as

$$Pr\left[\frac{\hat{y}_i - E(\hat{y}_i)}{\sqrt{\{var(\hat{y}_i)\}}} \geq \frac{-E(\hat{y}_i)}{\sqrt{\{var(\hat{y}_i)\}}}\right] \geq p_i$$

$$\text{or, } Pr\left[\frac{\hat{y}_i - E(\hat{y}_i)}{\sqrt{\{var(\hat{y}_i)\}}} \leq \frac{-E(\hat{y}_i)}{\sqrt{\{var(\hat{y}_i)\}}}\right] \leq 1 - p_i \tag{21}$$

where $\frac{\hat{y}_i - E(\hat{y}_i)}{\sqrt{\{var(\hat{y}_i)\}}}$ is a standard normal variate.
Using the notion of distribution function defined in (18), the deterministic equivalents of the constraints in (21) can be obtained as

$$\frac{-E(\hat{y}_i)}{\sqrt{\{var(\hat{y}_i)\}}} \leq F^{-1}(1 - p_i)$$

$$\text{or, } E(\hat{y}_i) + F^{-1}(1 - p_i)\sqrt{\{var(\hat{y}_i)\}} \geq 0, \quad i = 1,\ 2, \ldots,\ m_1 \tag{22}$$

Here, it is to be followed that the expression in (9) is quadratic in nature.
Proceeding in an analogous way, the equivalent deterministic nonlinear constraints of the chance constraints in (3) with $\leq$ type restriction can be obtained as

$$E(\hat{y}_i) + F^{-1}(p_i)\sqrt{\{var(\hat{y}_i)\}} \leq 0, \quad i = (m_1 + 1),\ (m_1 + 2), \ldots,\ m \tag{23}$$

Note 1: It is to be noted that if only $\hat{b}_i$ $(i = 1,\ 2, \ldots,\ m)$ are random in nature, then the constraints in (22) and (23) would be linear in form.

Now, to solve the problem by using linear goal programming (LGP) method, it is necessary to transform the quadratic constraints in (22) and (23) to their linear forms. Here, instead of using the traditional variable transformation approach [32], the widely used first-order *Taylor series approximation* method [41] can be employed to make an ease of solving the problem by using LGP methodology.

2.4 Use of Taylor Series Approximation

The expressions in (22) and (23) can be successively recast as:

$$E(\hat{y}_i) + f^{-1}(1 - p_i)\sqrt{var(\hat{y}_i)} = F_{1i}(\boldsymbol{X}), \quad i = 1, 2, \ldots, m_1 \tag{24}$$

$$\text{and, } E(\hat{y}_i) + f^{-1}(p_i)\sqrt{var\ (\hat{y}_i)} = F_{2i}(\boldsymbol{X}), \quad i = (m_1 + 1), (m_1 + 2), \ldots, m. \tag{25}$$

Since the defined constraints set are quadratic in form, the first-order *Taylor series approximation*, the simplest form of it, is sufficient to consider in the linearization process.

The *Taylor series approximation* is presented in the following steps:

Step 1: Determine $\boldsymbol{X}^* = (x_1^*, x_2^*, \ldots, x_n^*)$, where $\boldsymbol{X}^*$ is the initial approximate solution. Here, any one of the individual best solution of the objectives can be considered as the initial solution $\boldsymbol{X}^*$.

Step 2: Transform the function $F_{ki}(\boldsymbol{X})$ by using first order of *Taylor series expansion* as

$$F_{ki}(x_1, x_2, \ldots, x_n) = F_{ki}(x_1^*, x_2^*, \ldots, x_n^*) + \sum_{j=1}^{n} (x_j - x_j^*) \frac{\partial F_{ki}(x_1^*, x_2^*, \ldots, x_n^*)}{\partial x_j} \tag{26}$$

Step 3: Design the constraint sets in (24) and (25) by using the expression in (26) as:

$$F_{1i}(\boldsymbol{X}^*) + \sum_{j=1}^{n} (x_j - x_j^*) \frac{\partial F_{1i}(\boldsymbol{X}^*)}{\partial x_j} \geq 0 \tag{27}$$

$$\text{and,} \quad F_{2i}(\boldsymbol{X}^*) + \sum_{j=1}^{n} (x_j - x_j^*) \frac{\partial F_{2i}(\boldsymbol{X}^*)}{\partial x_j} \leq 0 \tag{28}$$

It is to be noted here that constraints sets defined in (27) and (28) are linear forms of the constraints sets in (24) and (25), respectively.

Then, in IVGP model formulation of the problem, the plan-valued objectives are to be transformed into planned interval goals by introducing target intervals to each of them.

2.5 Construction of Planned Interval Goals

In the decision environment, worst and best solutions of *k-th* objective that are associated with lower and upper limits of its planned interval in terms of their minimum and maximum outputs can be obtained as T_{kU} and T_{kL}, respectively.

Then, from the viewpoint of achieving an objective value with certain tolerance limits for satisfaction of DM, the target interval associated with achievement of *k-th* objective, which lies between the corresponding worst and best solutions can be considered as $\left[t_k^L, t_k^U\right]$, where $T_{kL} \leq t_k^L < t_k^U \leq T_{kU}, k = 1,\ 2, \ldots, K$.

The planned interval goals appear as [25]:

$$\left[\sum_{j=1}^{n} a_{kj}^L x_j + \alpha_k^L, \sum_{j=1}^{n} a_{kj}^U x_j + \alpha_k^U\right] = \left[t_k^L,\ t_k^U\right], \quad k \in K_1 \tag{29}$$

$$\left[\frac{\sum_{j=1}^{n} c_{kj}^{L} x_j + \mu_k^L}{\sum_{j=1}^{n} b_{kj}^{U} x_j + \rho_k^U}, \frac{\sum_{j=1}^{n} c_{kj}^{U} x_j + \mu_k^U}{\sum_{j=1}^{n} b_{kj}^{L} x_j + \rho_k^L,}\right] = \left[t_k^L, \ t_k^U\right], \quad k \in K_2, \tag{30}$$

Then, following interval arithmetic rules, the linear form of expression in (30) can be obtained as:

$$\left[\sum_{j=1}^{n}\left(c_{kj}^{L} - t_k^L b_{kj}^U\right) x_j + \left(\mu_k^L - t_k^U \rho_k^U\right), \sum_{j=1}^{n}\left(c_{kj}^{U} - t_k^U b_{kj}^L\right) x_j + \left(\mu_k^U - t_k^L \rho_k^U\right)\right] = [0, 0], \ k \in K_2 \tag{31}$$

Now, there are different versions of formulating IVGP models [25] that are mainly concerned with construction of goal achievement functions (called regret functions) for minimization of deviational variables associated with the defined goals. The notion of minimization of possible regrets from the optimistic point of view of DM is considered here as a simplest and promising one in the present decision-making context. However, in the process of formulating IVGP model, each of the defined goals in (29) and (31) are converted into equivalent two (flexible) standard goals by assigning the corresponding lower- and upper-bounds individually as the aspiration levels by using the interval arithmetic technique and introducing under- and over-deviational variables to each of them.

2.6 Conversion of Planned Interval Goals into Standard Goals

The standard goals associated with the goal expressions in (29) and (31) appear as:

$$\sum_{j=1}^{n} c_{kj}^{U} x_j + \alpha_k^U + d_{kL}^{-} - d_{kL}^{+} = t_k^L, \quad k \in K_1 \tag{32}$$

$$\sum_{j=1}^{n} c_{kj}^{L} x_j + \alpha_k^L + d_{kU}^{-} - d_{kU}^{+} = t_k^U, \quad k \in K_1 \tag{33}$$

$$\sum_{j=1}^{n}\left(c_{kj}^{U} - t_k^U b_{kj}^L\right) x_j + \left(\mu_k^U - t_k^L \rho_k^U\right) + d_{kL}^{-} - d_{kL}^{+} = 0, \quad k \in K_2 \tag{34}$$

$$\sum_{j=1}^{n}\left(c_{kj}^{L} - t_k^L b_{kj}^U\right) x_j + \left(\mu_k^L - t_k^U \rho_k^U\right) + d_{kU}^{-} - d_{kU}^{+} = 0, \quad k \in K_2 \tag{35}$$

where $\left(d_{kL}^{-}, d_{kU}^{-}\right)$ and $\left(d_{kL}^{+}, d_{kU}^{+}\right) \geq 0$ with $d_{kL}^{-} \cdot d_{kL}^{+} = 0$ and $d_{kU}^{-} \cdot d_{kU}^{+} = 0$, $(k = 1, 2, \ldots, K)$, represent the under- and over-deviational variables, respectively, associated with the respective goals.

Now, formulation of the GP model the proposed problem is presented in Sect. 3.

3 GP Model Formulation

In the decision-making context, it can easily be realized to the fact that the objective of achieving the goal values within their specified ranges means simultaneous minimization of under- and over-deviational variables associated with the goals in (32)–(35) to the extent possible in the decision environment. In the literature of GP, the two commonly used approaches are *minsum* GP [4] and *minmax* GP [42]. However, an intuitive idea of using GP is to take the convex combination of *minsum* GP and *minmax* GP formulations called extended GP (EGP) [43] in order to make a reasonable balance of achieving the decision with regard to take aggregated achievement of goals generated by using both the approaches simultaneously.

The GP model of the problem can be presented as [32]:

Find $\boldsymbol{X}$ so as to:

$$Minimize\ Z = \mu \sum_{k=1}^{K} w_k \left(d_{kL}^{-} + d_{kU}^{+}\right) + (1 - \mu)V,$$

and satisfy the goal expressions in (32)–(35) subject to the system constraints in (27) and (28) and

$$\left(d_{kL}^{-} + d_{kU}^{+}\right) - V < 0, \tag{36}$$

where Z represents the achievement functions, $V = \underset{k}{max}\ \left(d_{kL}^{-} + d_{kU}^{+}\right)$, and '*max*' stands for maximum.

To illustrate the effective use of the proposed approach, a numerical example is solved in the Sect. 4.

4 Numerical Example

A chance-constrained problem with two objectives having interval coefficients is considered as follows:

Table 1 Means and variances of the model parameters

Variable:	g_{11}	g_{12}	g_{13}	h_1	g_{21}	g_{22}	g_{23}	h_2
Mean:	2	4	6	9	3	5	7	20
Variance:	0.5	1	2.5	4	1.5	2	2.7	5

$$\begin{aligned}
&\text{Find } \boldsymbol{X}(x_1, x_2, x_3) \text{ so as to :}\\
&\mathit{Maximize}\, T_1 : [3,\ 5]x_1 + [11,\ 12]x_2 + [3,\ 6]x_3 + [3,\ 3]\\
&\mathit{Maximize}\, T_2 : \frac{[3,\ 11]x_1 + [15,\ 17]x_2 + [0.5,\ 3]x_3 + [4,\ 4]}{[6,\ 8]x_1 + [3,\ 5]x_2 + [2,\ 3]x_3 + [5,\ 5]}\\
&\text{subject to } Pr[g_{11}x_1 + g_{12}x_2 + g_{13}x_3 \geq h_1] \geq 0.90,\\
&\qquad Pr[g_{21}x_1 + g_{22}x_2 + g_{23}x_3 \leq h_2] \geq 0.85,
\end{aligned} \tag{37}$$

where $g_{ij}(i = 1,\ 2;\quad j = 1,\ 2,\ 3), h_1$ and h_2 are independent normal random variables.

In the decision situation, the means and variances of the system parameters are presented in Table 1.

In the model formulation, first using the data in Table 1 and following the expressions in (27) and (28), the deterministic equivalents of the chance constraints in (37) are successively determined as:

$$\begin{aligned}
&2x_1 + 4x_2 + 6x_3 - 1.281\left(0.5x_1^2 + x_2^2 + 2.5x_3^2 + 4\right)^{\frac{1}{2}} - 9 \geq 0,\\
&3x_1 + 5x_2 + 7x_3 + 1.036\left(1.5x_1^2 + 2x_2^2 + 2.7x_3^2 + 5\right)^{\frac{1}{2}} - 20 \leq 0
\end{aligned} \tag{38}$$

Then, *linear approximations* of both the constraints about the point (0, 0, 0) as an initial one are taken into account in the process of solving the problem.

The constraints in linear form are obtained as:

$$\begin{aligned}
&1.77x_1 + 3.57x_2 + 4.87x_3 - 10.84 \geq 0,\\
&3.46x_1 + 5.62x_2 + 7.83x_3 - 18.45 \leq 0.
\end{aligned} \tag{39}$$

Now, interval programming formulation of the problem is discussed as follows. Using the expressions in (15) and (16), the plan-valued objectives are obtained as

$$T_1(\boldsymbol{X}) = [3x_1 + 11x_2 + 3x_3 + 3,\ 5x_1 + 12x_2 + 6x_3 + 3] \tag{40}$$

$$T_2(\boldsymbol{X}) = \left[\frac{3x_1 + 15x_2 + 0.5x_3 + 4}{8x_1 + 5x_2 + 2x_3 + 5},\ \frac{4x_1 + 17x_2 + 3x_3 + 4}{6x_1 + 3x_2 + 3x_3 + 5}\right]. \tag{41}$$

Then, following the proposed procedure, the individual worst and best solutions of the objectives in (40) and (41) are obtained as $(T_{1L}, T_{1U}) = (15.38,\ 33.53)$ and $(T_{2L}, T_{2U}) = (0.80,\ 3.10)$, respectively.

In the decision process, the target intervals of the objectives in (40) and (41) are considered [17.78, 30.53] and [0.91, 2.75], respectively.

Now, using interval arithmetic rules and employing the proposed method, the standard linear goals of the model are obtained as follows.

$$\begin{aligned}
&5x_1 + 12x_2 + 6x_3 + d_{1L}^- - d_{1L}^+ = 14.78,\\
&3x_1 + 11x_2 + 3x_3 + d_{1U}^- - d_{1U}^+ = 27.53,\\
&-5.50x_1 + 8.75x_2 - 2.5x_3 + d_{2L}^- - d_{2L}^+ = 9.75,\\
&-4.28x_1 + 10.45x_2 - 1.32x_3 + d_{2U}^- - d_{2U}^+ = 0.55,
\end{aligned} \tag{42}$$

Then, following the procedure, the executable GP model of the problem is obtained as follows.

Find $\boldsymbol{X}(x_1,\ x_2)$ so as to:

$$Minimize\ Z = \mu\left[w_1\left(d_{1L}^- + d_{1U}^+\right) + w_2\left(d_{2L}^- + d_{2U}^+\right)\right] + (1-\mu)V$$

and satisfy the goal expressions in (42) subject to the system constraints in (39),

$$\text{and}\ d_{1L}^- + d_{1U}^+ \le V,\ d_{2L}^- + d_{2U}^+ \le V, \tag{43}$$

where $\left(d_{kL}^-, d_{kU}^-\right)$ and $\left(d_{kL}^+, d_{kU}^+\right) \ge 0$ with $d_{kL}^- \cdot d_{kL}^+ = 0$ and $d_{kU}^- \cdot d_{kU}^+ = 0$, $(k = 1,\ 2)$.

Now, for simplicity and without loss of generality, equal weights are assigned to all the goals for their achievement, i.e., $w_k = 0.50$, for $k = 1, 2$, and $\mu = 0.5$ are introduced in the decision-making context. The *Software* LINGO (ver. 12.0) is used to solve the problem.

The resultant decision is $(x_1,\ x_2,\ x_3) = (0, 1.48, 0.98)$.

The achieved objective values in interval form are found as:

$T_1 = [19.28, 27.22]$ and $T_2 = [0.91, 1.28]$.

The result shows that a satisfactory solution is reached for achievement of goals within their specified target intervals in the decision environment.

Note 2: The truncation errors occur for linearization of constraints are successively found as 0.06 and 0.04 % with regard to accuracy of satisfying the associated constraints in the decision situation. Thus, the tolerance values of arriving at optimality for the use of such linear constraints are quite acceptable in the uncertain environment.

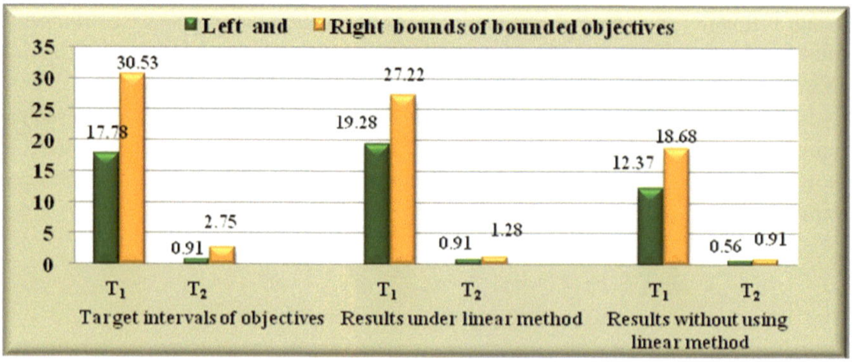

Fig. 5 Graphical representations of achieved objective values under the two approaches

4.1 Illustration for Performance Comparison

To expound the potential use of proposed method, execution of the problem is made without linearization of the fractional goals in the same decision environment.

Here, the solution is found as $(x_1,\ x_2,\ x_3) = (0, 0.3055, 2.0019)$ with objective values $T_1 = [12.37, 18.68]$ and $T_2 = [0.560, 0.091]$.
The results achieved under the two approaches are shown via the bar diagram presented in Fig. 5.

The graph shows that the decision achieved under the proposed approach is more acceptable than the conventional one with regard to achievement of values of the goals within the target intervals specified in the decision situation.

Remark In an uncertain environment, when model parameters follow other distributions [7] rather than normal one, like Poisson, Weibull distributions, etc., then the corresponding means and variances may not always agree with actual distribution patterns of random parameters. In such a case, computational complexity arises [46] with consideration of deterministic conversions of chance constraints in the decision situation. Here, stochastic simulation technique [47] can be effectively used to fit the chance parameters in the solution search process, which may be an extension in future study.

Note 3: It is worthy to note that, although interval programming has appeared as a pragmatic tool to solve real-world MODM problems, sometimes it becomes hard to specify the exact values of both lower- and upper-bounds of interval parameters owing to highly uncertain information about them in complex real-world decision situations. Again, in many practical problems, the lower and upper bounds of some interval parameters can rarely be acquired as deterministic in an uncertain environment. Consequently, robustness of an optimization process should have to be enhanced with the fluctuation of boundaries for searching better decision in inexact environment. The conceptual mathematical frame for solving such type of uncertain programming problems was introduced by Inuiguchi and Sakawa [48] in the past. The literature on uncertain programming for solving complex real-life problems has also been well discussed by Liu [49]. Here, one potential approach for better accounting for integrated uncertainties of parameters of a model might be the hybridization of interval programming method with incorporation of stochastic/ fuzzy distribution of bounds of interval parameters, and that can be effectively introduced to MODM problems with highly complex and uncertain information in managerial decision-making situations.

Here, it may be mentioned that, although the bounds of interval parameters are stochastically/fuzzily described, the elements within intervals are regarded as certain numbers. Although, the modeling of a real-world problem with such characteristics has been discussed by Nie et al. [50] in the past, study in the area is at an early stage in the literature of MODM in uncertain environment. However, the use of such an approach would be effective to establish a compromise between optimality and stability of the system and thereby potential solution can be generated for efficient management practices in uncertain decision making world.

5 Conclusions and Scope for Future Research

The main merit of using the proposed method is that all possible instances of uncertain data can be expressed as intervals, and interval information can be directly communicated in the process of solving problems without any distributional information that are required in conventional FP and SP methods. The potential use of the approach is illustrated by numerical example and the obtained solution is compared with solution of the problem obtained by using the conventional IVGP approach studied previously.

Further, the main advantage of the proposed approach is that the computational load [32] occurs for the uses of traditional linearization approaches to nonlinear functions does not arise here owing to the efficient use of interval arithmetic operation rules, and approximation error arises for the use of linearization technique is also negligible. In future study, the proposed approach can be extended to solve hierarchical decentralized decision problems [51] with interval parameter sets. However, it is hoped that approach presented here will open up new directions of research on IVGP for its actual implementation to real-world MODM problems in inexact environment.

Acknowledgments The authors would like to thank the Editors and anonymous Reviewer for their useful suggestions for improving the quality of presentation of the paper.

References

1. Steuer, R.E.: Multiple Criteria Optimization: Theory Computation and Applications. Wiley Publication, New York (1956)
2. Miettinen, M.K.: Nonlinear Multiobjective Optimization. Kluwer Academic Publishers, Boston (1999)
3. Hwang, C.-L., Masud, A.S.M.: Multiple Objective Decision Making- Methods and Applications. Springer, Berlin (1979)
4. Cohon, L.: Multiobjective Programming and Planning Courier. Dover Publications, New York (2004)
5. Ignizio, J.P.: Goal Programming and Extensions. Lexington Books, Massachusetts (1976)
6. Simon, H.A.: Administrative Behavior. Free Press, New York (1945)
7. King, J.K., Wallace, S.W.: Modeling with Stochastic Programming. Springer, New York (2012). Science
8. Charnes, A., Cooper, W.W.: Chance-constrained programming. Manag. Sci. **6**, 73–79 (1959)
9. Liu, B.: Dependent-chance programming: a class of stochastic optimization. Comput. Math. Appl. **34**, 89–104 (1997)
10. Zanakis, S.H., Evans, J.R., Vazacopoulos, A.A.: Heuristic methods and applications: a categorized survey. Euro. J. Oper. Res. **43**, 88–110 (1989)
11. Sengupta, J.K.: Stochastic Programming Methods and Applications. North-Holland Publishing Company, Amsterdam (1972)
12. Pal, B.B., Moitra, B.N.: A goal programming procedure for solving problems with multiple fuzzy goals using dynamic programming. Euro. J. Oper. Res. **144**, 480–491 (2003)
13. Dubois, D., Prade, H.: Fundamentals of Fuzzy Sets. Kluwer Academic Publication, Boston (2000)
14. Zimmermann, H.-J.: Fuzzy Sets Decision Making and Expert Systems. Kluwer Academic Publishers, Boston (1987)
15. Zadeh, L.A.: Fuzzy Sets. Information and Control **8**, 338–353 (1965)
16. Pal, B.B.: GA based FGP for resource allocation in farming. In: Wang, J. (ed.) Encyclopedia of Business Analytics and Optimization. IGI Global, USA, 5, pp. 1013–1029 (2014)
17. Kumar, M., Pal, B.B.: FGP model for emission-economic power dispatch. In: Wang, J. (ed.) Encyclopedia of Business Analytics and Optimization. IGI Global, USA, 5, pp. 936–950 (2014)
18. Slowinski, R.: A multicriteria fuzzy linear programming method for water supply system development planning. Fuzzy Sets Syst. **19**, 217–237 (1986)
19. Jiang, C., Han, X., Liu, G.R., Liu, G.P.: A nonlinear interval number programming method for uncertain optimization problems. Euro. J. Oper. Res. **188**, 1–13 (2008)
20. Bitran, G.R.: Linear multiobjective problems with interval coefficients. Manag. Sci. **26**, 694–706 (1980)
21. Moore, R.E., Kearfort, R.B., Cloud, M.J.: Introduction to Interval Analysis. Society for Industrial and Applied Mathematics (SIAM), Philadelphia (2009)
22. Huang, G.H., Baetz, B.W., Patry, G.G.: A grey fuzzy linear programming approach for municipal solid waste management planning under uncertainty. Civ. Eng. Syst. **10**, 123–146 (1993)
23. Moore, R.E.: Interval Analysis. Prentice-Hall, New Jersey (1966)
24. Hansen, E.R.: Global Optimization Using Interval Analysis. Marcel Dekker Inc. and Sun Microsystem Inc., Pub., New York (2003)
25. Inuiguchi, M., Kume, Y.: Goal programming problems with interval coefficients and target intervals. Euro. J. Oper. Res. **52**, 345–361 (1991)
26. Oliviera, C., Antunes, C.H.: Multiple objective linear programming models with interval coefficients-an illustrated overview. Euro. J. Oper. Res. **118**, 1434–1463 (2007)
27. Ida, M.: Interval multiobjective programming and mobile robot path planning. In: Mohammadian, M. (ed.) New Frontiers in Computational Intelligence and It's Applications. IOS Press, Amsterdam (2000)
28. Ida, M.: Portfolio selection problem interval coefficients. Appl. Math. Lett. **16**, 709–713 (2003)

29. Inuiguchi, M., Mizoshita, F.: Qualitative and Quantitative Data Envelopment Analysis with Interval Data. Integrated Uncertainty Management and Applications. Advances in Intelligent and Soft Computing, 68, pp. 163–174 (2010)
30. Pal, B.B., Kumar, M., Sen, S.: A Priority Based Interval-Valued Goal Programming Approach for Land Utilization Planning in Agricultural System: A Case Study. IEEE Explore, 1–9 (2010)
31. Pal, B.B., Biswas, P., Mukhopadhyay, A.: Using genetic algorithm to goal programming model of solving economic-environmental electric power generation problem with interval-valued target goals. Mathematical Modeling and Scientific Computation Communications in Computer and Information Science, pp. 156–169. Springer, Berlin (2012)
32. Pal, B.B., Kumar, M., Sen, S.: A priority based goal programming method for solving academic personnel planning problems with interval-valued resource goals in university management system. Int. J. Appl. Manag. Sci. **4**, 284–312 (2012)
33. Holland, J.H.: Genetic algorithms and optimal allocation of trials. SIAM J. Comput. **2**, 88–105 (1973)
34. Pal, B.B., Chakraborti, D., Biswas, P.: A genetic algorithm based goal programming method for solving patrol manpower deployment planning problems with interval-valued resource goals in traffic management system: a case study. IEEE Xplore Digit. Libr. 61–69 (2009)
35. Michalewicz, Z.: Genetic Algorithms + Data Structures = Evolution Programs, 3rd edn. Spriger, Berlin (1996)
36. Awerbach, S., Ecker, J.G., Wallace, W.A.: A note: hidden nonlinearities in the application of goal programming. Manag. Sci. **22**, 918–920 (1976)
37. Steuer, R.E.: Algorithm for linear programming problems with interval objectives function coefficients. Manag. Sci. **26**, 333–348 (1981)
38. Pal, B.B., Moitra, B.N., Sen, S.: A linear goal programming approach to multiobjective fractional programming with interval parameter sets. Int. J. Math. Oper. Res. **3**, 697–714 (2011)
39. Pal, B.B., Kumar, M.: Interval goal programming for economic-environmental power generation dispatch problems. IEEE Xplore Digit. Libr. 1–8 (2013)
40. Feiring, B.R., Sastri, T., Sim, L.S.M.: A stochastic programming model for water resource planning. Math. Comp. Mod. **27**, 1–7 (1998)
41. Pal, B.B., Sen, S., Kumar, M.: A Fuzzy goal programming approach for solving multiobjective fuzzy stochastic programming problems IEEE Xplore Digit. Libr. 239–243 (2009)
42. Romero, C.: Handbook of Critical Issues in Goal Programming. Pergamon Press, Oxford (1991)
43. Pal, B.B., Kumar, M.: Extended goal programming approach with interval data uncertainty for resource allocation in farm planning: a case study. Advances in Intelligent Systems and Computing, pp. 639–651. Springer, New York (2014)
44. Dawood, H.: Theories of Interval Arithmetic: Mathematical Foundations and Applications. Lap Lambert Academic Publishing, Saarbrücken (2011)
45. Kulpa, Z.: Diagrammatic representation for interval arithmetic. Linear Algebra Appl. **324**(1–3), 55–80 (2001)
46. Blumenfeld, D.: Operations Research Calculations Handbook. CRC Press, New York (2010)
47. Pal, B.B., Gupta, S.: A genetic algorithm approach to fuzzy goal programming formulation of chance constrained problems using stochastic simulation. IEEE Xplore Digit. Libr. 187–192 (2009)
48. Inuiguchi, M., Sakawa, M.: Robust optimization under softness in a fuzzy linear programming problem. Int. J. Approx. Res. **18**, 21–34 (1998)
49. Liu, B.D.: Theory and Practice of Uncertain Programming. Springer, Berlin (2009)
50. Nie, X.H., Huang, G.H., Li, Y.P., Liu, L.: IFRP: a hybrid interval-parameter fuzzy robust programming approach for waste management planning under uncertainty. J. Env. Manag. **84**, 1–11 (2007)
51. Sakawa, M., Nishizaki, I.: Cooperative and Non-cooperative Multi-Level Programming. Springer, New York (2009)

Correction to: Dealing with Uncertainty: From Rough Sets to Interactive Rough-Granular Computing

Andrzej Jankowski, Andrzej Skowron and Roman Swiniarski

Correction to: Chapter "Dealing with Uncertainty: From Rough Sets to Interactive Rough-Granular Computing" in: M.K. Chakraborty et al. (eds.), *Facets of Uncertainties and Applications*, Springer Proceedings in Mathematics & Statistics 125, https://doi.org/10.1007/978-81-322-2301-6_2

In the original version of the book, the grant numbers approved by Polish National Science Centre (NCN), mentioned in the acknowledgement section of Chapter 2 has been revised from "DEC-2011/01/D /ST6/ 06981, DEC-2012/05/B/ST6/03215, and DEC-2013/09/B/ST6/01568" to "DEC-2011/01/D /ST6/ 06981 and DEC-2012/05/B/ST6/03215", which is a belated correction. The correction has been updated in the chapter.

The updated version of this chapter can be found at
https://doi.org/10.1007/978-81-322-2301-6_2

M.K. Chakraborty et al. (eds.), *Facets of Uncertainties and Applications*,
Springer Proceedings in Mathematics & Statistics 125,
https://doi.org/10.1007/978-81-322-2301-6_26